Symmetry and Symmetry-Breaking in Fluid Dynamics

Symmetry and Symmetry-Breaking in Fluid Dynamics

Guest Editors

Andrzej Herczyński
Roberto Zenit

MDPI

Basel • Beijing • Wuhan • Barcelona • Belgrade • Novi Sad • Cluj • Manchester

Guest Editors
Andrzej Herczyński
Department of Physics
Boston College
Chestnut Hill, MA
United States

Roberto Zenit
Department of Engineering
Brown University
Providence, RI
United States

Editorial Office
MDPI AG
Grosspeteranlage 5
4052 Basel, Switzerland

This is a reprint of articles from the Special Issue published online in the open access journal *Symmetry* (ISSN 2073-8994) (available at: www.mdpi.com/journal/symmetry/special_issues/Symmetry_Symmetry_Breaking_Fluid_Dynamics).

For citation purposes, cite each article independently as indicated on the article page online and using the guide below:

Lastname, A.A.; Lastname, B.B. Article Title. *Journal Name* **Year**, *Volume Number*, Page Range.

ISBN 978-3-7258-1864-8 (Hbk)
ISBN 978-3-7258-1863-1 (PDF)
https://doi.org/10.3390/books978-3-7258-1863-1

Contents

About the Editors . . . . **vii**

Andrzej Herczyński and Roberto Zenit
Symmetry and Symmetry-Breaking in Fluid Dynamics
Reprinted from: *Symmetry* **2024**, *16*, 621, doi:10.3390/sym16050621 . . . . **1**

Alireza, Hooshanginejad and Sunghwan jung
Buoyancy-Marangoni Fingering of a Miscible Spreading Drop
Reprinted from: *Symmetry* **2022**, *14*, 425, doi:10.3390/sym14020425 . . . . **4**

M. G. E. Roach and Herbert Huppert
Spreading or Contraction of an Axisymmetric Viscous Drop between Two Plates in a Rapidly Rotating Frame
Reprinted from: *Symmetry* **2022**, *14*, 2488, doi:10.3390/sym14122488 . . . . **13**

Miles M. P. Couchman, Davis J. Evans and John W. M. Bush
The Stability of a Hydrodynamic Bravais Lattice
Reprinted from: *Symmetry* **2022**, *14*, 1524, doi:10.3390/sym14081524 . . . . **22**

Paweł Sznajder, Bogdan Cichocki and Maria Ekiel-Jeżewska
Lack of Plasma-like Screening Mechanism in Sedimentation of a Non-Brownian Suspension
Reprinted from: *Symmetry* **2022**, *14*, 63, doi:10.3390/sym14010063 . . . . **41**

Neil M. Ribe, Pierre-Thomas Brun and Basile Audoly
Symmetry and Asymmetry in the Fluid Mechanical Sewing Machine
Reprinted from: *Symmetry* **2022**, *14*, 772, doi:10.3390/sym14040772 . . . . **51**

Maciej Lisicki, Łukasz Adamowicz, Andrzej Herczyński and Henry Keith Moffatt
Viscous Thread Falling on a Spinning Surface
Reprinted from: *Symmetry* **2022**, *14*, 1550, doi:10.3390/sym14081550 . . . . **68**

Krzysztof Andrzej Mizerski
Possible Role of Non-Stationarity of Magnetohydrodynamic Turbulence in Understanding of Geomagnetic Excursions
Reprinted from: *Symmetry* **2021**, *13*, 1881, doi:10.3390/sym13101881 . . . . **78**

Zhanbin Lu and Moshe Matalon
Anchored and Lifted Diffusion Flames Supported by Symmetric and Asymmetric Edge Flames
Reprinted from: *Symmetry* **2023**, *15*, 1547, doi:10.3390/sym15081547 . . . . **98**

Yuli D. Chashechkin
Discrete and Continuous Symmetries of Stratified Flows Past a Sphere
Reprinted from: *Symmetry* **2022**, *14*, 1278, doi:10.3390/sym14061278 . . . . **126**

Jean-Luc Thiffeault
Moving Forward by Shaking Sideways
Reprinted from: *Symmetry* **2022**, *14*, 620, doi:10.3390/sym14030620 . . . . **150**

Hai An, Zhenyu Hu, Haozhe Pan and Po Yang
Numerical Investigation of the Automatic Air Intake Drag Reduction Strut Based on the Venturi Effect
Reprinted from: *Symmetry* **2022**, *14*, 367, doi:10.3390/sym14020367 . . . . **158**

About the Editors

Andrzej Herczyński

Andrzej Herczyński's research has been in fluid dynamics and applied mathematics, including linear waves in complex environments. He has also been interested in the connections between the natural sciences, visual arts, and music.

Roberto Zenit

Roberto Zenit's research projects span across diverse areas of fluid mechanics, including two-phase and biological flows. Although his work is mainly experimental, he has also conducted numerical simulations and theoretical studies.

symmetry

MDPI

Editorial

Symmetry and Symmetry-Breaking in Fluid Dynamics

Andrzej Herczyński [1,]*and Roberto Zenit [2]

[1] Department of Physics, Boston College, Chestnut Hill, MA 02467, USA
[2] School of Engineering, Brown University, Providence, RI 02912, USA; roberto_zenit@brown.edu
* Correspondence: andrzej@bc.edu

It may seem that the heading of this Special Issue of *Symmetry*—though narrower than the famous all-inclusive title of an essay by Jean-Paul Sartre, *Being and Nothingness*—encompasses most, if not all, fluid phenomena. While we did endeavor to represent a broad range of flows, the primary aim was to provoke new questions and insights to those wherein the role of symmetry, a lack thereof, or transition away or toward symmetry are key to understanding them.

We thus anticipated attracting contributions spanning different scales and levels of complexity, in either two or three dimensions, including symmetry-breaking instabilities, symmetry-imposing boundary conditions, and flows that are symmetric partially, locally, or intermittently. Our hope was that a collection of such varied topics and results may suggest new approaches to fluid phenomena, whether well studied already or less familiar.

The eleven papers collected in this Special Issue are indeed diverse: in two or three dimensions, fundamental and applied, at macro and micro scales, elucidating flows that are symmetric, symmetry-breaking, and intermittently symmetric, and in uniform and stratified fluids. The phenomena addressed include single-drop dynamics, hydrodynamic lattices, viscous flows, interfacial instability, magneto-hydrodynamics, turbulence, swimming dynamics, sedimentation, and diffusion flames.

Several papers—contributions on spreading and rotating drops, viscous jets, the dynamo problem, and diffusion flames—illustrate phenomena wherein a flow ceases to be symmetrical when it transitions from one regime to another, an indication of the non-linear nature of the Navier–Stokes equations. The solutions can jump from symmetric to non-symmetric only when the balance of forces (viscous, inertial, surface tension) is broken, leading to a new state.

The collection opens with three papers [1–3] concerning drops but in highly diverse settings. The contribution from the research group of Sunghwan (Sunny) Jung [1] at Cornell University presents new experiments exploring the spreading of a water droplet on a bath of glycerol–water solutions. In this scenario, the outward buoyancy competes with the inward forcing due to Marangoni and viscous effects and a quasi-symmetric fingering pattern develops. A paper by Roach and Huppert [2] from the University of Cambridge takes up the problem of axisymmetric drop expanding between two rotating discs at low Reynolds numbers, and also the effect of squeezing the droplet, which has a similar solution. In the latter case, critical parameters are determined when the system breaks symmetry. The third contribution in this trio, by John Bush and his collaborators at MIT [3], considers the stability of a two-dimensional lattices, square and triangular, of oil droplets bouncing in synchrony on the surface of the same liquid.

The next paper, by Sznajder, Cichocki and Ekiel-Jeżewska [4] from the Polish Academy of Science and the University of Warsaw, analyzes the sedimentation of particles in the vanishing Reynolds number limit. Their results, based on BBGKY hierarchy derived from the Liouville equation, suggest the breaking of translational symmetry of the system due to plasma-like screening.

The next two papers [5,6] concern viscous jets falling on a moving substrate, symmetry-breaking extensions of the classical, axis-symmetric coiling instability. The first of these

Citation: Herczyński, A.; Zenit, R. Symmetry and Symmetry-Breaking in Fluid Dynamics. *Symmetry* **2024**, *16*, 621. https://doi.org/10.3390/sym16050621

Received: 7 May 2024
Accepted: 11 May 2024
Published: 17 May 2024

contributions, by Neil Ribe from CNRS and collaborators [5], revisits the version of the problem wherein the substrate is a translating belt, the so-called fluid mechanical sewing machine. As the name suggests, a rich variety of trace patterns can be obtained depending on the relative speed of the belt and the fluid at the contact point, such as overlapping loops, separated loops, meanders, figures of eight, and other forms. In the second, experimental contribution from Lisicki, Adamowicz, Herczyński, and Moffatt [6], the jet falls on a spinning surface at various radial distances. Similar patterns emerge in this case, but their center-line symmetry is subtly broken by centrifugal effects.

Three papers [7–9] bring the lens of symmetry/symmetry-breaking to somewhat more specialized problems. The magnetohydrodynamic turbulence in the Earth's core, the dynamo problem, is the subject of a paper by Krzysztof Mizerski [7] from the Polish Academy of Science. The focus here is on the asymmetric flow due to turbulent wave fields. The role of both symmetric and asymmetric edge flames in stabilizing diffusion flames is analyzed in a paper by Lu and Matalon [8]. The asymmetry is caused not by the geometry of the mixing zone, but by the unequal fuel and oxidizer Lewis numbers. The third paper in this group, a review article by Yuli Chashechkin [9] from the Russian Academy of Science, concerns stratified flow past a sphere, which has been visualized using the schlieren method. The images reveal the presence of very fine ligaments that, at higher velocities, introduce asymmetric features to the flow in the wake.

The last two papers [10,11] address propulsion in water, but in very different contexts: the first is a fundamental investigation of a surprising swimming mechanism; the second, an engineering design for reducing drag in high-speed vessels. Jean-Luc Thiffeault from the University of Wisconsin-Madison offers [10] a simple mechanical "toy" model for a micro swimmer, which shakes from side to side but nevertheless moves forward. The effect depends on forcing that also exerts a torque and, crucially, on some form of friction. A numerical investigation of a drag-reducing mechanism for a surface ship is carried out in the concluding paper by a group of researchers from Harbin Engineering University led by Hai An [11]. The paper proposes a novel design of a partially submerged strut attached to the vehicle, with the air intake above the water line and air outlet below, producing a bubbly flow, which reduces skin friction.

It is our hope that this small collection of papers may serve to illustrate a remarkably wide range of fluid dynamical phenomena wherein consideration of symmetry proves worthwhile, and perhaps also to encourage such considerations. After all, as Blaise Pascal remarked in his *Pensée*, "symmetry is what we see at a glance".

Conflicts of Interest: The authors declare no conflict of interest.

References

1. Hooshanginejad, A.; Jung, S. Buoyancy-Marangoni Fingering of a Miscible Spreading Drop. *Symmetry* **2022**, *14*, 425. [CrossRef]
2. Roach, M.G.; Huppert, H.E. Spreading and Contraction of an Axisymmetric Viscous Drop between Two Plates in a Rapidly Rotating Frame. *Symmetry* **2022**, *14*, 2488. [CrossRef]
3. Couchman, M.M.; Evans, D.J.; Bush, J.W. The stability of a Hydrodynamic Bravais Lattice. *Symmetry* **2022**, *14*, 1524. [CrossRef]
4. Sznajder, P.; Cichocki, B.; Ekiel-Jeżewska, M. Lack of Plasma-like Screening Mechanism in Sedimentation of a Non-Brownian Suspension. *Symmetry* **2022**, *14*, 63. [CrossRef]
5. Ribe, N.M.; Brun, P.T.; Audoly, B. Symmetry and asymmetry in the fluid mechanical sewing machine. *Symmetry* **2022**, *14*, 772. [CrossRef]
6. Lisicki, M.; Adamowicz, Ł.; Herczyński, A.; Moffatt, H.K. Viscous thread falling on a spinning surface. *Symmetry* **2022**, *14*, 1550. [CrossRef]
7. Mizerski, K.A. Possible Role of Non-Stationarity of Magnetohydrodynamic Turbulence in Understanding of Geomagnetic Excursions. *Symmetry* **2021**, *13*, 1881. [CrossRef]
8. Lu, Z.; Matalon, M. Anchored and Lifted Diffusion Flames Supported by Symmetric and Asymmetric Edge Flames. *Symmetry* **2023**, *15*, 1547. [CrossRef]
9. Chashechkin, Y.D. Discrete and Continuous Symmetries of Stratified Flows Past a Sphere. *Symmetry* **2022**, *14*, 1278. [CrossRef]

10. Thiffeault, J.L. Moving Forward by Shaking Sideways. *Symmetry* **2022**, *14*, 620. [CrossRef]
11. An, H.; Hu, Z.; Pan, H.; Yang, P. Numerical Investigation of the Automatic Air Intake Drag Reduction Strut Base on the Venturi Effect. *Symmetry* **2022**, *14*, 367. [CrossRef]

symmetry

Article

Buoyancy-Marangoni Fingering of a Miscible Spreading Drop

Alireza Hooshanginejad * and Sunghwan Jung *

Department of Biological and Environmental Engineering, Cornell University, Ithaca, NY 14853, USA
* Correspondence: hooshanginejad@cornell.edu (A.H.); sunnyjsh@cornell.edu (S.J.)

Abstract: We experimentally investigate the interfacial instability that emerges when a water droplet is deposited on a bath of glycerol-water solution. Despite the absence of surface tension to stabilize short-wavelength undulations, we observe finite-size fingers that grow and pinch off from the drop. We show that the fingering patterns formed in the experiments resultes from a balance between the outward buoyancy effect and inward Marangoni flow. This induced Marangoni flow inhibits small perturbations and acts as an effective surface tension on the miscible interface of the spreading drop. To characterize the final size and shape of the drop, we perform systematic experiments by varying the drop volume and the glycerol-water volume fraction. In addition, we have developed scaling arguments for the drop's final radius using key physical forces, and show that the final wavelength is inversely proportional to the Bond number.

Keywords: Marangoni flow; interfacial instability; miscible interface

Citation: Hooshanginejad, A.; Jung, S. Buoyancy-Marangoni Fingering of a Miscible Spreading Drop. *Symmetry* **2022**, *14*, 425. https://doi.org/10.3390/sym14020425

Academic Editors: Andrzej Herczyński and Roberto Zenit

Received: 15 January 2022
Accepted: 16 February 2022
Published: 21 February 2022

1. Introduction

The spreading of a drop over a liquid layer is a fundamental problem of great importance in many natural and industrial problems, such as oil spills [1], polymer processing [2], shaping optical lenses [3,4], and lab-on-a-chip devices [5]. The mechanics of a drop spreading on an immiscible liquid layer under buoyancy have been studied extensively (e.g., [6–8]), as two immiscible fluids create a clear interface to be characterized.

Marangoni stress (i.e., stress induced by a gradient in surface tension) can be used to manipulate the spreading dynamics of drops [9–11]. The physical picture of drop spreading is more intricate when mixing comes into play in miscible fluids [12–14]. Marangoni spreading of a miscible drop has been studied both in the presence of surfactants [15,16] and with lower surface tension of the drop [14,17–19]. In the latter case, the outward Marangoni flow often causes fingering instabilities [20–22]. Despite numerous previous studies, the dynamics of a drop on a miscible liquid layer is still poorly characterized due to the complexity of visualizations [14]. A better understanding of this topic would benefit various applications in the areas of renewable energy [23] and biomaterials [24].

Although a number of studies have investigated the spreading of miscible drops under outward Marangoni flow, very few studies exist that focus on an inward Marangoni flow. Tan and Thoroddsen [25] first reported that when a drop of water impacts a bath of glycerol, the drop forms flower-like fingering patterns. They showed that the combined effect of an inward Marangoni flow with the impact inertia leads to an intermediate circle drained from the water. Similarly, when a liquid drop of calcium chloride is deposited on a bath of sodium silicate with higher surface tension, fingering patterns emerge at the growing front, whereas a spiral channel forms in the center [26].

In this study, we present a new instability caused by the interactions between buoyancy and Marangoni effects. We conducted a series of experiments, depositing a water drop on a millimetric layer of glycerol. Unlike [25], we observed interfacial fingers pinching off at the growing front. In addition, we observed an outer shallow layer separated from a deeper inner drop from the top-down view. We also performed experiments with different volume fractions of a glycerol-water mixture to observe its effect on the instability. The paper is organized as follows. In Section 2, we describe our experimental procedure, along with our observations. In Section 3 we present the results from image processing of the data and scaling arguments justifying our observations. Finally, in Section 4 our findings and potential future studies are discussed.

2. Experiments

We gently deposit a drop of deionized water with volume V on a bath of glycerol-water solution that is 5 mm deep, as illustrated in Figure 1a. To remove any effects of inertia, the drops are slowly deposited using a pipette on the glycerol-water surface. The volume fraction of glycerol in the glycerol-water solution is denoted by ϕ_g. To make the glycerol-water mixture, we prepare a total volume of 200 mL for each trial. First, we collect the required volumes of water and pure glycerol in two different containers. Then, we mix and stir the two liquids for 15–20 min until the solution turnes from a semi-transparent color to a completely transparent color. The solution is then left for at least 4–5 h before stirring again for another 5 min. We establish this procedure to ensure the homogeneity of the glycerol and water mixture. The water surface tension with air, and the glycerol-water solution's surface tension with air, density, and viscosity are denoted by γ_{wa}, γ_{ga}, ρ_g, and μ_g, respectively. We note that water and glycerol are miscible. In addition, $\gamma_{ga} < \gamma_{wa}$. To express the glycerol-water solution density, we use Equation (3) in [27]. To find the value for the solution's viscosity, we use Equation (6) and Figure 1 in [28]. To calculate the solution's surface tension, we use the interpolation of Table 4 in [29] after converting the volume fraction of glycerol to a mass fraction. All variables are evaluated at the room temperature (i.e., T = 21 C). We record the top view of the drop as it spread on the glycerol-water solution, as shown in Figure 1b, using a digital camera (Nikon 7500) and a macro lens (Nikkor 105 mm). A small amount of food dye is added to the water for clearer visualization.

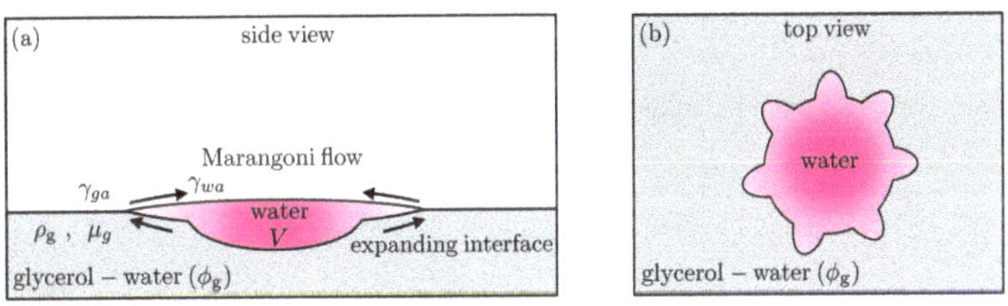

Figure 1. (**a**) Side-view and (**b**) top-view schematics of the experiments.

Figure 2 shows snapshots of a 500 µL water drop spreading on a bath of pure glycerol (i.e., $\phi_g = 0.999$). As the drop of water starts spreading on glycerol, the symmetry of the growing contact line is broken, and interfacial undulations emerge, as shown in Figure 2i. The instantaneous outer radius of the drop is denoted as $r_o(t)$, whereas the mean outer radius is denoted as $\bar{r}_o$. The perturbations continue to grow as indicated in Figure 2ii, with a certain wave number, n, that changes with time. At the early stage of spreading, the Marangoni flow from glycerol to the water drop is strong. As the perturbations grow, a number of discrete finger-like undulations form at the interface, and continue to evolve in the radial direction, as shown in Figure 2iii and its inset. In addition, the internal flows induced by the Marangoni stress at the mixing front begin to divide the drop into two regions of an inner zone and an outer zone. The inset of Figure 2v demonstrates the borderline separating the two regions with a mean radius $\bar{r}_i(t)$. The dark borderline is indicative of either a depth change, which suggests that the inner zone is deeper than the outer zone, or an abrupt concentration gradient for glycerol in water. We note that the glycerol solution has a higher density than the water drop. Hence, separation of the drop into an inner and outer zone is presumably due to the Rayleigh–Taylor instability, which happens when the glycerol solution is flowing over the water drop under Marangoni effects.

Confirming this hypothesis requires further investigation of visualizing the side-view of the drop, which will be carried out in a future study.

As the fingers grow, the centerline of each finger is filled with the glycerol-water mixture, as shown in the inset of Figure 2iv. However, the fingers still continue to grow radially outward (see the inset of Figure 2iv) until they pinch off from the miscible interface. After the pinch-off, the separated fingers start to expand along the azimuthal direction, as shown in Figure 2v. A ring of separated fingers continues to gradually move away from the center even after the interface stops spreading, as shown in Figure 2v and its inset. After the pinch-off event, the Marangoni flow weakens noticeably; therefore, the drop's shape remains unchanged. At this stage, dark radial lines are seen in the outer zone, as shown in Figure 2v, which is indicative of variations in either depth or glycerol concentration in the azimuthal direction. These variations are presumably induced by the vortical Marangoni flows as water mixes with glycerol at the interface, as illustrated in the inset of Figure 2. The right inset of Figure 2v shows that the final shape of the drop's outer interface maintains its cusp-like fingers remaining from the finger pinch-off event due to the absence of interfacial tension between water and the glycerol solution. The mean outer radius and mean inner radius for the final shape of the drop are denoted as R_o and R_i, respectively. In addition, the wave number and wave length of the final drop shape are denoted as N and λ, respectively.

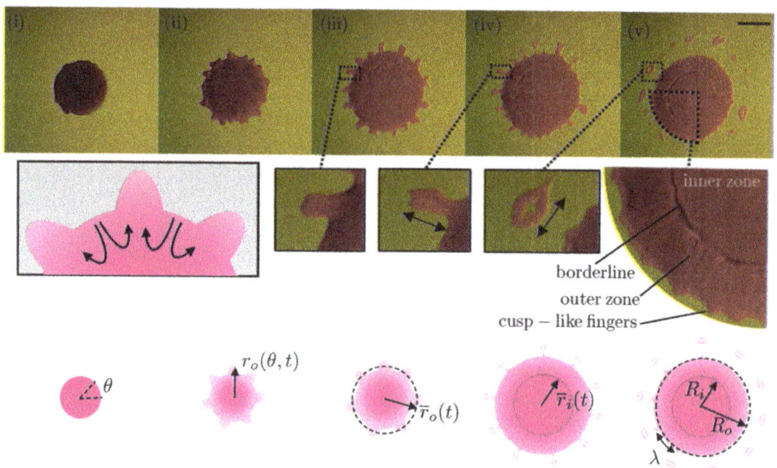

Figure 2. Sequential snapshots of a water drop spreading on glycerol ($V = 500$ µL, and $\phi_g = 0.999$) at **(i)** $t = 10$ s, **(ii)** $t = 20$ s, **(iii)** $t = 30$ s, **(iv)** $t = 40$ s, and **(v)** $t = 50$ s, where the inner zone is fully formed with a clear borderline. The scale bar shows 1 cm. The insets in the second row represent zoomed-in views of a single finger. The schematics in the third row illustrate the key characteristics of the spreading mechanism, including the outer interface undulations, the emergence of the inner zone, and the fingers pinch-off event.

3. Results

In this section, we first discuss the effects of two parameters (V and ϕ_g) on the growth of the fingers. We then discuss the final size and wavelength of the drop in the form of scaling arguments and dimensionless parameters.

Figure 3a shows the time evolution of r_o along the azimuthal direction, θ, starting from when the drop is completely deposited on the liquid layer. As indicated in Figure 3a, the fingers grow over time until the drop reaches its final shape with a certain wavenumber. Figure 3b shows $\overline{r}_o$ and $\overline{r}_i$ over the same timespan. At $t \simeq 35$ s the inner zone emerges when the outer interface has nearly reached its final size, as shown in Figure 3b. Here, the size of the inner zone does not change noticeably. In addition, we note that the outer interface spreading scales as $t^{1/7}$, which is in the same range as the scalings reported for the dynamic wetting of a viscous liquid on a smooth solid surface [30–33]. Finally, Figure 3c shows the dominant wavenumber based on the FFT results of the interface temporal shape. As seen in Figure 3c, the wavenumber in the experiments increases with time until the drop reaches its final shape.

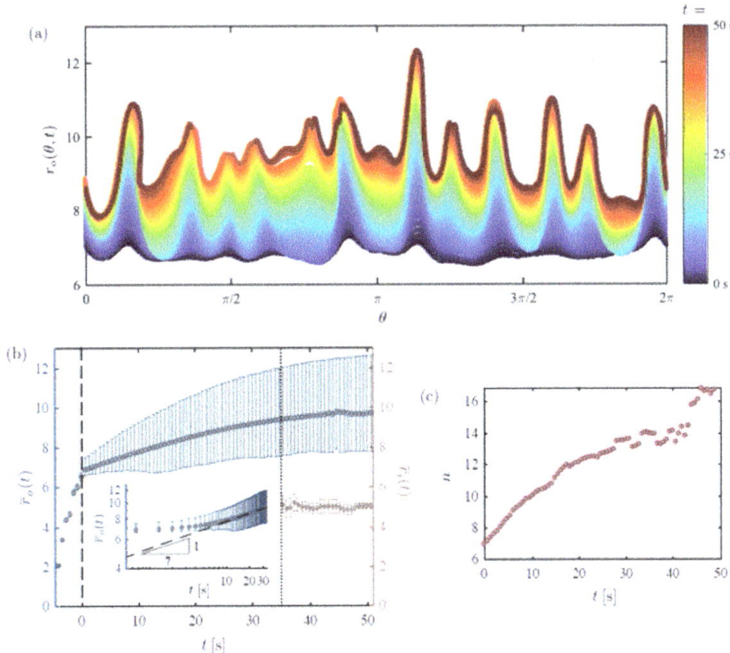

Figure 3. (**a**) Time evolution of r_o vs. θ for $V = 200$ µL, and $\phi_g = 0.999$; (**b**) $\bar{r}_o$ vs. t on the left vertical axis, and $\bar{r}_i$ vs. t on the right vertical axis for $V = 200$ µL, and $\phi_g = 0.999$. The inset plot shows the log-log plot of $\bar{r}_o$ scaling as $t^{1/7}$. (**c**) n vs. t for $V = 200$ µL, and $\phi_g = 0.999$.

3.1. Effect of the Drop Volume, V

We systematically changed the volume of the deposited water drop in the range of V = 100–600 µL on pure glycerol (i.e., $\phi = 0.999$) to characterize the effect of the drop volume on the final radius and shape of the drop. Figure 4a(i–iii) shows the final shapes of three different drops with $V = 200$, 400, and 600 µL, respectively. Figure 4b shows how the final outer radius, R_o, and the inner radius, R_i, of the drop increase with the volume, V. Notably, $R_o - R_i$ remains nearly constant even with different V. We also extracted the dominant wavenumber, N, which increases monotonically with volume V, as shown in Figure 4c. Notably, the corresponding wavelength, namely, $\lambda = 2\pi R_o/N$, decreases with increasing V (see the inset of Figure 4c).

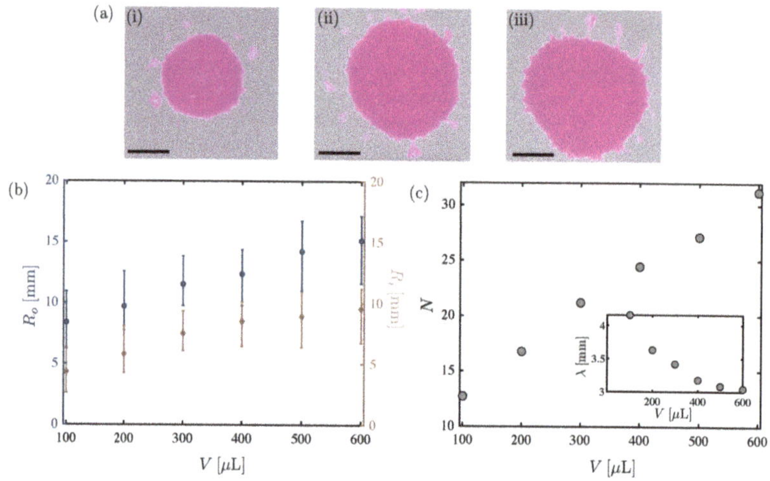

Figure 4. (**a**) The final drop shape for (i) $V = 200$ µL, (ii) $V = 400$ µL, and (iii) $V = 600$ µL on pure glycerol (i.e., $\phi_g = 0.999$). The scale bars show 1 cm. (**b**) R_o vs. V on the left vertical axis, and R_i vs. V on the right vertical axis for $\phi_g = 0.999$. The error bars represent 3 trials. (**c**) N vs. V for $\phi_g = 0.999$. The inset shows $\lambda = 2\pi R_o/N$ vs. V.

3.2. *Effect of the Glycerol Volume Fraction, ϕ_g*

We systematically changed the volume fraction of the glycerol solution in the range of $\phi_g = 0.8$–0.999 to characterize the Marangoni effects and the effects of buoyancy and viscosity on the final size and shape of the drop. With decreasing ϕ_g, both the density difference and the surface tension difference between the water drop and the glycerol-water solution decrease. Therefore, the outward buoyance force and inward Marangoni stress of the drop get suppressed. The values of ρ, μ_g, and γ_{ga} for varying ϕ_g are extracted from [27–29], and listed in the supplemental material. In addition, the corresponding molar fraction of glycerol in the mix, χ_g, is included for varying ϕ_g values in the supplemental material.

As ϕ_g decreases, the shape of the fingers changes from the cusp-like fingers, shown in Figure 2a, to rounded fingers, shown in Figure 5a(iv–v) at $\phi_g = 0.9$, until the outer interface becomes stable at $\phi_g = 0.8$, as shown in Figure 5b. However, groove-like patterns are observed to grow in the middle of the water drop at $\phi_g = 0.8$, as shown in sequential snapshots in Figure 5b. Figure 5c shows that R_o and R_i decrease as ϕ_g increases. Note that the inner zone forms closer to the outer interface with decreasing ϕ_g until $R_o = R_i$, when $\phi_g \leq 0.85$. Finally, Figure 5d shows that N increases and λ decreases as ϕ_g increases. This suggests that the Marangoni stress is more effective in stabilizing short-wavelength perturbations; therefore, inhibiting the Marangoni stress (i.e., lower ϕ_g) yields smaller λ.

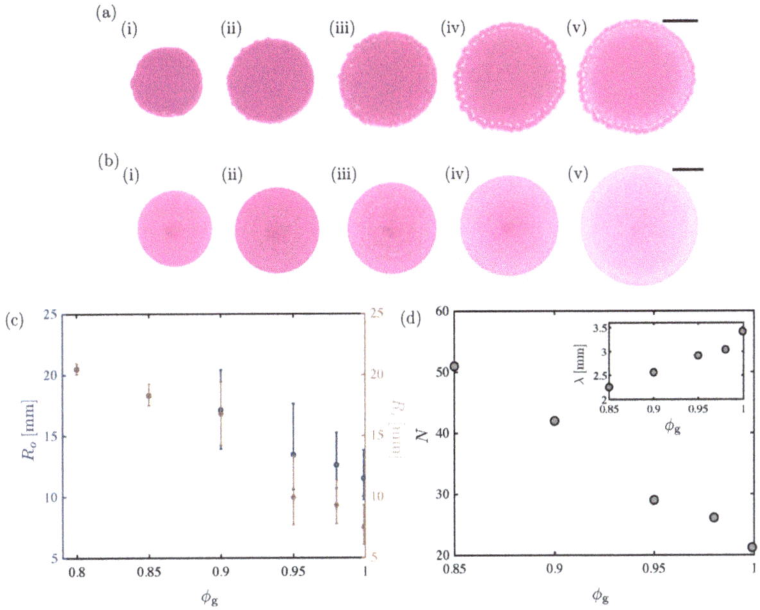

Figure 5. (**a**) Snapshots for $V = 300\ \mu L$, and $\phi_g = 0.9$ at (i) $t = 10$ s, (ii) $t = 20$ s, (iii) $t = 30$ s, (iv) $t = 40$ s, and (v) $t = 50$ s. The scale bars show 1 cm. (**b**) Snapshots for $V = 300\ \mu L$, and $\phi_g = 0.8$ at (i) $t = 10$ s, (ii) $t = 20$ s, (iii) $t = 30$ s, (iv) $t = 40$ s, and (v) $t = 50$ s. The scale bars show 1 cm. (**c**) R_o vs. ϕ_g on the left vertical axis, and R_i vs. ϕ_g on the right vertical axis for $V = 300\ \mu L$. The error bars represent 3 trials. (**d**) N vs. ϕ_g for $V = 300\ \mu L$. The inset shows $\lambda = 2\pi R_o/N$ vs. ϕ_g.

3.3. Scaling Arguments

Here we consider the forces acting on the drop as it spreads over the fluid interface. The stress from surface tension scales as $\Delta\gamma/H$, whereas the viscous force scales as $\mu_g u/H$, where $\Delta\gamma = \gamma_{wa} - \gamma_{ga}$, u is the characteristic speed, and H denotes the characteristic height of the drop, as indicated in Figure 6a. Then, balancing the two forces yields a modified capillary number in the form of $\text{Ca} = \mu_g u/\Delta\gamma$. Using the mean velocity of the outer interface, the capillary number ranged from 0.009–0.04 in our experiments. Therefore, neglecting viscous forces gives buoyancy and Marangoni stress as the driving force and resisting force, respectively. The force balance yields

$$\Delta\gamma/H^2 \sim g\Delta\rho, \tag{1}$$

where $\Delta\rho$ denotes the difference between the glycerol solution density and the water density. Substituting $H \sim V/R_o^2$ in Equation (1) yields

$$\begin{aligned} \Delta\gamma R_o^4/V^2 &\sim g\Delta\rho \\ R_o &\sim \left(\frac{g\Delta\rho}{\Delta\gamma}V^2\right)^{1/4}. \end{aligned} \tag{2}$$

Figure 6b shows the plot of R_o versus $\left(g\Delta\rho V^2/\Delta\gamma\right)^{1/4}$ for all cases of varying V and ϕ_g. Despite the changes in V, $\Delta\gamma$, and $\Delta\rho$, all experimental data show a good agreement with the scaling for R_o from Equation (2).

In addition, we nondimensionalize λ by means of the capillary length, $l_c = \sqrt{\Delta\gamma/(\rho_g g)}$. Figure 6c shows that the non-dimensional wavelength is inversely proportional to the Bond number as $\lambda/l_c \sim \tilde{Bo}^{-1}$, where $\tilde{Bo} = \Delta\rho g V^{2/3}/\Delta\gamma$ is the Bond number. Qualitatively, it is expected that the wavelength increases with the Marangoni stress and inversely with buoyancy.

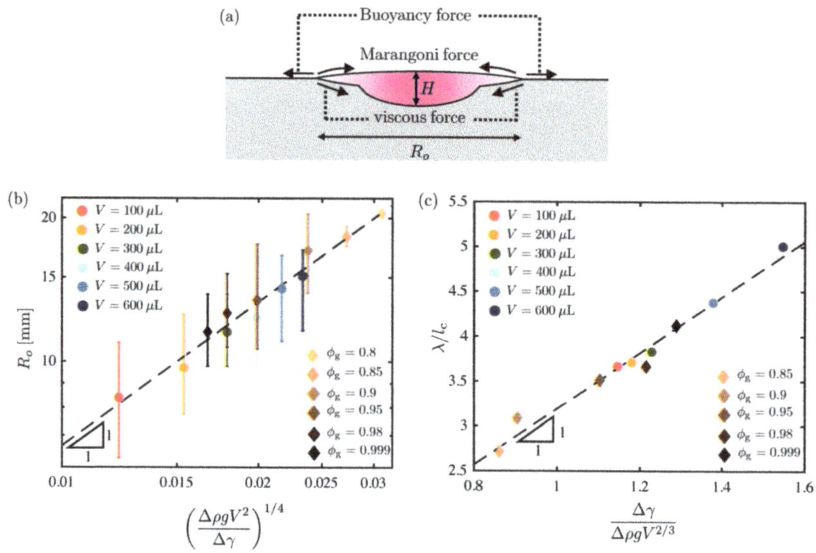

Figure 6. (**a**) Force balance schematic. (**b**) R_o vs. $\left(g\Delta\rho V^2/\Delta\gamma\right)^{1/4}$ for all cases of varying V and ϕ_g. (**c**) λ vs. $\tilde{Bo}^{-1} = \Delta\gamma/\Delta\rho g V^{2/3}$ for all cases of varying V and ϕ_g.

4. Discussion and Conclusions

In this paper, we have presented a new form of instability when a less viscous and less buoyant drop spreads over a millimetric miscible liquid layer that has lower surface tension than the drop. As the glycerol solution mixes with water at the moving front, an inward Marangoni flow is generated against a spreading flow under buoyancy effects. Therefore, the glycerol solution flows along the free surface under the Marangoni flow. As a high concentration of glycerol solution flows over the water drop, it becomes unstable due to buoyancy effects. Therefore, there must be a characteristic length arising from the interplay between the Marangoni flow and the occurrence of Rayleigh–Taylor instability. This characteristic length hints at the formation of the inner zone, separated by the observed borderline in the experiments, which is indicative of either a depth gradient or concentration gradient. In addition, we showed that after the outer interface stops spreading, the fingers continue to grow until they pinch off from the miscible interface. After the pinch-off, the fingers continue moving radially outward and slowly mix with the glycerol solution. In the final stages, the Marangoni flow is noticeably weakened; therefore, the drop's outer interface maintains its cusp-like patterns.

We observed that both the outer and inner radii, R_0 and R_i, grow with a similar trend regardless of the drop volume. In addition, the total number of fingers in the final shape increases monotonically with increasing droplet volume, whereas the final wavelength decreases. These results showed that buoyancy shortens the wavelength, thereby increasing the wavenumber. Furthermore, we showed that by adding water to glycerol and suppressing the inward Marangoni flow, the borderline between the two zones approaches the outer interface until it is not observed below $\phi_g \simeq 0.85$. In addition, the final wavenumber increases with a decreasing volume fraction, whereas the final wavelength decreases. In other words, inhibiting the Marangoni flow yields a smaller wavelength. Therefore, the results of our two parametric experiments suggest that buoyancy destabilizes the interface as the driving force, and the Marangoni stress stabilizes small undulations as the resisting force. By balancing the two opposite forces, we incorporated scaling laws for the growing outer interface, which were in good agreement with the experiments. In addition, we found that the final wavelength is inversely proportional to the modified bond number for all cases.

Despite the similarities between our system and that of [25], we observed new behaviors, such as the pinch-off of fingers, the formation of inner and outer zones, and new inner streak patterns that emerged for lower glycerol concentrations. These differences are presumably due to the negligible effect of inertia in the present study as opposed to that in [25], in which drops impacted a fluid bath. The radial motion of the fingers after the pinch-off event is reminiscent of the atomization process of drops as a water-IPA drop spreads on an oil bath [11]. However, we note that in the case of a water-IPA drop on an oil bath, outward Marangoni flow induced by IPA evaporation results in fragmentation, whereas in the current system, the inward Marangoni flow induces the fragmentation of the fingers. Further analysis is required to gain a better understanding of the underlying mechanism in order to rationalize the wavelength. More details related to the current miscible instability will be investigated in a future study.

Author Contributions: A.H. conceived the idea and performed experiments. A.H. and S.J. analyzed data and wrote the manuscript. All authors have read and agreed to the published version of the manuscript.

Funding: This work was supported by the National Science Foundation (NSF) under Grant No. CBET-1919753.

Informed Consent Statement: Not Applicable.

Data Availability Statement: The data is accessible through https://osf.io/czw93/, accessed on 15 February 2022. DOI:10.17605/OSF.IO/CZW93.

Acknowledgments: The authors would like to thank O. K. Matar, H. A. Stone, I. Bischofberger, and J. W. M. Bush for fruitful discussions.

Conflicts of Interest: The authors declare no conflict of interest.

References

1. Hoult, D.P. Oil Spreading on the Sea. *Annu. Rev. Fluid Mech.* **1972**, *4*, 341–368. [CrossRef]
2. Plochocki, A.P.; Dagli, S.S.; Andrews, R.D. The interface in binary mixtures of polymers containing a corresponding block copolymer: Effects of industrial mixing processes and of coalescence. *Polym. Eng. Sci.* **1990**, *30*, 741–752. [CrossRef]
3. Cruz-Campa, J.L.; Okandan, M.; Busse, M.L.; Nielson, G.N. Microlens rapid prototyping technique with capability for wide variation in lens diameter and focal length. *Microelectron. Eng.* **2010**, *87*, 2376–2381. [CrossRef]
4. Frumkin, V.; Bercovici, M. Fluidic shaping of optical components. *Flow* **2021**, *1*, E2. [CrossRef]
5. Stone, H.A.; Stroock, A.D.; Ajdari, A. Engineering Flows in Small Devices: Microfluidics Toward a Lab-on-a-Chip. *Annu. Rev. Fluid Mech.* **2004**, *36*, 381–411. [CrossRef]
6. Fraaije, J.G.E.M.; Cazabat, A.M. Dynamics of spreading on a liquid substrate. *J. Colloid Interface Sci.* **1989**, *133*, 452–460. [CrossRef]
7. Brochard-Wyart, F.; Debrégeas, G.; de Gennes, P.G. Spreading of viscous droplets on a non viscous liquid. *Colloid Polym. Sci.* **1996**, *274*, 70–72. [CrossRef]
8. Bacri, L.; Debrégeas, G. Experimental Study of the Spreading of a Viscous Droplet on a Nonviscous Liquid. *Langmuir* **1996**, *12*, 6708–6711. [CrossRef]
9. Brochard, F. Motions of droplets on solid surfaces induced by chemical or thermal gradients. *Langmuir* **1989**, *5*, 432–438. [CrossRef]
10. Carles, P.; Cazabat, A.M. Spreading of oil drops under a solvent vapor: Influence of Marangoni effect. In *Surfactants and Macromolecules: Self-Assembly at Interfaces and in Bulk*; Lindman, B., Rosenholm, J.B., Stenius, P., Eds.; Steinkopff: Darmstadt, Germany, 1990; pp. 76–81.
11. Keiser, L.; Bense, H.; Colinet, P.; Bico, J.; Reyssat, E. Marangoni Bursting: Evaporation-Induced Emulsification of Binary Mixtures on a Liquid Layer. *Phys. Rev. Lett.* **2017**, *118*, 074504. [CrossRef]
12. Thoroddsen, S.T.; Takehara, K. The coalescence cascade of a drop. *Phys. Fluids* **2000**, *12*, 1265–1267. [CrossRef]
13. Thoroddsen, S.T.; Qian, B.; Etoh, T.G.; Takehara, K. The initial coalescence of miscible drops. *Phys. Fluids* **2007**, *19*, 072110. [CrossRef]
14. Kim, H.; Muller, K.; Shardt, O.; Afkhami, S.; Stone, H.A. Solutal Marangoni flows of miscible liquids drive transport without surface contamination. *Nat. Phys.* **2017**, *13*, 1105–1110. [CrossRef]

15. Warner, M.R.E.; Craster, R.V.; Matar, O.K. Fingering phenomena associated with insoluble surfactant spreading on thin liquid films. *J. Fluid Mech.* **2004**, *510*, 169–200. [CrossRef]
16. Stocker, R.; Bush, J.W.M. Spontaneous oscillations of a sessile lens. *J. Fluid Mech.* **2007**, *583*, 465–475. doi: 10.1017/S0022112007005940. [CrossRef]
17. Ruckenstein, E.; Smigelschi, O.; Suciu, D.G. A steady dissolving drop method for studying the pure Marangoni effect. *Chem. Eng. Sci.* **1970**, *25*, 1249–1254. [CrossRef]
18. Walters, D.A. A stable transitional flow pattern in the surface tension driven spreading of ethanol-water solutions. *Langmuir* **1990**, *6*, 991–994. [CrossRef]
19. Kim, H.; Lee, J.; Kim, T.H.; Kim, H.Y. Spontaneous Marangoni Mixing of Miscible Liquids at a Liquid–Liquid–Air Contact Line. *Langmuir* **2015**, *31*, 8726–8731. [CrossRef]
20. Marmur, A.; Lelah, M.D. The Spreading of Aqueous Surfactant Solutions on Glass. *Chem. Eng. Commun.* **1981**, *13*, 133–143. [CrossRef]
21. Troian, S.M.; Wu, X.L.; Safran, S.A. Fingering instability in thin wetting films. *Phys. Rev. Lett.* **1989**, *62*, 1496–1499. [CrossRef]
22. Troian, S.M.; Herbolzheimer, E.; Safran, S.A. Model for the fingering instability of spreading surfactant drops. *Phys. Rev. Lett.* **1990**, *65*, 333–336. [CrossRef] [PubMed]
23. Li, Y.X.; Khan, M.I.; Gowda, R.J.P.; Ali, A.; Farooq, S.; Chu, Y.M.; Khan, S.U. Dynamics of aluminum oxide and copper hybrid nanofluid in nonlinear mixed Marangoni convective flow with entropy generation: Applications to renewable energy. *Chin. J. Phys.* **2021**, *73*, 275–287. [CrossRef]
24. Song, Y.Q.; Waqas, H.; Al-Khaled, K.; Farooq, U.; Ullah Khan, S.; Ijaz Khan, M.; Chu, Y.M.; Qayyum, S. Bioconvection analysis for Sutterby nanofluid over an axially stretched cylinder with melting heat transfer and variable thermal features: A Marangoni and solutal model. *Alex. Eng. J.* **2021**, *60*, 4663–4675. [CrossRef]
25. Tan, E.; Thoroddsen, S.T. Marangoni instability of two liquids mixing at a free surface. *Phys. Fluids* **1998**, *10*, 3038–3040. [CrossRef]
26. Hussein, S.; Maselko, J.; Pantaleone, J.T. Growing a Chemical Garden at the Air–Fluid Interface. *Langmuir* **2016**, *32*, 706–711. [CrossRef] [PubMed]
27. Volk, A.; Kähler, C.J. Density model for aqueous glycerol solutions. *Exp. Fluids* **2018**, *59*, 75. [CrossRef]
28. Cheng, N.S. Formula for the Viscosity of a Glycerol—Water Mixture. *Ind. Eng. Chem. Res.* **2008**, *47*, 3285–3288. [CrossRef]
29. Takamura, K.; Fischer, H.; Morrow, N.R. Physical properties of aqueous glycerol solutions. *J. Pet. Sci. Eng.* **2012**, *98–99*, 50–60. [CrossRef]
30. Huppert, H.E. Flow and instability of a viscous current down a slope. *Nature* **1982**, *300*, 427–429. [CrossRef]
31. Ehrhard, P. Experiments on isothermal and non-isothermal spreading. *J. Fluid Mech.* **1993**, *257*, 463–483. doi: 10.1017/S0022112093003167. [CrossRef]
32. McHale, G.; Newton, M.I.; Rowan, S.M.; Banerjee, M. The spreading of small viscous stripes of oil. *J. Phys. D Appl. Phys.* **1995**, *28*, 1925–1929. [CrossRef]
33. Kim, S.J.; Fezzaa, K.; An, J.; Sun, T.; Jung, S. Capillary spreading of contact line over a sinking sphere. *Appl. Phys. Lett.* **2017**, *111*, 134102. [CrossRef]

symmetry

Article

Spreading or Contraction of an Axisymmetric Viscous Drop between Two Plates in a Rapidly Rotating Frame

M. G. E. Roach [1] and Herbert E. Huppert [2,*]

[1] Emmanuel College, Cambridge CB2 3AP, UK
[2] Institute of Theoretical Geophysics, King's College, Cambridge CB2 1ST, UK
* Correspondence: heh1@cam.ac.uk

Abstract: We analyse the motions of a axisymmetric drop expanding between two rotating discs. We restrict to the case of a highly viscous fluid and a rapid rate of rotation. Therefore, we make modelling assumptions following from both a low Reynolds number and a low Rossby number. We investigate both the squeezing problem, where the top disc is pushed down on the drop; and the contraction problem, where the top plate is pulled away from the drop. Both problems have similar solutions to the non-rotating case but we find that the rotation term in the contraction problem allows a critical rotation rate that prevents the plates from moving apart. This exists because pressure in the fluid layer is lowered by the rotation and thus there is a suction effect between the two plates which promotes adhesion. We also complete the linear instability analysis of the squeezing problem and determine the critical values where the system shifts from symmetrical to asymmetrical.

Keywords: Hele-Shaw cell; rotation; viscous; squeezing; contraction

Citation: Roach, M.G.E.; Huppert, H.E. Spreading or Contraction of an Axisymmetric Viscous Drop between Two Plates in a Rapidly Rotating Frame. *Symmetry* **2022**, *14*, 2488. https://doi.org/10.3390/sym14122488

Academic Editors: Andrzej Herczynski, Tomohiro Inagaki and Roberto Zenit

Received: 21 August 2022
Accepted: 11 November 2022
Published: 24 November 2022

1. Introduction

The study of expanding viscous drops has seen a lot of development in the last few decades. A good outline of this field can be found in Huppert (2006) [1], beginning with analysis from von Kármán (1940) [2]. Recent developments from Ward (2006) [3] and Moffatt et al. (2021) [4] have focused on the case of a drop expanding in the thin layer between two horizontal plates (a Hele-Shaw cell). These papers have fully described the motion and shown that the squeezing problem is stable under linear perturbations. Investigation by Gay (2002) [5] has demonstrated that the contraction problem is unstable: air is drawn in as the radius of the viscous drop decreases and, as the two interact, a viscous fingering instability develops and thus the symmetry breaks down. This is an example of a Saffman-Taylor instability (Saffman and Taylor, 1958 [6]). The rotating contraction problem is also expected to dominated by a Saffman-Taylor instability, despite the additional complexities caused by rotation but this has not previously been investigated; this paper is original in its inclusion of rotation in both the squeezing and contraction problems. Both of these rotating problems have great relevance to the adhesion industry which, in turn, has applications to the construction, transportation and machinery manufacture industries, among others, as detailed in Dinte and Sylvester (2017) [7]. The viscous fluid we consider throughout this paper can be used to model an adhesive, which Dinte and Sylvester (2017) [7] define "as a mixture in a liquid or semi-liquid state, capable to join permanently to surfaces, by an adhesive process". In particular, the theory of this paper is most relevant to adhesively-bonded lap joints which were analytically investigated by Her (1999) [8]. It was demonstrated by de Bruyne and Houwink (1952) [9] that such a joint between the ends of two coaxial cylinders will break at high loads when subjected to torsion (by applying torque in opposite directions of rotation to the two cylinders). We can investigate an identical joint, but in the case that the cylinders rotate in the same direction, using the theory of this paper. In industry, such joints are used for torque transmission and have applications in aircraft, space structures and robot arms as outlined by Choi and Lee (1994) [10]. It is demonstrated in this paper that, by rotating such cylinders at a sufficiently fast rate, the adhesive layer between them can pull them closer together, even when the external forces are pulling them apart. We explicitly identify the minimum rotation rate required to achieve this effect.

2. Background

The system we are analysing consists of two plates which are a vertical distance h apart. Between the two plates is a volume of fluid V with surface tension γ, uniform density ρ and uniform dynamic viscosity μ. The two plates are rotated about a vertical axis through the centre of the fluid drop at a rate of rotation $\mathbf{\Omega} = \Omega \mathbf{e}_z$ and a force $\mathbf{F} = -F\mathbf{e}_z$ acts vertically and uniformly on the top plate. The fluid begins in a cylindrical drop of radius a_0 and the initial vertical distance between the plates is h_0. We make the modelling assumptions that the height of the fluid drop does not vary radially and that the front of the fluid drop remains vertical at all times. We choose to work in a coordinate system with origin on the bottom plate directly below initial centre of the fluid drop. The pressure field in the fluid is denoted by p and the velocity field by $\mathbf{u}$.

We investigate this system by following a similar mathematical analysis to Moffatt et al. (2021) [4]. In contrast, we work in a rapidly rotating frame, more specifically at low Rossby number. The axis of rotation is vertical and passes through the centre of the drop.

We consider a reference frame rotating with the plates at constant angular frequency $\mathbf{\Omega}$. This gives the same no-slip boundary conditions as the initial problem investigated by Moffatt et al. (2021) [4]; as well as adding Coriolis and centrifugal terms to the Navier-Stokes equations. We consider low Rossby number, so the centrifugal force dominates over both the Coriolis and advective effects, leaving us with the governing equation

$$\rho \frac{\partial \mathbf{u}}{\partial t} + \rho \mathbf{\Omega} \times (\mathbf{\Omega} \times \mathbf{x}) = -\nabla p + \rho \mathbf{g} + \mu \nabla^2 \mathbf{u}. \tag{1}$$

3. The Pseudo-Pressure Field

We now observe the following identity from vector calculus, as in Tritton (1988) [11]

$$\mathbf{\Omega} \times (\mathbf{\Omega} \times \mathbf{x}) = \nabla(-\frac{1}{2}|\mathbf{\Omega} \times \mathbf{x}|^2). \tag{2}$$

Hence, we can rewrite the governing equation as

$$\rho \frac{\partial \mathbf{u}}{\partial t} = -\nabla P + \rho \mathbf{g} + \mu \nabla^2 \mathbf{u}, \tag{3}$$

$$\text{where } P = p - \frac{1}{2}\rho |\mathbf{\Omega} \times \mathbf{x}|^2. \tag{4}$$

Since we are working in the case where the axis of rotation passes vertically through the centre of the drop, $\mathbf{\Omega} = \Omega \mathbf{e}_z$ and $\mathbf{x} = r\mathbf{e}_r + z\mathbf{e}_z$. Hence

$$P = p - \frac{1}{2}\rho \Omega^2 r^2. \tag{5}$$

We can also use the pseudo-pressure field to rewrite another of the key equations from Moffatt et al. (2021) [4]. Standard lubrication theory indicates that the pressure is independent of z, i.e., $p = p(r,t)$, and the radial component of velocity, $u = u(r,z,t)$, satisfies $|\partial u/\partial z| >> |\partial u/\partial r|$ and obeys:

$$\frac{\partial p}{\partial r} = \mu \frac{\partial^2 u}{\partial y^2} + \rho \Omega^2 r, \tag{6}$$

which we can rewrite as

$$\frac{\partial P}{\partial r} = \mu \frac{\partial^2 u}{\partial z^2}. \tag{7}$$

It is also worth noting that

$$p = p(r,t) \Rightarrow P = P(r,t). \tag{8}$$

Hence, the problem now closely resembles that investigated by Moffatt et al. (2021) [4] but with this pseudo-pressure field, $P(r,t)$, in the place of $p(r,t)$. The only difference being that, on the boundary $r = a$, we have the condition $P = p_a - \frac{1}{2}\rho\Omega^2 a^2 - 2\gamma/h$, where γ is the surface tension. This result can be recovered from Ungarish and Huppert (1998) [12] by directly considering the pressure field and neglecting the gravitational term due to both the thin layer approximation and low Rossby number.

4. Analysis of Basic State

Using the no slip boundary conditions, $u(r,0,t) = u(r,h,t) = 0$, we can solve for $u(r,z,t)$ in terms of the pressure field

$$u(r,z,t) = \frac{1}{2\mu}\frac{\partial P}{\partial r} z(z-h). \tag{9}$$

Thus, averaging over the depth, we find that:

$$\bar{u}(r,t) \equiv \frac{1}{h}\int_0^h u(r,z,t)dz = -\frac{h^2}{12\mu}\frac{\partial P}{\partial r}, \tag{10}$$

$$\text{i.e., } \bar{u}(r,t) = -\frac{h^2}{12\mu}\left(\frac{\partial p}{\partial r} - \rho\Omega^2 r\right). \tag{11}$$

Proceeding with the analysis, we find a corresponding result to the Reynolds equation

$$\nabla^2 P = \frac{12\mu}{h^3}\frac{dh}{dt} \tag{12}$$

$$\Rightarrow \nabla^2 p = \frac{12\mu}{h^3}\frac{dh}{dt} + 2\rho\Omega^2, \tag{13}$$

with boundary condition $p(a,t) = p_a - 2\gamma/h$. We determine the solution as

$$p(r,t) = \left(\frac{3\mu}{h^3}\frac{dh}{dt} + \frac{1}{2}\rho\Omega^2\right)\left(r^2 - a^2\right) + p_a - 2\gamma/h. \tag{14}$$

We balance forces on the upper plate to obtain:

$$F = 2\pi\int_0^a (p - p_a) r dr \tag{15}$$

and arrive at the nonlinear ODE

$$F = \frac{3\mu V^2}{8\pi}\frac{d}{dt}\left(\frac{1}{h^4}\right) - \left(\frac{\rho\Omega^2 V^2}{4\pi} + 2\gamma V\right)\frac{1}{h^2}. \tag{16}$$

5. The Squeezing Problem

In this section, we focus on the case $F > 0$. We use a change of variables to rewrite (16) as:

$$\frac{dX^2}{dT} = 1 + \lambda X, \;\; X(0) = 1, \tag{17}$$

$$\text{where} \begin{cases} X = h_0^2/h^2 = a^4/a_0^4 \\ T = t/t_0 \\ t_0 = 3\mu V^2/(8\pi F h_0^4) \\ \lambda = \frac{\rho\Omega^2 V^2}{4\pi F h_0^2} + \frac{2\gamma V}{F h_0^2} = \frac{\pi\rho\Omega^2 a_0^4}{4F} + \frac{2\pi^2 a_0^4 \gamma}{VF}. \end{cases} \tag{18}$$

This is very similar to the ODE found by Moffatt et al. (2021) [4] for the corresponding problem in the non-rotating case; however, the parameter λ has now been increased by a quantity proportional to Ω^2. We can solve the equation to determine the implicit solution

$$T = \frac{2}{\lambda^2}\left[\lambda(X-1) - \log\left(\frac{1+\lambda X}{1+\lambda}\right)\right]. \tag{19}$$

6. The Contraction Problem

We find a more interesting difference when we investigate the contraction problem, $F < 0$. We can, again, use a change of variables to rewrite the ODE as

$$\frac{dX^2}{dT} = -1 + |\lambda|X, \quad X(0) = 1, \tag{20}$$

$$\text{where} \begin{cases} X = h_0^2/h^2 = a^4/a_0^4 \\ T = t/|t_0| \\ t_0 = 3\mu V^2/(8\pi F h_0^4) \\ \lambda = \frac{\rho\Omega^2 V^2}{4\pi F h_0^2} + \frac{2\gamma V}{F h_0^2} = \frac{\pi\rho\Omega^2 a_0^4}{4F} + \frac{2\pi^2 a_0^4 \gamma}{VF}. \end{cases} \tag{21}$$

But we now note that the system can be forced to remain at $X = 1$ (i.e., $h = h_0, a = a_0$) if we choose Ω such that $\lambda = -1$. The critical value of Ω which achieves this is

$$\Omega_c = \left(\frac{4|F|}{\pi\rho a_0^4} - \frac{8\pi\gamma}{\rho V}\right)^{\frac{1}{2}}. \tag{22}$$

This indicates that, by spinning the plates at $\Omega = \Omega_c$, we can prevent the contraction force from pulling the plates apart. If $\Omega < \Omega_c$, then the plates pull apart; if $\Omega > \Omega_c$, then the plates are pushed together. We note that Ω_c does not necessarily exist: if $a_0 > a_c = \left(\frac{V|F|}{2\pi^2\gamma}\right)^{\frac{1}{4}}$, then $|\lambda| > 1$ and hence the plates are always pushed together.

7. Stability of Squeezing Problem

Other differences from the non-rotating case arise when we carry out a linear stability analysis. As noted in the introduction, the contraction problem, $F < 0$, is expected to develop complicated Saffman-Taylor instabilities at the interface between the air and the fluid. Therefore, we will focus on the squeezing problem, $F > 0$. We consider perturbing the boundary by:

$$\tilde{a}(\theta, t) = a(t) + \varepsilon\alpha(t)cos(n\theta), \tag{23}$$

$$\text{where } 0 < \varepsilon << 1 \text{ and } n \geq 2 \text{ is an integer,} \tag{24}$$

and we consider a perturbed pressure field given by

$$\tilde{p}(r, \theta, t) = p(r, t) + \varepsilon p_1(r, t)cos(n\theta). \tag{25}$$

As in the paper by Moffatt et al. (2021) [4], we find $\nabla^2[p_1 cos(n\theta)] = 0$ and hence $p_1(r, t) = k(t)r^n$. Thus, we have

$$\tilde{p}(r, \theta, t) = p(r, t) + \varepsilon k(t) r^n cos(n\theta). \tag{26}$$

We now find $k(t)$, using the condition $\tilde{p}(\tilde{a},\theta,t) = p_a - 2\gamma/h$,

$$k(t) = -a(t)\alpha(t)\frac{1}{a(t)^n}cos(n\theta)\left(\frac{6\mu}{h(t)^3} + \rho\Omega^2\right). \tag{27}$$

Hence:

$$\tilde{p}(r,\theta,t) = \left[r^2 - a^2 - 2\varepsilon a\alpha\left(\frac{r}{a}\right)^n cos(n\theta)\right]\left[\frac{3\mu}{h^3}\frac{dh}{dt} + \frac{1}{2}\rho\Omega^2\right] + p_a - 2\gamma/h \tag{28}$$

and

$$\frac{\partial\tilde{p}}{\partial r} = \left[r - n\varepsilon\alpha\left(\frac{r}{a}\right)^{n-1} cos(n\theta)\right]\left[\frac{6\mu}{h^3}\frac{dh}{dt} + \rho\Omega^2\right]. \tag{29}$$

We recall (11) and observe that $\tilde{u}(\tilde{a},t) = \frac{\partial\tilde{a}}{\partial t} = \frac{da}{dt} + \varepsilon\frac{d\alpha}{dt}cos(n\theta)$. Hence, evaluating $\frac{\partial p}{\partial r}$ at $r = \tilde{a}$, we obtain

$$\frac{da}{dt} + \varepsilon\frac{d\alpha}{dt}cos(n\theta) = -\frac{a}{2h}\frac{dh}{dt} + \varepsilon\alpha cos(n\theta)\left(\frac{1}{2h}\frac{dh}{dt}(n-1) + \frac{\rho\Omega^2h^2}{12\mu}n\right) + O\left(\varepsilon^2\right). \tag{30}$$

Therefore, as in Moffatt et al. (2021) [4], at leading order we simply see a statement of the conservation of $V = \pi a^2 h$. At order ε, we find

$$\frac{d\alpha}{dt} = \alpha\left[\frac{1}{2h}\frac{dh}{dt}(n-1) + \frac{\rho\Omega^2h^2}{12\mu}n\right]. \tag{31}$$

This equation governs the stability of the basic state. It is different to the corresponding equation in Moffatt et al. (2021) [4] due to the Ω term. Hence:

$$|\alpha| \text{ is decreasing (i.e., the n}^{th}\text{ mode is stable)} \Leftrightarrow \frac{1}{2h}\frac{dh}{dt}(n-1) + \frac{\rho\Omega^2h^2}{12\mu}n < 0. \tag{32}$$

We can rewrite this condition using X and T as

$$\text{n}^{th}\text{ mode is stable} \Leftrightarrow \frac{dX}{dT} > \frac{n}{n-1}\frac{\rho\Omega^2h_0^2t_0}{3\mu}. \tag{33}$$

Equation (17) tells us that $\frac{dX}{dT} = \frac{1}{2X} + \frac{\lambda}{2}$, where $\lambda = \frac{\rho\Omega^2V^2}{4\pi Fh_0^2} + \frac{2\gamma V}{Fh_0^2}$. We also recall that $t_0 = \frac{3\mu V^2}{8\pi Fh_0^4}$ and, hence , we can rewrite the stability condition as

$$\text{n}^{th}\text{ mode is stable} \Leftrightarrow h^2 > \frac{1}{n-1}\frac{\rho\Omega^2V^2}{4\pi F} - \frac{2\gamma V}{F}. \tag{34}$$

Therefore, we can see that the system is always stable if $\gamma \geq \frac{\rho\Omega^2 V}{8\pi}$; equivalently, the system is always stable if $\Omega \leq \Omega_s = \left(\frac{8\pi\gamma}{\rho V}\right)^{\frac{1}{2}}$. Otherwise, we can rewrite (34) in terms of the radius of the drop as:

$$\text{n}^{th}\text{ mode is stable} \Leftrightarrow \frac{V^2}{\pi^2a^4} > \frac{1}{n-1}\frac{\rho\Omega^2V^2}{4\pi F} - \frac{2\gamma V}{F}, \tag{35}$$

which we can rearrange as

$$\text{n}^{th}\text{ mode is stable} \Leftrightarrow a < \left(\frac{1}{n-1}\frac{\pi\rho\Omega^2}{4F} - \frac{2\pi^2\gamma}{VF}\right)^{-\frac{1}{4}}. \tag{36}$$

Hence, if we let $a_n = \left(\frac{1}{n-1}\frac{\pi\rho\Omega^2}{4F} - \frac{2\pi^2\gamma}{VF}\right)^{-\frac{1}{4}}$, then the nth mode is stable until the drop's radius reaches the critical value a_n.

These values are increasing as n increases so we observe that if $a \leq a_2$, then the drop is symmetrical and stable under any perturbation. Therefore, a_2 is the critical radius of the drop at which the system becomes unstable, which we can explicitly write as

$$a_2 = \left(\frac{\pi\rho\Omega^2}{4F} - \frac{2\pi^2\gamma}{VF}\right)^{-\frac{1}{4}} = \left(\frac{4VF}{\pi\rho\Omega^2 V - 8\pi^2\gamma}\right)^{\frac{1}{4}}. \tag{37}$$

The above results are consistent with the results found by Moffatt et al. (2021) [4] because in their analysis $\Omega = 0$ so the condition $\Omega \leq \Omega_s$ is satisfied and hence the motion is always stable. These results also incorporate surface tension which is crucial for calculating Ω_s but is less relevant in the non-rotating case.

We also observe that a_2 is decreasing as a function of Ω so the system becomes unstable at a smaller critical radius when it is rotated at a faster rate.

8. Instability of Contraction Problem

If we use the same linear perturbation model as in the investigation of the squeezing problem, then again we arrive at (32). There is a difference when we rewrite this equation in terms of X and T (because $T = t/|t_0| = -t/t_0$) so we find that

$$\text{the system is stable} \Leftrightarrow \frac{dX}{dT} > -\frac{n}{n-1}\frac{\rho\Omega^2 h_0^2 t_0}{3\mu}. \tag{38}$$

Equation (20) tells us that $\frac{dX}{dT} = -\frac{1}{2X} - \frac{\lambda}{2}$, where $\lambda = \frac{\rho\Omega^2 V^2}{4\pi F h_0^2} + \frac{2\gamma V}{F h_0^2}$. We also recall that $t_0 = \frac{3\mu V^2}{8\pi F h_0^4}$ and, hence, we can rewrite the stability condition as

$$\text{n}^{th} \text{ mode is stable} \Leftrightarrow h^2 < \frac{2\gamma V}{|F|} - \frac{1}{n-1}\frac{\rho\Omega^2 V^2}{4\pi|F|}. \tag{39}$$

Therefore, we can see that the system is always unstable if $\gamma \leq \frac{\rho\Omega^2 V}{8\pi}$; equivalently, the system is always unstable if: $\Omega \geq \Omega_s = \left(\frac{8\pi\gamma}{\rho V}\right)^{\frac{1}{2}}$. Otherwise, we can rewrite (39) in terms of the radius of the drop as:

$$\text{n}^{th} \text{ mode is stable} \Leftrightarrow \frac{V^2}{\pi^2 a^4} < \frac{2\gamma V}{|F|} - \frac{1}{n-1}\frac{\rho\Omega^2 V^2}{4\pi|F|}, \tag{40}$$

which we can rearrange as

$$\text{n}^{th} \text{ mode is stable} \Leftrightarrow a > \left(\frac{2\pi^2\gamma}{V|F|} - \frac{1}{n-1}\frac{\pi\rho\Omega^2}{4|F|}\right)^{-\frac{1}{4}}. \tag{41}$$

We recall $a_n = \left(\frac{1}{n-1}\frac{\pi\rho\Omega^2}{4F} - \frac{2\pi^2\gamma}{VF}\right)^{-\frac{1}{4}} = \left(\frac{2\pi^2\gamma}{V|F|} - \frac{1}{n-1}\frac{\pi\rho\Omega^2}{4|F|}\right)^{-\frac{1}{4}}$; thus the nth mode is unstable if $a \leq a_n$ but it is stable if $a > a_n$.

As before, a_n is increasing in n and we can calculate $\lim_{n\to\infty} a_n = \left(\frac{V|F|}{2\pi^2\gamma}\right)^{\frac{1}{4}} = a_c$ thus, if $\Omega < \Omega_s$ and $a \geq a_c$, then the system is stable. At first glance, this may appear to be inconsistent with Saffman and Taylor (1958) [6]. However, we recall that, if $a_0 \geq a_c$, then the plates are not pulled apart, hence no Saffman-Taylor instabilities develop. Indeed, there is an equivalence between the statements $a_0 \geq a_c$ and $a \geq a_c$ because $a_0 \geq a_c \Leftrightarrow a \geq a_0$ (as the plates are pushed together).

In the case of non-rotating plates, the necessary and sufficient condition for stability simply reduces to $a_0 \geq a_c$ which is equivalent to the plates not pulling apart.

9. Numerical Values

In this section we will use the following parameters to calculate the predicted critical values [these were the experimental values used by Moffatt et al. (2021) [4]].

$$\rho = 1.41 \times 10^3 \text{ kg m}^{-3}, \quad \mu = 64.3 \text{ kg s}^{-1}\text{m}^{-1}, \quad V = 5 \times 10^{-6} \text{ m}^3, \tag{42}$$

$$|F| = 11.04 \text{ kgms}^{-2}, \quad \gamma = 0.07 \text{ kg s}^{-2}, \quad a_0 = 30.72 \times 10^{-3} \text{ m}. \tag{43}$$

Therefore, we can calculate the critical values identified earlier in the paper.

$$\Omega_c = 105 \text{ s}^{-1} = 999 \text{ rpm}, \tag{44}$$

$$\Omega_s = 15.8 \text{ s}^{-1} = 151 \text{ rpm}, \tag{45}$$

$$a_c = 0.0795 \text{ m} = 79.5 \text{ mm} \tag{46}$$

These values give a useful insight into how plausible it would be to stabilise the motion of an expanding drop and to force two contracting plates together. Using these values we can also calculate a_2 for a range of values of Ω and accordingly construct phase diagram showing the different regions where the system is either stable (and will remain axisymmetric) or unstable (and will not) for the squeezing problem. The result of this is shown in Figure 1.

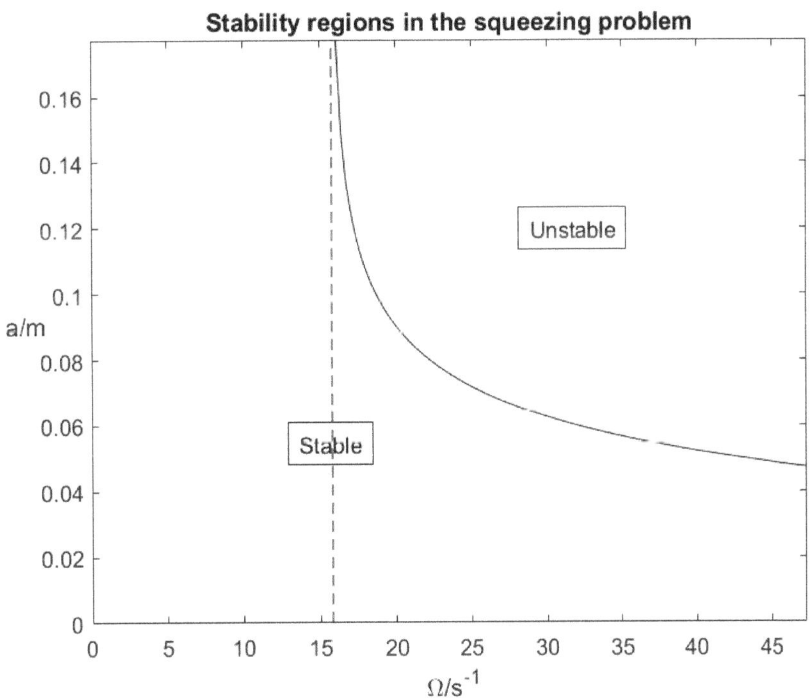

Figure 1. Regions of stability in the $\Omega - a$ phase plane under a squeezing force.

We can also plot the corresponding regions for the contraction problem, which is shown in Figure 2.

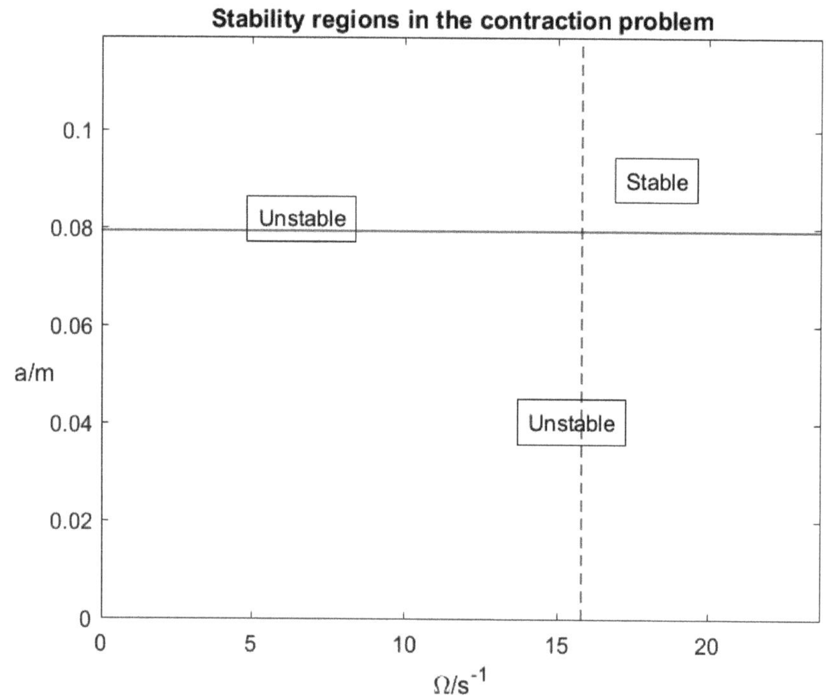

Figure 2. Regions of stability in the $\Omega - a$ phase plane under a contraction force.

10. Conclusions

We have identified the key equations for an axisymmetric viscous drop under compressing and contracting forces in a rotating system and the critical values at which the symmetry breaks down. In particular, we have shown that, under a squeezing force, the rotating viscous drop maintains its axisymmetry unless $\Omega \geq \Omega_s$ and the radius grows to be larger than a_2, where:

$$\Omega_s = \left(\frac{8\pi\gamma}{\rho V}\right)^{\frac{1}{2}}, \tag{47}$$

and:

$$a_2 = \left(\frac{4VF}{\pi\rho\Omega^2 V - 8\pi^2\gamma}\right)^{\frac{1}{4}}; \tag{48}$$

whereas, under a contraction force, symmetry breaks down if $\Omega < \Omega_s$ or if the drop starts at an initial radius less than

$$a_c = \left(\frac{V|F|}{2\pi^2\gamma}\right)^{\frac{1}{4}}. \tag{49}$$

We have also found that the critical rotation rate which prevents a contracting force from pulling apart two plates joined by a viscous drop is:

$$\Omega_c = \left(\frac{4|F|}{\pi\rho a_0^4} - \frac{8\pi\gamma}{\rho V}\right)^{\frac{1}{2}}, \tag{50}$$

which exists given that $a_0 \leq a_c$. Further research may be done in a laboratory to experimentally confirm these results.

Author Contributions: Formal analysis, writing—original draft preparation: M.G.E.R.; Conceptualization, supervision, writing—review and editing, correspondence: H.E.H. All authors have read and agreed to the published version of the manuscript.

Funding: This research received no external funding.

Institutional Review Board Statement: Not applicable.

Informed Consent Statement: Not applicable.

Data Availability Statement: The data used in the Numerical Values section can be found in Moffatt et al. (2021) [4]. DOI:10.1017/jfm.2021.668

Acknowledgments: Let us firstly extend our thanks to a very talented group of summer undergraduate students: A. Cox, M. Liu, M. Loncar, J. Saville and O. Wilson. It was a pleasure to work alongside them. We must also thank H.K. Moffatt for taking the time to read an earlier draft of this article and helping us to progress. Lastly, we would like to give our dearest thanks to King's College, Cambridge for providing us with a workspace and the Braithwaite Batty Fund for providing an accommodation subsidy.

Conflicts of Interest: The authors declare no conflict of interest.

References

1. Huppert, H.E. Gravity Currents: A personal perspective. *J. Fluid Mech.* **2006**, *554*, 299–322. [CrossRef]
2. von Kármán, T. The engineer grapples with nonlinear problems. *Bull. Am. Math. Soc.* **1940**, *46*, 615–683. [CrossRef]
3. Ward, T. Radial spreading of a viscous drop between parallel-plane surfaces. *Phys. Fluids* **2006**, *354*, 816–824. [CrossRef]
4. Moffatt, H.K.; Guest, H.; Huppert, H.E. Spreading or contraction of viscous drops between plates: Single, multiple or annular drops. *J. Fluid Mech.* **2021**, *925*, A26. [CrossRef]
5. Gay, C. Stickiness—Some fundamentals of adhesion. *Integr. Comp. Biol.* **2002**, *42*, 1123–1126. [CrossRef]
6. Saffman, P.G.; Taylor, G.I. The penetration of a fluid into a porous medium or Hele–Shaw cell containing a more viscous liquid. *Proc. R. Soc. Lond.* **1958**, *A245*, 312–329.
7. Dinte, E.; Sylvester, B. Adhesives: Applications and Recent Advances. In *Applied Adhesive Bonding in Science and Technology*; InTech Rijeka: Rijeka, Croatia, 2017
8. Her, S.-C. Stress analysis of adhesively-bonded lap joints. *Compos. Struct.* **1999**, *47*, 673–678. [CrossRef]
9. de Bruyne, N.A.; Houwink, R. *Adhesion and Adhesives*; Elsevier Publishing Co. Ltd.: Amsterdam, The Netherlands; Cleaver-Hume Press: London, UK, 1952.
10. Choi, J.H.; Lee, D.G. The Torque Transmission Capabilities of the Adhesively-Bonded Tubular Single Lap Joint and the Double Lap Joint. *J. Adhes.* **1994**, *44*, 197–212. [CrossRef]
11. Tritton, D.J. *Physical Fluid Dynamics*, 2nd ed.; Oxford Science Publications: Oxford, UK, 1988.
12. Ungarish, M.; Huppert, H.E. The effects of rotation on axisymmetric gravity currents. *J. Fluid Mech.* **1998**, *362*, 17–51. [CrossRef]

symmetry

MDPI

Article

The Stability of a Hydrodynamic Bravais Lattice

Miles M. P. Couchman [1], Davis J. Evans [2] and John W. M. Bush [2,*]

[1] Department of Applied Mathematics and Theoretical Physics, University of Cambridge, Cambridge CB3 0WA, UK; mc2277@cam.ac.uk
[2] Department of Mathematics, Massachusetts Institute of Technology, Cambridge, MA 02139, USA; djevans@mit.edu
* Correspondence: bush@math.mit.edu

Abstract: We present the results of a theoretical investigation of the stability and collective vibrations of a two-dimensional hydrodynamic lattice comprised of millimetric droplets bouncing on the surface of a vibrating liquid bath. We derive the linearized equations of motion describing the dynamics of a generic Bravais lattice, as encompasses all possible tilings of parallelograms in an infinite plane-filling array. Focusing on square and triangular lattice geometries, we demonstrate that for relatively low driving accelerations of the bath, only a subset of inter-drop spacings exist for which stable lattices may be achieved. The range of stable spacings is prescribed by the structure of the underlying wavefield. As the driving acceleration is increased progressively, the initially stationary lattices destabilize into coherent oscillatory motion. Our analysis yields both the instability threshold and the wavevector and polarization of the most unstable vibrational mode. The non-Markovian nature of the droplet dynamics renders the stability analysis of the hydrodynamic lattice more rich and subtle than that of its solid state counterpart.

Keywords: bouncing droplets; Faraday waves; lattice instability; normal-mode analysis; phonons

Citation: Couchman, M.M.P.; Evans, D.J.; Bush, J.W.M. The Stability of a Hydrodynamic Bravais Lattice. *Symmetry* **2022**, *14*, 1524. https://doi.org/10.3390/sym14081524

Academic Editor: Andrzej Herczyński

Received: 20 June 2022
Accepted: 21 July 2022
Published: 26 July 2022

1. Introduction

The collective vibrations of atoms within a crystal lattice, referred to as phonons, are of fundamental interest in materials science and solid-state physics, and govern bulk properties of the crystal such as the heat capacity, thermal and electrical conductivity, and elastic modulus [1–4]. Complementary insight into lattice dynamics is often gained through consideration of macroscopic analog systems, including mass-and-spring networks [5,6] and more complex metamaterials [7,8]. Numerous hydrodynamic analogues of crystals have been explored, from Bragg's packed bubbles at an air–liquid interface [9] to more recent studies of vibrations in lattices of colloidal particles [10–14] and microfluidic droplets [15–18]. We here consider theoretically the dynamics of a two-dimensional lattice comprised of millimetric droplets that bounce on the surface of a vibrating liquid bath and are coupled through an underlying Faraday wavefield.

A fluid bath vibrating vertically with acceleration $\gamma \sin(2\pi f t)$ will destabilize into a subharmonic field of standing Faraday waves, characterized by period $T_F = 2/f$ and wavelength λ_F (as is related to $\omega_F = \pi f$ through the standard water-wave dispersion relation), when the vibrational acceleration γ exceeds a critical value γ_F known as the Faraday threshold [19,20]. Below γ_F, but above the bouncing threshold $\gamma_B < \gamma_F$, a millimetric droplet may bounce indefinitely on the bath's surface, generating a spatially extended, temporally decaying wavefield at each impact [21,22]. As γ is gradually increased beyond γ_B, the droplet's bouncing period increases until eventually becoming twice that of the vibrational driving, thus achieving resonance with the Faraday wavefield and prompting a substantial increase of the system's wave energy. In this period-doubled bouncing regime, droplets may bounce in either low- or high-energy states, as distinguished by bouncing amplitude and denoted by $(2,1)^1$ and $(2,1)^2$, respectively [23]. Such period-doubled drops may bounce either in phase or out of phase with respect to each other. For $\gamma > \gamma_W$, the walking threshold, a single droplet destabilizes from stationary bouncing into steady horizontal motion, propelled through a resonant interaction with its own wavefield [24,25]. As γ approaches γ_F, the persistence time of the waves increases and the droplet's dynamics become more strongly influenced by its past trajectory, and the system has heightened 'path memory' [26,27]. The resulting non-Markovian nature of the droplet dynamics gives rise to numerous features reminiscent of quantum systems; consequently, this system has provided the basis for the burgeoning field of hydrodynamic quantum analogues [28–31].

Multiple droplets may organize into static and dynamic bound states by virtue of their shared wavefield. Specifically, the static bound states of droplet pairs [32], free rings [33], and radially confined rings (modelling a one-dimensional periodic lattice) [34] have inter-drop spacings related to the Faraday wavelength λ_F. As the vibrational acceleration is increased, droplet pairs destabilize into either in-line oscillations, or orbital or promenading (side by side) motion, depending on the droplet size and inter-drop distance [32]. Radially confined rings destabilize into out-of-phase angular oscillations or propagating soliton-like waves [34], and free rings exhibit additional radial vibrational modes [33]. The stability and resonant oscillations of forced chains with a free end have also been considered theoretically [35]. Pairs of identical walking drops may form dynamic bound states consisting of either orbital [24,25,36,37] or promenading motion [38,39].

A variety of two-dimensional bound structures have been studied, including the rotational [40] and translational instabilities of droplet aggregates [41]. Eddi et al. [42] constructed eight of the eleven possible Archimedean tilings of the plane, some of which required tuning the relative bouncing phase of neighbouring droplets. Eddi et al. [43] considered square and triangular lattices and observed the emergence of coherent modes of oscillation, a hydrodynamic analog of phonons, beyond a critical vibrational acceleration. Only one lattice spacing was considered for each geometry and a full characterization of the emergent oscillations as a function of the lattice spacing was not undertaken. Edge effects were seen to influence the observed lattice dynamics owing to their finite size. To rationalize their experimental observations, Eddi et al. [43] proposed a phenomenological, one-dimensional model in which each droplet was connected to its nearest neighbours via an effective spring force proportional to the wave amplitude. Notably, their model did not include an explicit waveform or the influence of the system memory, which precluded a quantitative characterization of the lattice stability.

A detailed linear stability analysis of an infinite one-dimensional droplet lattice was performed by Thomson et al. [44], using the stroboscopic model of Oza et al. [45], in order to rationalize the observed dynamics of a periodic droplet chain [34]. In this configuration, only certain lattice spacings were found to remain stable below the instability threshold of a single drop γ_W, and the lattice subsequently destabilized via either super- or sub-critical Hopf bifurcations as the driving acceleration was increased. This linear stability analysis was then extended to investigate both weakly nonlinear oscillations and solitary waves [46].

We here build upon the work of Eddi et al. [43] and Thomson et al. [44] by developing a theoretical framework for studying the stability and dynamics of two-dimensional droplet lattices, which one expects to exhibit a richer set of instabilities than their one-dimensional counterparts. Specifically, we consider the Bravais lattice, a theoretical construct used in solid-state physics to describe regular crystalline structures. The defining feature of the Bravais lattice is its discrete translational symmetry, which allows its lattice points to be expressed as integer multiples of two basis vectors. The Bravais lattice thus appears to be identical from each constituent lattice point [4]. In two dimensions, five possible geometries satisfy this required symmetry, specifically, square, triangular, rectangular, centred rectangular, and oblique lattices.

In Section 2, we use the theoretical model of Couchman et al. [32] to derive the dispersion relation governing the stability of a generic Bravais lattice. In Section 3, we focus on the square and triangular geometries, which admit analytical simplifications due to their rotational symmetry. For these geometries, we predict the stability threshold and mode of vibrational instability that emerge as the bath's vibrational acceleration is increased progressively. Our theoretical predictions are compared to the experimental results of Eddi et al. [43]. In Section 4, we summarize our results and propose future avenues of investigation.

2. Normal Mode Analysis

In this section, we characterize the linear stability of a Bravais droplet lattice. In Section 2.1, we review the variable-phase stroboscopic model of Couchman et al. [32] for the dynamics of multiple interacting droplets, which provides the basis of our analysis. Definitions of relevant variables and parameters are provided in Table 1. In Section 2.2, we then characterize the base state of the Bravais lattice and, in Section 2.3, perturb the base state to derive the dispersion relation governing the lattice's stability and normal modes of vibration.

Table 1. Definitions of relevant variables and parameters.

Symbol	Definition
Fluid	
ρ, σ, ν_e	Density, surface tension, effective kinematic viscosity [47]
$f, T_F = 2/f, \lambda_F, k_F = 2\pi/\lambda_F$	Bath driving frequency, Faraday period, wavelength, wavenumber
$\gamma, \gamma_W, \gamma^*, \gamma_F$	Peak driving acceleration of bath, walking threshold of single drop, lattice instability threshold, Faraday threshold
Trajectory equation	
$\boldsymbol{x} = (x, y), t, g$	Horizontal position, time, gravitational acceleration
$R, m = 4\pi\rho R^3/3$	Droplet radius, mass
$h(\boldsymbol{x}, t)$	Wave amplitude strobed at bouncing period T_F
$f(r)$	Wave kernel
$T_d = 1/(\nu_e k_F^2)$	Wave decay timescale [47]
$M_e = \frac{T_d}{T_F(1-\gamma/\gamma_F)}$	Memory parameter
$A = \sqrt{\frac{\nu_e T_F}{2\pi}} \frac{mgk_F^3}{3k_F^2\sigma + \rho g}$	Wave-amplitude coefficient
$\alpha = \frac{\epsilon^2}{2\nu_e(1+2\epsilon^2)}, \epsilon = \frac{2\pi f \rho \nu_e k_F}{3k_F^2\sigma + \rho g}$	Spatio-temporal damping coefficient, viscosity induced wavenumber correction [48]
$\zeta = \frac{2}{k_F}\sqrt{\frac{\alpha}{T_F M_e}}$	Non-dimensional spatial-damping coefficient
$D = 0.17mg\sqrt{\frac{\rho R}{\sigma}} + 6\pi R\mu_a, \mu_a$	Horizontal drag coefficient [47], air viscosity
$\kappa = \frac{m}{T_F D}$	Non-dimensional droplet mass
$\beta = \frac{mgAT_F k_F^2}{D}$	Non-dimensional wave-force coefficient
$\mathcal{S}, \mathcal{C}$	Impact phase parameters (see Appendix A) [32]
Lattice	
$\boldsymbol{a}, \boldsymbol{b}$	Basis vectors defining geometry of Bravais lattice
$\boldsymbol{d}_{mn} = m\boldsymbol{a} + n\boldsymbol{b}$	Horizontal position of droplets in base lattice, $(m, n) \in \mathbb{Z}$
$\boldsymbol{k}, \boldsymbol{\xi}, \omega_{\boldsymbol{k}}$	Wave vector, polarization vector, complex-valued frequency of vibrational mode

2.1. Review of Theoretical Model

Throughout this work, we choose physical parameters for the drop and bath liquids corresponding to silicone oil with kinematic viscosity $\nu = 20$ cSt, density $\rho = 949$ kg m^{-3}, and surface tension $\sigma = 20.6 \times 10^{-3}$ N m^{-1}, which have been widely used in experimental and theoretical studies of droplet–droplet interactions [32,33,37,39]. In a deep bath vibrating vertically at $f = 80$ Hz, such a fluid will be characterized by a Faraday wavelength $\lambda_F \approx 4.75$ mm and Faraday threshold $\gamma_F \approx 4.25g$, where g denotes the gravitational acceleration. We assume that all droplets have radius $R = 0.36$ mm and bounce in phase with each other at the Faraday period T_F in the higher-energy $(2,1)^2$ mode, as assumed in the prior analysis of free rings [33]. An example of a triangular droplet lattice constructed in the laboratory is shown in Figure 1a.

The trajectory equation for walking droplets was developed by Moláček and Bush [47]. The 'stroboscopic' trajectory equation of Oza et al. [45] is a simplification thereof, so called because it effectively eliminates the vertical droplet motion from consideration by averaging the drop dynamics over a bouncing period, and so describes the horizontal dynamics visible in the laboratory when the system is strobed at the Faraday period [49]. We here employ the variable-phase stroboscopic model of Couchman et al. [32], an extension of the stroboscopic model that accounts for variations in the phase of impact between the droplet and bath. Consideration of such variations has been found to be necessary in order to rationalize the observed stability of multi-drop systems [48], including bound droplet pairs [32] and rings [33].

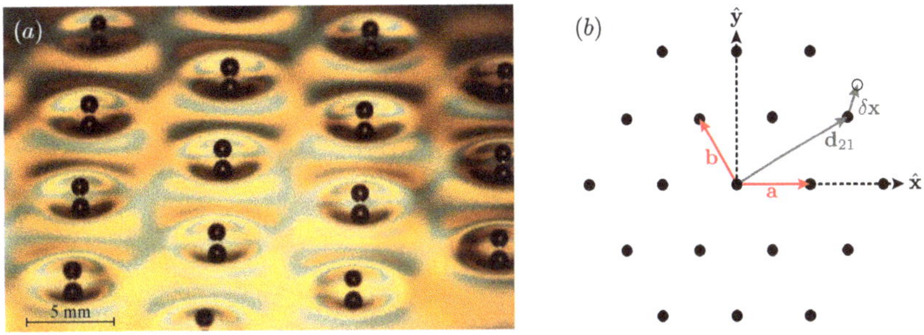

Figure 1. (**a**) An oblique view of a triangular lattice of millimetric droplets bouncing on the surface of a vertically vibrating liquid bath. The droplets are coupled by a shared wavefield, as is visualized by illuminating the bath with a striped pattern of coloured light [49]. We note that beneath each drop, the drop's reflection from the bath surface is visible. (**b**) A Bravais lattice is generated via enumeration of the points $\boldsymbol{d}_{mn} = m\boldsymbol{a} + n\boldsymbol{b}$, $(m, n) \in \mathbb{Z}$, with point $(m, n) = (2, 1)$ highlighted for illustration. The basis vectors describing the triangular lattice shown here are $\boldsymbol{a} = a(1, 0)$ and $\boldsymbol{b} = a(-1/2, \sqrt{3}/2)$, where a has units of length and sets the lattice spacing. We study the response of the lattice to perturbations $\delta\boldsymbol{x}$ away from its base state.

The variable-phase stroboscopic model predicts that the horizontal positions, $\boldsymbol{x}_{mn}$, of interacting drops of mass m in the lattice evolve according to

$$m\ddot{\boldsymbol{x}}_{mn} + D\dot{\boldsymbol{x}}_{mn} = -mg\mathcal{C}_{mn}\nabla h(\boldsymbol{x}_{mn}, t), \tag{1}$$

where overdots denote time derivatives, and the wavefield h is described by

$$h(\boldsymbol{x}, t) = A\sum_{pq}\int_{-\infty}^{t} \mathcal{S}_{pq} f(k_F|\boldsymbol{x} - \boldsymbol{x}_{pq}(s)|)e^{-(t-s)/(T_F M_e)}ds, \tag{2}$$

with wave kernel

$$f(r) = J_0(r)\left(1 + (K_1(\zeta r)\zeta r - 1)e^{-r^{-2}}\right), \tag{3}$$

where J_0 and K_1 denote Bessel functions of the first kind and modified second kind, respectively.

In Equation (1), drop motion is driven by a wave force, proportional to the local gradient of the underlying wavefield h, and resisted by a linear drag force with coefficient D. The wavefield, defined in Equation (2), is modelled as the superposition of waves of spatial form $f(r)$ generated by each droplet along its past trajectory, as summed over all of the droplets in the lattice. The wave kernel $f(r)$ is based on the wave model of Moláček and Bush [47] but more accurately captures the experimentally observed far-field decay of the wavefield [50], and so the long-range interactions between droplets [32,48,50,51]. For $r < 1$, $f(r)$ is well approximated by $J_0(r)$, but then decays more rapidly than $J_0(r)$ for $r > 1$, as prescribed by the spatial damping factor ζ which is defined in terms of fluid parameters in Table 1. The memory timescale $T_F M_e \sim (1 - \gamma/\gamma_F)^{-1}$ appearing in Equation (2) characterizes the temporal decay of the waves, with higher values of the memory parameter M_e indicating waves that decay more slowly and thus have a greater influence on the lattice's evolution.

The phase factors $\mathcal{S}$ and $\mathcal{C}$ capture the phase shift between the resonant oscillations of the drop and wavefield, which may vary as a function of the system parameters. Such a phase shift influences both the wave amplitude generated by each droplet at impact (as captured by $\mathcal{S}$) and the horizontal wave force imparted to the droplet (as captured by $\mathcal{C}$). Couchman et al. [32] determined the dependence of these phase factors on the droplet radius, local wave height, and vibrational acceleration. They also found that, for drops in the $(2,1)^2$ bouncing mode [52] commonly used in experiments and treated here, neglecting variations in these phase factors may lead to the prediction that bound states destabilize below the walking threshold of a single droplet γ_W, a prediction at odds with experimental observations of drop–drop interactions [32–34,37,39]. Further details concerning the phase parameters $\mathcal{S}$ and $\mathcal{C}$, and their explicit functional forms, may be found in Appendix A.

By introducing the following non-dimensional variables $\bar{x} = k_F x, \bar{h} = h/A, \bar{t} = t/T_F$ for horizontal position, wave height, and time, respectively, the governing Equations (1)–(2) take the following non-dimensional form:

$$\kappa \ddot{\bar{x}}_{mn} + \dot{\bar{x}}_{mn} = -\beta \mathcal{C}_{mn} \nabla \bar{h}(\bar{x}_{mn}, \bar{t}), \tag{4}$$

$$\bar{h}(\bar{x}, \bar{t}) = \sum_{pq} \int_{-\infty}^{\bar{t}} \mathcal{S}_{pq} f(|\bar{x} - \bar{x}_{pq}(\bar{s})|) e^{-(\bar{t}-\bar{s})/M_e} d\bar{s}. \tag{5}$$

Expressions for the non-dimensional mass κ and waveforce coefficient β are given in Table 1. For the remainder of the paper, we drop the overbars denoting non-dimensional variables for the sake of notational simplicity.

2.2. Base State of the Bravais Lattice

The unperturbed Bravais lattice has droplets located at horizontal positions

$$\boldsymbol{d}_{mn} = m\boldsymbol{a} + n\boldsymbol{b}, \tag{6}$$

where m and n are integers and $\boldsymbol{a}$ and $\boldsymbol{b}$ are the basis vectors that define the lattice geometry, commonly referred to as 'primitive vectors' (see Figure 1b). We first demonstrate that a generic Bravais lattice is a stationary solution to Equation (4). Substituting $\boldsymbol{x}_{mn}(t) = \boldsymbol{d}_{mn}$ into Equation (5) yields the wavefield for the Bravais lattice,

$$h(\boldsymbol{x}, t) = M_e \sum_{pq} \mathcal{S}(\gamma, h(\boldsymbol{d}_{pq})) f(|\boldsymbol{x} - \boldsymbol{d}_{pq}|). \tag{7}$$

We note that the lattice symmetry ensures that all drops encounter the same wave height $h(\boldsymbol{d}_{pq}) = h_0$, so their phase factors are likewise identical, $\mathcal{S}(h_0) = \mathcal{S}_0$. Evaluating Equation (7) at the lattice points yields an implicit expression for the local wave amplitude,

$$h_0 = M_e \mathcal{S}(\gamma, h_0) \sum_{pq} f(|\boldsymbol{d}_{pq}|). \tag{8}$$

Making use of the functional form for $\mathcal{S}$ presented in Equation (A2), one may solve the implicit Equation (8) numerically to obtain h_0. Having obtained h_0, the phase parameters $\mathcal{S}_0 = \mathcal{S}(\gamma, h_0)$ and $\mathcal{C}_0 = \mathcal{C}(\gamma, h_0)$ may be computed using Equations (A2) and (A3), respectively. In Section 3, it will be shown that the values $\mathcal{S}_0$ and $\mathcal{C}_0$ depend on the initial geometry of the Bravais lattice, which in turn strongly influence the lattice stability.

Having solved for the base state wavefield, it is immediately evident that $\boldsymbol{x}_{mn}(t) = \boldsymbol{d}_{mn}$ satisfies Equation (4) when $\nabla h(\boldsymbol{d}_{mn}, t) = 0$, which may be written explicitly as

$$\sum_{pq} f'(|\boldsymbol{d}_{pq}|) \frac{\boldsymbol{d}_{pq}}{|\boldsymbol{d}_{pq}|} = \boldsymbol{0}. \tag{9}$$

Physically, Equation (9) signifies that the net waveforce on each droplet must vanish, or equivalently that the slope of the wavefield beneath each droplet is zero. Noting that $\boldsymbol{d}_{(-p,-q)} = -\boldsymbol{d}_{(p,q)}$, the pairwise terms (p,q) and $(-p,-q)$ cancel in Equation (9), leaving only the $(p,q) = (0,0)$ term which vanishes because the wave kernel is even and so $f'(0) = 0$ (see Equation (3)). It is thus apparent that, by virtue of its translational symmetry, any Bravais lattice is a stationary solution of the trajectory equation.

2.3. Dispersion Relation of the Perturbed Lattice

We may now assess the stability of the Bravais lattice by studying the growth of small perturbations to the base state, $\boldsymbol{x}_{mn}(t) = \boldsymbol{d}_{mn} + \varepsilon\delta\boldsymbol{x}_{mn}(t)$, where $\varepsilon \ll 1$. Substituting this perturbation into Equation (4), using the equilibrium condition (9), and expanding to first order in ε, yields the following linearized equations of motion governing the perturbations $\delta\boldsymbol{x}_{mn}(t)$,

$$\kappa\delta\ddot{\boldsymbol{x}}_{mn} + \delta\dot{\boldsymbol{x}}_{mn} = -\varphi\sum_{pq}\int_{-\infty}^{t}\mathbf{H}_f(\boldsymbol{d}_{pq})\cdot(\delta\boldsymbol{x}_{mn}(t) - \delta\boldsymbol{x}_{pq}(s))e^{-(t-s)/M_e}ds, \quad (10)$$

where

$$\varphi = \beta\mathcal{S}_0\mathcal{C}_0, \quad (11)$$

and $\mathbf{H}_f$ denotes the Hessian matrix of the radially symmetric wave kernel f (Equation (3)),

$$\mathbf{H}_f(\boldsymbol{x}) = \begin{bmatrix} \frac{\partial^2 f}{\partial x^2} & \frac{\partial^2 f}{\partial x\partial y} \\ \frac{\partial^2 f}{\partial x\partial y} & \frac{\partial^2 f}{\partial y^2} \end{bmatrix}_{\boldsymbol{x}}. \quad (12)$$

The elements of $\mathbf{H}_f$ may be expressed explicitly as

$$\frac{\partial^2 f}{\partial x^2}(\boldsymbol{x}) = \frac{f''(r)}{r^2}x^2 + \frac{f'(r)}{r^3}y^2, \quad (13)$$

$$\frac{\partial^2 f}{\partial y^2}(\boldsymbol{x}) = \frac{f''(r)}{r^2}y^2 + \frac{f'(r)}{r^3}x^2, \quad (14)$$

$$\frac{\partial^2 f}{\partial x\partial y}(\boldsymbol{x}) = \left(\frac{f''(r)}{r^2} - \frac{f'(r)}{r^3}\right)xy, \quad (15)$$

where $\boldsymbol{x} = (x, y)$ and $r = |\boldsymbol{x}|$.

In deriving Equation (10), we note that variations in the phase parameters around the base values $\mathcal{S}_0$, $\mathcal{C}_0$ are of $\mathcal{O}(\varepsilon^2)$ and so can be neglected. To see this, note that

$$\mathcal{S}(\gamma, h(\boldsymbol{d}_{mn} + \epsilon\delta\boldsymbol{x}_{mn})) = \mathcal{S}\left(\gamma, h(\boldsymbol{d}_{mn}) + \varepsilon\nabla h(\boldsymbol{d}_{mn})\cdot\delta\boldsymbol{x}_{mn} + \mathcal{O}\left(\varepsilon^2\right)\right), \quad (16)$$

but $\nabla h(\boldsymbol{d}_{mn}) = 0$ in the base state (Equation (9)). Therefore, the impact phase parameters only enter our linear analysis through their base values $\mathcal{S}_0$, $\mathcal{C}_0$, which influence the resulting stability of the lattice through the parameter φ (Equation (11)) that scales the waveforce in Equation (10).

To derive the dispersion relation governing the linear stability of the lattice, we substitute the following normal mode perturbation into Equation (10)

$$\delta\boldsymbol{x}_{mn} = \boldsymbol{\xi}e^{i\boldsymbol{k}\cdot\boldsymbol{d}_{mn} + \omega_{\boldsymbol{k}}t} + \text{c.c.}, \quad (17)$$

where c.c. denotes the complex conjugate of the preceding exponential term. Equation (17) represents a plane wave characterized by wavevector $\boldsymbol{k}$, polarization vector $\boldsymbol{\xi}$, and complex frequency $\omega_{\boldsymbol{k}}$. The real and imaginary components of $\omega_{\boldsymbol{k}}$ represent the mode's growth rate and oscillation frequency, respectively. Making use of the following integral, $\int_{-\infty}^{t} e^{\omega s}e^{-(t-s)/M_e}ds = M_e e^{\omega t}/(1 + M_e\omega)$, we thus obtain the dispersion relation

$$\underbrace{\left[\left(\kappa\omega_{\boldsymbol{k}}^2 + \omega_{\boldsymbol{k}}\right)\mathbf{I} + \varphi M_e\left(\mathbf{H}^{(0)} - \frac{\mathbf{H}^{(\boldsymbol{k})}}{1 + M_e\omega_{\boldsymbol{k}}}\right)\right]}_{\mathbf{D}_{\boldsymbol{k}}(\omega_{\boldsymbol{k}})}\boldsymbol{\xi} = \mathbf{0}, \quad (18)$$

where $\mathbf{I}$ denotes the identity matrix and

$$\mathbf{H}^{(k)} = 2\sum_{pq} \mathbf{H}_f(\boldsymbol{d}_{pq}) \cos(\boldsymbol{k} \cdot \boldsymbol{d}_{pq}) \tag{19}$$

may be recognized as the elementwise discrete cosine transform of the Hessian matrix over the lattice. For $\boldsymbol{k} = \mathbf{0}$, $\mathbf{H}^{(0)}$ is the Hessian matrix of the net wavefield at the origin, as characterizes the wavefield's local curvature.

Non-trivial solutions ($\boldsymbol{\xi} \neq \mathbf{0}$) to the dispersion relation (18) exist when

$$\det \mathbf{D}_{\boldsymbol{k}}(\omega_{\boldsymbol{k}}) = 0, \tag{20}$$

which takes the form of a sextic polynomial in the complex-valued frequency $\omega_{\boldsymbol{k}}$ and may be solved numerically. If a wavevector $\boldsymbol{k}$ exists such that any root $\omega_{\boldsymbol{k}}^*$ of Equation (20) has a positive real part, then the associated mode will grow and the lattice will destabilize. If no such $\boldsymbol{k}$ exist, the lattice remains stable. For an initially stable lattice, as the vibrational forcing γ is increased, the instability threshold γ^* is reached when there exists at least one $\boldsymbol{k}^*$ such that $\mathrm{Re}(\omega_{\boldsymbol{k}}^*) \geq 0$. The polarization vector $\boldsymbol{\xi}^*$ associated with the unstable mode $\boldsymbol{k}^*$ may then be found by solving $\mathbf{D}_{\boldsymbol{k}}(\omega_{\boldsymbol{k}}^*)\boldsymbol{\xi}^* = \mathbf{0}$ (Equation (18)). We note that the polarization vector $\boldsymbol{\xi}$ must be real-valued as $\mathbf{D}_{\boldsymbol{k}}$ is a symmetric matrix.

It is noteworthy that, owing to the discrete translational symmetry of the Bravais lattice, not all wavevectors $\boldsymbol{k}$ produce physically distinguishable oscillations when substituted into the normal mode ansatz (17). For example, consider a square lattice characterized by $\boldsymbol{d}_{mn} = a(m\hat{\boldsymbol{x}} + n\hat{\boldsymbol{y}})$ and two wavevectors $\boldsymbol{k} = (k_x, k_y)$ and $\boldsymbol{k}' = (k_x + 2\pi/a, k_y + 2\pi/a)$. Substituting $\boldsymbol{k}'$ into Equation (17) yields

$$\delta\boldsymbol{x}_{mn} = \boldsymbol{\xi} e^{2\pi i(m+n)} e^{i\boldsymbol{k}\cdot\boldsymbol{d}_{mn} + \omega_{\boldsymbol{k}} t} = \boldsymbol{\xi} e^{i\boldsymbol{k}\cdot\boldsymbol{d}_{mn} + \omega_{\boldsymbol{k}} t}, \tag{21}$$

an oscillation characterized by wavevector $\boldsymbol{k}$. Therefore, $\boldsymbol{k}'$ and $\boldsymbol{k}$ result in the same physical oscillation. It thus suffices to consider only wavevectors $\boldsymbol{k}$ in the lattice's so-called Brillouin zone, defined as the smallest set of $\boldsymbol{k}$ required to describe all distinguishable vibrations of the discrete lattice [4]. The Brillouin zones for the square and triangular geometries considered in Section 3 are illustrated in Figure 2, and the procedure for generating the Brillouin zone for a generic Bravais lattice may be found in standard reference texts on solid-state physics [4].

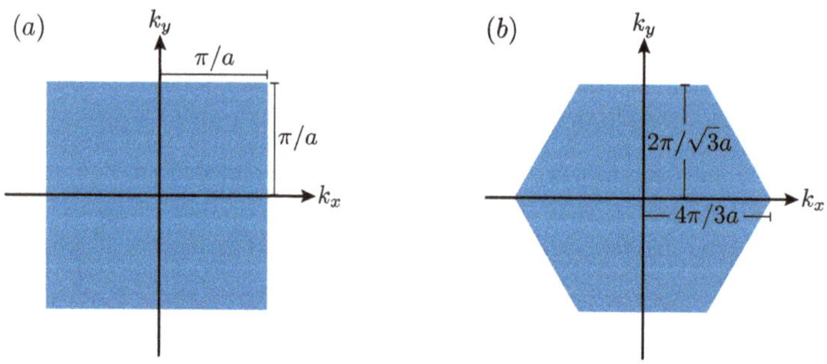

Figure 2. The extent of the Brillouin zone in wavevector space (k_x, k_y) for (**a**) square and (**b**) triangular Bravais lattices with lattice spacing a [4]. Wavevectors $\boldsymbol{k}'$ outside the Brillouin zone are paired with an equivalent $\boldsymbol{k}$ within the Brillouin zone such that $\boldsymbol{k}'$ and $\boldsymbol{k}$ yield the same physical vibration once substituted into the normal mode ansatz (17), a result of the discrete translational symmetry of the lattice. In characterizing lattice stability, it thus suffices to consider only modes with wavevectors $\boldsymbol{k}$ within the Brillouin zone.

3. Results

3.1. Square and Triangular Lattices

We first consider the stability of square and triangular lattices, as are characterized by a single inter-drop spacing a with droplets at the base state positions

$$\boldsymbol{d}_{mn}^{(\mathrm{square})} = a(m\hat{\boldsymbol{x}} + n\hat{\boldsymbol{y}}), \tag{22}$$

$$\boldsymbol{d}_{mn}^{(\mathrm{triangle})} = a\left(m\hat{\boldsymbol{x}} + (n/2)\left(-\hat{\boldsymbol{x}} + \sqrt{3}\hat{\boldsymbol{y}}\right)\right), \tag{23}$$

respectively. As is demonstrated in Appendix B, the four- and six-fold symmetry of the square and triangular geometries, respectively, result in the Hessian matrix for the net wavefield $\mathbf{H}^{(0)}$ (Equation (19), with $\boldsymbol{k} = \mathbf{0}$) reducing to a scalar multiple of the identity matrix $\mathbf{I}$,

$$\mathbf{H}^{(0)} = \eta_0 \mathbf{I}, \tag{24}$$

where η_0 is the unique (repeated) eigenvalue of $\mathbf{H}^{(0)}$. Geometrically, Equation (24) signifies that the curvature of the local wavefield beneath each droplet in the square and triangular base lattices is isotropic, having the same value in all horizontal directions. Using Equation (24), the general dispersion relation (18) reduces to the simpler eigenvalue problem,

$$\mathbf{H}^{(k)}\boldsymbol{\xi} = \left[\frac{1}{\varphi M_e}(1 + M_e\omega_k)\left(\kappa\omega_k^2 + \omega_k + \varphi\eta_0 M_e\right)\right]\boldsymbol{\xi}, \tag{25}$$

which admits non-trivial solutions when $\boldsymbol{\xi}$ is an eigenvector of $\mathbf{H}^{(k)}$ with corresponding eigenvalue η_k. When such is the case, one may equate η_k with the bracketed term on the right-hand side of Equation (25), yielding

$$\mathcal{P}_1(\omega_k) \equiv \kappa\omega_k^3 + \left(1 + \frac{\kappa}{M_e}\right)\omega_k^2 + \left(\frac{1}{M_e} + M_e\varphi\eta_0\right)\omega_k + \varphi(\eta_0 - \eta_k) = 0, \tag{26}$$

which may be solved for the three ω_k associated with each of the two eigenvalues η_k. Since $\kappa > 0$ and $\mathcal{P}_1(\omega_k) \to \infty$ as $\omega_k \to \infty$, the intermediate value theorem guarantees a positive real root whenever $\mathcal{P}_1(0) < 0$, i.e., $\varphi(\eta_0 - \eta_k) < 0$. We thus deduce that a lattice is always unstable if there are any wavevectors $\boldsymbol{k}$ such that either eigenvalue η_k of $\mathbf{H}^{(k)}$ is greater than η_0.

Based on Equation (26), we now have a simple procedure for assessing whether a square or triangular lattice is unstable at the initial vibrational acceleration $\gamma = 0.7\gamma_F$, which is below the walking threshold of a single drop γ_W and corresponds approximately to the lowest γ at which the droplets bounce in a period-doubled mode [52]. Specifically, we compute the eigenvalues η_0 and η_k of $\mathbf{H}^{(k)}$ (Equation (19)) numerically, noting that there are two η_k for each wave vector $\boldsymbol{k}$ in the Brillouin zone, and plot their dependence on the lattice spacing a for both the square (Figure 3a) and triangular (Figure 4a) lattice geometries. Intervals of a where $\eta_k > \eta_0$ for all $\boldsymbol{k}$ are shaded in red, indicating that the lattice is already unstable at $\gamma = 0.7\gamma_F$. We observe that for both the square and triangular geometries, there are only discrete intervals of a for which the lattice is initially stable, which roughly correspond to geometries in which drops bounce in minima of the local wave amplitude (see Figures 3e and 4e). We note that this behaviour is in accord with previous studies reporting that bound states are most stable when each drop bounces in a minimum of the net wavefield produced by its neighbours [32,33]. Thomson et al. [44] found similar swaths of initial instability as a function of the lattice spacing for infinite one-dimensional droplet chains at low vibrational accelerations, which they referred to as 'geometric instabilities'.

For the set of lattices that are initially stable at $\gamma = 0.7\gamma_F$, we may then determine the instability threshold $\gamma^* > 0.7\gamma_F$ at which destabilization occurs. At γ^*, the real part of ω_k will vanish and the imaginary part will correspond to the frequency of the destabilizing oscillation. Substituting $\omega_{\boldsymbol{k}} = i\Omega$ ($\Omega \in \mathbb{R}$) into Equation (26), separating the complex polynomial into real and imaginary parts, and eliminating Ω yields

$$\mathcal{P}_{\boldsymbol{k}}(M_e) \equiv \varphi\eta_0 M_e^3 + \varphi\kappa\eta_{\boldsymbol{k}} M_e^2 + M_e + \kappa = 0. \tag{27}$$

Noting that $\kappa > 0$, $\varphi > 0$ and $\eta_0 < 0$, we observe that $\mathcal{P}_{\boldsymbol{k}}(M_e) = 0$ has at least one positive real root, the minimum of which corresponds to the memory M_e^* at which the lattice destabilizes. Because $\mathcal{Q}_{\boldsymbol{k}}(M_e) \equiv \mathcal{P}_{\boldsymbol{k}}(M_e) - CM_e^2 < \mathcal{P}_{\boldsymbol{k}}(M_e)$ for every non-zero M_e provided $C > 0$, we infer that $\mathcal{Q}_{\boldsymbol{k}}$ must have a root at a lower M_e than does $\mathcal{P}_{\boldsymbol{k}}$. Furthermore, if there is a wavevector $\boldsymbol{k}'$ such that $\eta_{\boldsymbol{k}'} < \eta_{\boldsymbol{k}}$, then since $\mathcal{P}_{\boldsymbol{k}}(M_e) - \mathcal{P}_{\boldsymbol{k}'}(M_e) = \varphi\kappa M_e^2(\eta_{\boldsymbol{k}} - \eta_{\boldsymbol{k}'}) > 0$, the smallest real positive root of $\mathcal{P}_{\boldsymbol{k}'}(M_e)$ must be smaller than that of $\mathcal{P}_{\boldsymbol{k}}(M_e)$. We thus conclude that the wave vector $\boldsymbol{k}$ that goes unstable at the lowest memory value M_e, corresponds to that $\boldsymbol{k}$ with the minimum eigenvalue $\eta_{\boldsymbol{k}}$.

We may thus gain additional insight from the curves $\eta_{\boldsymbol{k}}$ plotted in Figures 3a and 4a for the square and triangular geometries, respectively. Namely, in the initially stable intervals of a, the lowermost curve $\eta_{\boldsymbol{k}}$ corresponds to the destabilizing vibrational wavemode $\boldsymbol{k}$ that will emerge at the instability threshold γ^*. In Figures 3 and 4, we plot the instability thresholds γ^* (panel b), and the magnitude and direction of the wavevector characterizing this destabilizing mode (panels c and d, respectively) as a function of the lattice spacing a. In all cases, we find that the polarization vector $\boldsymbol{\xi}$ is parallel to the wave vector $\boldsymbol{k}$, signifying that the lattices always destabilize into longitudinal, as opposed to transverse, oscillations.

For both the square and triangular geometries, the instability threshold of the lattice, γ^*, is almost always greater than the walking threshold γ_W of a single drop. The lattices are most stable when the constituent drops bounce in the deepest minima of the wavefield produced by their neighbours, as was the case for droplet pairs and rings [32,33]. This effect is a direct result of variations in the impact phase, with the product $\mathcal{S}_0\mathcal{C}_0$ (see Equations (10) and (11)) decreasing with decreasing local wave amplitude (see Figures 3e and 4e) and thus reducing the horizontal waveforce exerted on each droplet at impact. Conversely, we note that in their theoretical analysis of a one-dimensional lattice, Thomson et al. [44] predicted destabilization below γ_W, despite the fact that experimentally such lattices were found to remain stable above γ_W [34]. This mismatch followed from their assumption of a constant impact phase, and highlights the importance of accounting for variations in the vertical dynamics by using a variable-impact-phase model when considering droplet–droplet interactions. In Figures 3b and 4b we observe that γ^* approaches γ_W in the limit of large a, consistent with the droplets in the lattice becoming effectively uncoupled from their neighbours at sufficiently large distances.

In Figure 3c,d, we highlight two dominant modes of vibration for the square lattice, as are further illustrated in Figure 5. The first mode (green) corresponds to out-of-phase oscillations of neighbouring lattice planes along either the $\hat{x}$ or $\hat{y}$ directions, which are equivalent given the four-fold rotational symmetry of the square lattice. We note that Eddi et al. [43] observed a superposition of such oscillations in the $\hat{x}$ and $\hat{y}$ directions, resulting in the appearance of each droplet exhibiting roughly circular orbits around its base point (see Figure 4a of [43]). In magenta, we highlight a separate mode characterized by out-of-phase oscillations along 45 degree planes in the lattice with a wavelength of $\sqrt{2}a$, corresponding to the diagonal of a square cell. Apart from these two readily identifiable modes, the scaled wavevector $a|\boldsymbol{k}|/\pi$ (Figure 3c) varies approximately linearly with the scaled lattice spacing a/λ_F, thus representing an oscillation wavelength that is no longer solely governed by a, but is now also influenced by the Faraday wavelength λ_F that characterizes the underlying wavefield.

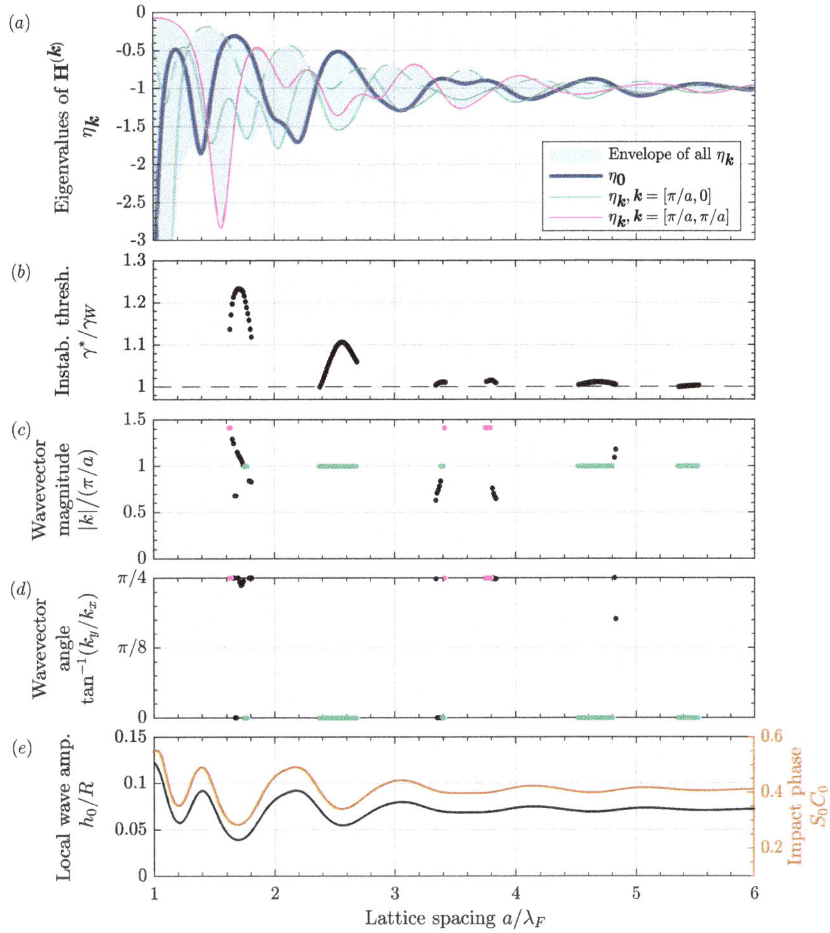

Figure 3. The square lattice. (**a**) The eigenvalues η_k of the matrix $\mathbf{H}^{(k)}$ (19), governing the lattice stability, are plotted as a function of the inter-drop spacing a, normalized by the Faraday wavelength λ_F. In intervals of a where η_0 is greater than all η_k, the lattice is initially stable at $\gamma/\gamma_F = 0.7$. Otherwise, the lattice is already unstable as denoted by red shading. Curves η_k for two common modes of instability are highlighted. There are two eigenvalues η_k for a single mode $\boldsymbol{k}$, denoted by dashed and solid lines of the same colour. We note that the two magenta curves are virtually indistinguishable. For a given a, the most negative η_k corresponds to the destabilizing wave mode. (**b**) The instability threshold γ^*, normalized by the walking threshold for a single drop γ_W, for initially stable spacings a. The corresponding wavevector magnitude $|\boldsymbol{k}|$, normalized by π/a (see Figure 2a), and wave angle with respect to the x-axis, are shown for the destabilizing mode in panels (**c**,**d**), respectively. Green and magenta points indicate spacings a where one of the two wave modes highlighted in panel (**a**) are found to be destabilizing. These modes are further illustrated in Figure 5. (**e**) The black curve indicates the dependence on lattice spacing of the local wave amplitude h_0 beneath each droplet in the base lattice, normalized by the drop radius R. The orange curve denotes the corresponding product of impact phases $\mathcal{S}_0\mathcal{C}_0$ that influences the lattice stability through φ in Equation (10).

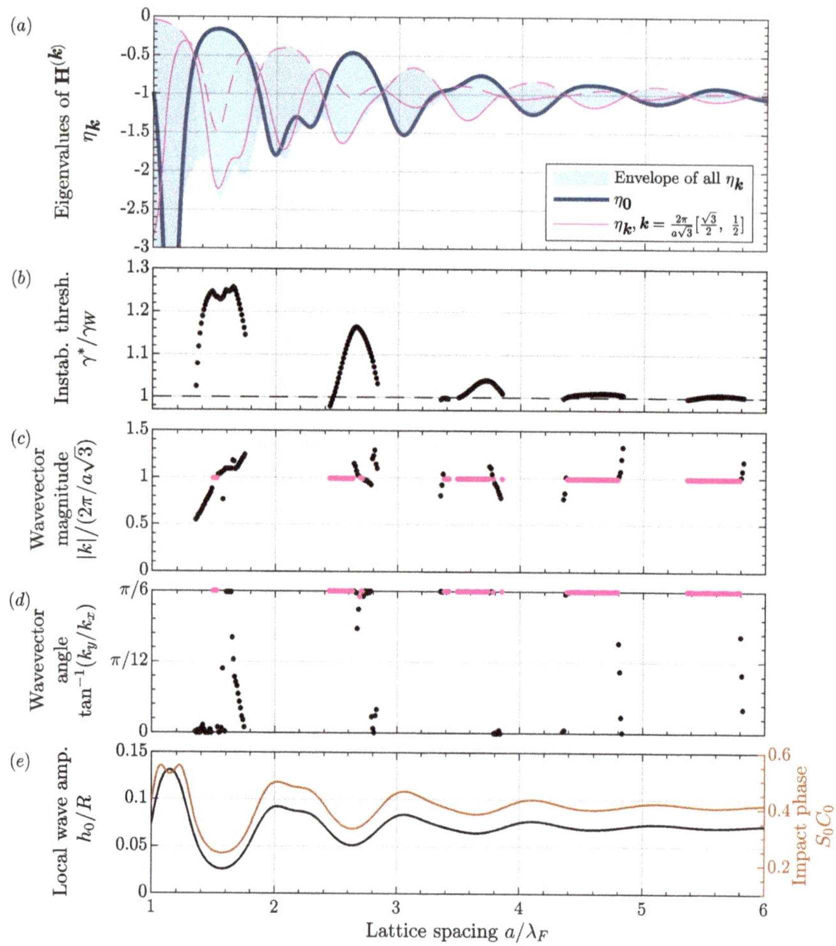

Figure 4. The triangular lattice. Quantities characterizing the stability of a triangular lattice are plotted as a function of the lattice spacing a, in the same manner as for the square lattice in Figure 3. The dominant mode of instability is highlighted in magenta in panels (**c**,**d**), and is illustrated in Figure 6.

In Figure 4c,d, we highlight the dominant mode of vibration for the triangular lattice (magenta) which, as illustrated in Figure 6, corresponds to out-of-phase oscillations of neighbouring lattice planes along the $\pi/6$ direction (or equivalently, the $\pi/6 + n(\pi/3)$, $n = (1, 2, \ldots, 5)$ directions given the six-fold rotational symmetry of the lattice). This mode is consistent with that reported by Eddi et al. [43] (see Figure 4b of [43]). We note that in their experiments, Eddi et al. [43] used different drop sizes and fluid parameters than considered here, so we cannot present a quantitative comparison between their results and our theoretical predictions.

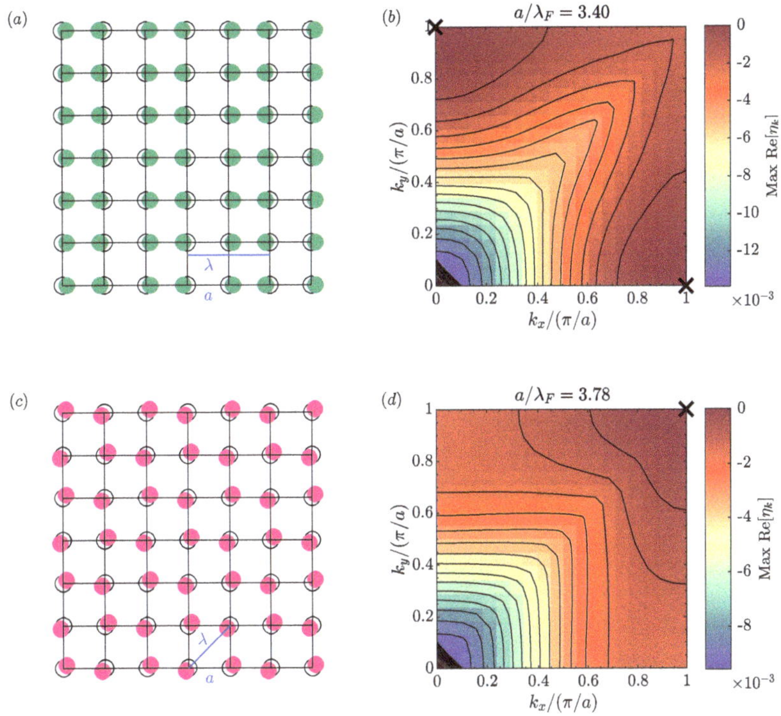

Figure 5. Common vibrational modes of the square lattice corresponding to longitudinal waves along high-symmetry lattice directions. (**a**) An illustration of the $\boldsymbol{k} = (\pi/a, 0)$ wave mode highlighted by green markers in Figure 3, corresponding to the out-of-phase oscillations of neighbouring planes along the $\hat{x}$ direction with wavelength $\lambda = 2a$. (**b**) The maximum real part of the eigenvalues η_k are shown for a lattice spacing $a/\lambda_F = 3.40$, with the most unstable modes $\boldsymbol{k} = (\pi/a, 0)$ and $(0, \pi/a)$ marked by crosses. (**c**,**d**) Analogous plots for the wave mode $\boldsymbol{k} = (\pi/a, \pi/a)$, highlighted by magenta markers in Figure 3, corresponding to the out-of-phase oscillations of neighbouring planes along the 45 degree diagonal, with wavelength $\lambda = \sqrt{2}a$.

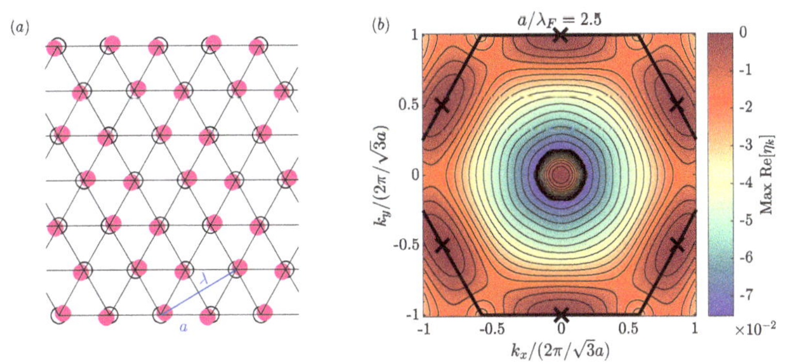

Figure 6. Common vibrational mode of the triangular lattice. (**a**) An illustration of the $\boldsymbol{k} = \frac{2\pi}{\sqrt{3}a}(\frac{\sqrt{3}}{2}, \frac{1}{2})$ wave mode highlighted by magenta markers in Figure 4, corresponding to the out-of-phase oscillations of neighbouring planes along a line 30 degrees to the horizontal, with wavelength $\lambda = \sqrt{3}a$. (**b**) The maximum real part of the eigenvalues η_k for the lattice spacing $a/\lambda_F = 2.5$. The Brillouin zone boundary is traced in black (see Figure 2b), and crosses indicate the most unstable modes: $\boldsymbol{k} = \frac{2\pi}{\sqrt{3}a}(\frac{\sqrt{3}}{2}, \frac{1}{2})$ and $\pi/3$ rotations thereof.

3.2. Geometric Instabilities in the Low-Memory Limit

We have demonstrated that, for the relatively low driving acceleration $\gamma = 0.7\gamma_F$, square and triangular lattices are already unstable for certain values of the lattice spacing a. Following Thomson et al. [44], we refer to such lattices as being 'geometrically unstable'. We proceed by demonstrating a method for predicting whether a generic Bravais lattice is geometrically unstable on the basis of the shape of the lattice-induced wavefield. In the low-memory limit $M_e \ll 1$, the dispersion relation (18) reduces to the eigenvalue problem

$$\left(\mathbf{H}^{(0)} - \mathbf{H}^{(\boldsymbol{k})}\right)\boldsymbol{\xi} = \left[-\frac{1}{\varphi M_e}\left(\kappa\omega_{\boldsymbol{k}}^2 + \omega_{\boldsymbol{k}}\right)\right]\boldsymbol{\xi}, \tag{28}$$

which admits non-trivial solutions when $\boldsymbol{\xi}$ is an eigenvector of the matrix $\left(\mathbf{H}^{(0)} - \mathbf{H}^{(\boldsymbol{k})}\right)$ with corresponding eigenvalue $\alpha_{\boldsymbol{k}}$. When such is the case, one may equate $\alpha_{\boldsymbol{k}}$ with the bracketed term on the right-hand side of Equation (28), yielding

$$\mathcal{P}_2(\omega_{\boldsymbol{k}}) \equiv \kappa\omega_{\boldsymbol{k}}^2 + \omega_{\boldsymbol{k}} + \varphi M_e \alpha_{\boldsymbol{k}} = 0, \tag{29}$$

which may be solved to obtain the two $\omega_{\boldsymbol{k}}$ associated with each of the two eigenvalues $\alpha_{\boldsymbol{k}}$. We note that $\mathcal{P}_2(\omega_{\boldsymbol{k}}) \to \infty$ as $\omega_{\boldsymbol{k}} \to \infty$. In the case $\alpha_{\boldsymbol{k}} < 0$, we also have $\mathcal{P}_2(0) < 0$, and so the intermediate value theorem guarantees the existence of a positive real root. Therefore, $\alpha_{\boldsymbol{k}} < 0$ is a sufficient condition for the geometric instability of a generic Bravais lattice; while we have derived this result in the low-memory limit, one expects that increasing the memory will tend to promote lattice destabilization. Thus, a lattice that is geometrically unstable in the low-memory limit should remain so at higher memories.

In Figure 7, we use the above criterion to determine the regions of initial stability of a rectangular lattice, as is described by two lattice spacings a and b. We note that the diagonal line $a = b$ in Figure 7 yields the intervals of stability shown in Figure 3b. As was the case for the square and triangular geometries, the pockets of stability are correlated with minima in the local wave amplitude. Having identified the regions of initial stability, the dispersion relation (18) could then be used to characterize the instability threshold and most unstable vibrational mode for each lattice.

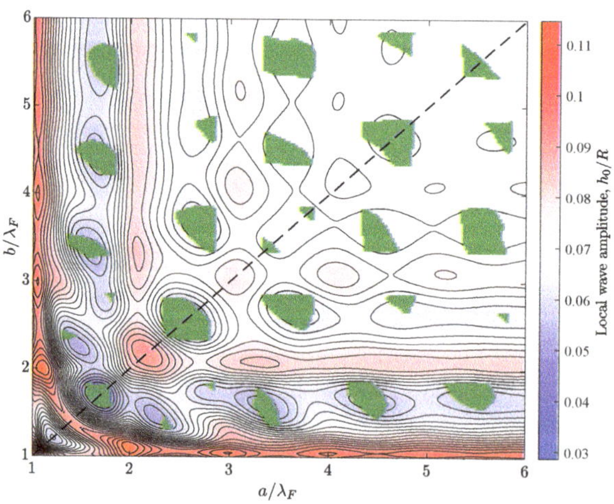

Figure 7. The rectangular lattice is parameterised by two lattice spacings, a and b. Regions of initial stability at low memory are shaded in green. All other lattices are geometrically unstable. The colourmap corresponds to the local wave amplitude beneath each drop h_0, normalized by the drop radius R. Stable regions are roughly correlated with minima in h_0. The diagonal line $a = b$ corresponds to the square lattice, where the regions of initial stability correspond to those presented in Figure 3.

4. Discussion

We have used the variable-phase stroboscopic model of Couchman et al. [32] to consider the linear stability of a group of droplets arranged in a Bravais lattice. All Bravais lattices are stationary solutions to the stroboscopic model, since the lattice's discrete translational symmetry ensures that the net waveforce on each droplet vanishes. By considering the response of the lattice to normal mode perturbations, our analysis yielded an implicit dispersion relation relating the vibrational mode's wavevector $\boldsymbol{k}$ to its complex frequency ω_k, as captures both the mode's growth rate and oscillation frequency. Particular attention was given to the square and triangular Bravais geometries, for which the rotational symmetry of the lattice allowed our general dispersion relation to be reduced to a form similar to that considered by Thomson et al. [44] in their investigation of a one-dimensional droplet chain. A distinctive feature of our analysis is the inclusion of variations in the drop's vertical motion, as are required to capture the stabilizing influence of droplet–droplet interactions that allow lattices to remain stable above the walking threshold of the individual constituent drops. This stabilizing phenomenon has also been reported for a variety of other bound states [32–34,37,39].

While all lattices are stationary solutions to the trajectory equation, not all are stable at low vibrational accelerations; indeed, the majority are geometrically unstable [44]. For the square and triangular lattices, we deduced a criterion for geometric instability solely in terms of the local wave curvature beneath each droplet. This criterion predicts discrete intervals of stability in the lattice spacing. These stable regions correspond roughly to geometries that minimize the local wave amplitude beneath each droplet, a feature also reported in bound droplet pairs [32], rings [33], and one-dimensional lattices [44]. By considering the low-memory limit of our dispersion relation, we deduced a similar criterion for assessing the geometric instability of a generic Bravais lattice, which we used to identify the initial regions of stability for the rectangular lattice.

Consistent with the experimental observations reported by Eddi et al. [43], increasing the memory causes an initially stable lattice to destabilize into phonon-like motions characterized by coherent small-amplitude oscillations. The most unstable modes were longitudinal waves in all cases and were usually aligned along high-symmetry directions of the lattice; shear modes only destabilize at higher vibrational forcings. We numerically computed the instability threshold of the most unstable modes, and found that they typically arise above the walking threshold γ_W for a single droplet, as highlights the stability imparted by neighbouring droplets. Furthermore, we found local maxima for the instability threshold at the centre of the initially stable regions, corresponding to minima in the local wave amplitude, a result not captured in the theoretical modelling of the one-dimensional lattice [44] where variable phase factors were neglected.

The hydrodynamic lattice exhibits certain features that are distinct from canonical models of phonons in crystalline lattices, such as those analysed by Blackman [53] and Montroll [54]. In these models, a discrete lattice of masses is connected to their nearest and next-nearest neighbours via a linear spring force. If we were to remove the damping term from our dispersion relation (18), and take the low-memory limit considered in Section 3.2, we would recover a dispersion relation of the same form as that of Blackman and Montroll, for which all modes are neutrally stable. For droplet lattices accessible in the laboratory, the damping term in the trajectory Equation (1) dominates the inertia term in the low-memory limit, resulting in overdamped oscillations. In our system, neutrally stable oscillations only arise when the stabilizing effects of damping precisely balance the destabilizing effects of memory.

In the low-memory limit, the potential landscape is a sum of wave kernels centred at each droplet of the lattice. This potential induces a linear spring force with the local curvature playing the role of the spring constant. Since the wave kernel is oscillatory, an individual droplet may contribute a negative curvature (analogously, a negative spring constant) to this sum, depending on the inter-droplet distance. This effect is not seen in a canonical mass-spring lattice, as the springs are only perturbed about their equilibrium lengths. The oscillatory wave kernel of the hydrodynamic lattice thus gives rise to much richer dynamics than arise in generic crystal lattices, including the discrete windows of initial stability.

Our linear theory for the hydrodynamic lattice cannot predict the amplitude of the emergent oscillations. A weakly nonlinear extension of our theory, which parallels that developed for a one-dimensional hydrodynamic lattice [46], could yield insight into the more complex oscillations reported by Eddi et al. [43]. The potential for further theoretical explorations abound. For example, one might consider how the introduction of defects, such as holes or an additional droplet, might modify the resulting lattice dynamics [55].

In solid-state physics, more complex lattice geometries arise, for example, in ionic lattices comprised of more than one type of atom. In such cases, the 'unit cell' constitutes the smallest non-repeating subcomponent of the lattice, and the structure may be defined in terms of a Bravais lattice of such unit cells. As is well established in the phonon literature, having more than one element per unit cell is a prerequisite for optical modes and band gaps [4,56], frequency ranges in which no vibrational states arise. In our study, we restricted our attention to the case where there is a single droplet per unit cell, and focused on relatively simple geometries. In the future, one might extend this framework to describe lattices tiled by unit cells containing multiple droplets, thereby describing arbitrary lattices in the plane. We note that for such arrangements, the equilibrium condition $\nabla h(\boldsymbol{d}_{mn}) = 0$ (Equation (9)) would not necessarily be satisfied for all arbitrary unit cells. Nevertheless, an extended theoretical framework might enable an investigation of the Archimedian tilings that Eddi et al. [42] were unable to access in the laboratory, such as the truncated hexagonal and great rhombi-trihexagonal lattices.

Author Contributions: Conceptualization, M.M.P.C. and J.W.M.B.; methodology, software validation, analysis, data curation, and visualization, M.M.P.C. and D.J.E.; writing—original draft preparation, M.M.P.C. and D.J.E.; writing—review and editing, M.M.P.C., D.J.E. and J.W.M.B.; resources, supervision, administration, and funding acquisition, J.W.M.B. All authors have read and agreed to the published version of the manuscript.

Funding: J.W.M.B. gratefully acknowledges the financial support of the NSF through grant CMMI-2154151.

Institutional Review Board Statement: Not applicable.

Informed Consent Statement: Not applicable.

Data Availability Statement: Not applicable.

Conflicts of Interest: The authors declare no conflict of interest.

Appendix A. Impact-Phase Parameters

The strobing of the trajectory equation of Moláček and Bush [47] at the vertical bouncing period T_F introduces two impact-phase parameters,

$$\mathcal{S} = \frac{\int_t^{t+T_F} F_N(t')\sin(\pi f t')dt'}{\int_t^{t+T_F} F_N(t')dt'}, \quad \mathcal{C} = \frac{\int_t^{t+T_F} F_N(t')\cos(\pi f t')dt'}{\int_t^{t+T_F} F_N(t')dt'}, \tag{A1}$$

that represent averages of the sine and cosine of the bath's phase of oscillation over the duration of droplet impact, weighted by the vertical contact force F_N exerted on the drop by the bath [47]. The impact phase directly scales both the wave amplitude generated at each impact (see Equation (2)) and the horizontal wave force imparted to the drop by the bath (see Equation (1)), as captured by $\mathcal{S}$ and $\mathcal{C}$, respectively. While the phase parameters are often combined into a constant fitting parameter $\sin\Phi$, modulations in a drop's impact phase have been shown to be critically important in accurately capturing the stability of bound droplet states [32,33,37,39]. Couchman et al. [32] thus derived functional forms for the dependencies of $\mathcal{S}$ and $\mathcal{C}$ on the bath's vibrational acceleration γ, the local wave amplitude beneath each drop $h_0 = h(\boldsymbol{x}_0, t)$, and the drop radius R. To ensure that our theoretical predictions may provide the best possible comparison with future experimental studies, we here include these variable-phase parameters in our analysis, noting that the product $\mathcal{SC}$ may be set to a constant for a simpler description of the system.

The functional dependencies of $\mathcal{S}$ and $\mathcal{C}$ on the bath's vibrational acceleration γ, local-wave amplitude h_0, and droplet radius R, are presented in Table 2 of Couchman et al. [32]. To obtain the explicit results presented in Section 3, we here focus our theoretical analysis on drops of radius $R = 0.36$ mm bouncing in a $(2,1)^2$ mode, as are typical parameters for experimental studies [32,33,37,39]. In this case, the impact phase functions take the form

$$\mathcal{S}(\gamma, h_0) = 1 - 1.32\exp\{-3.52(\gamma/g - 5.73h_0/R - 2)\}, \quad \text{(A2)}$$

$$\mathcal{C}(\gamma, h_0) = 1.98\exp\{-2.37(\gamma/g - 5.86h_0/R - 2)\}, \quad \text{(A3)}$$

where the gravitational acceleration g and drop radius R are used to non-dimensionalize γ and h_0, respectively.

Appendix B. Rotational Symmetries of the Square and Triangular Lattices

We here derive the simplification $\mathbf{H}^{(0)} = \eta_0\mathbf{I}$ (Equation (24)) which holds for the square and triangular lattices, due to their respective four- and six-fold symmetries (see Figure A1). We adopt the notation

$$c_{pq} \equiv x_{pq}/|\boldsymbol{d}_{pq}|, \;\; s_{pq} \equiv y_{pq}/|\boldsymbol{d}_{pq}|, \quad \text{(A4)}$$

where $\boldsymbol{d}_{pq} = (x_{pq}, y_{pq})$, and c_{pq} and s_{pq} thus represent the cosine and sine of the angle between the position vector $\boldsymbol{d}_{pq}$ and the x-axis. Equation (19), using $\boldsymbol{k} = 0$, thus yields

$$\mathbf{H}^{(0)} = 2\sum_{pq}\left[f''(|\boldsymbol{d}_{pq}|)\underbrace{\begin{bmatrix} c_{pq}^2 & c_{pq}s_{pq} \\ c_{pq}s_{pq} & s_{pq}^2 \end{bmatrix}}_{\mathbf{A}_1} + \frac{f'(|\boldsymbol{d}_{pq}|)}{|\boldsymbol{d}_{pq}|}\underbrace{\begin{bmatrix} s_{pq}^2 & -c_{pq}s_{pq} \\ -c_{pq}s_{pq} & c_{pq}^2 \end{bmatrix}}_{\mathbf{A}_2} \right]. \quad \text{(A5)}$$

For both the square and triangular geometries, we now demonstrate that the right-hand side of Equation (A5) is proportional to the identity matrix $\mathbf{I}$.

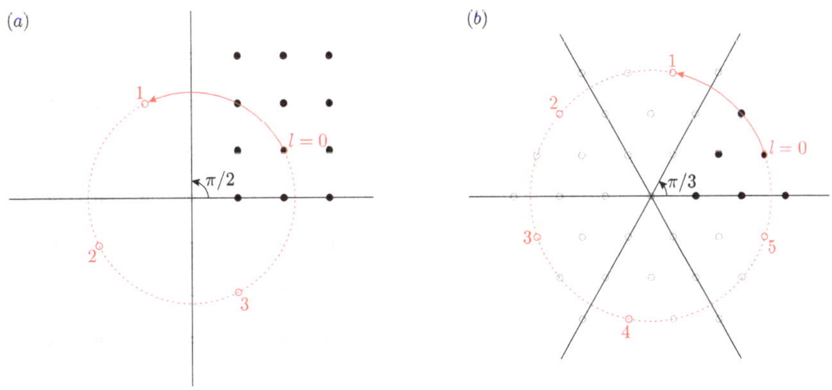

Figure A1. Square (**a**) and triangular (**b**) lattices exhibit four- and six-fold symmetry, respectively, and may be generated by rotating the bolded points around the origin in increments of $\pi/2$ and $\pi/3$.

Appendix B.1. Square Lattice

Consider performing the sums in Equation (A5) by grouping four points at a time, selected by rotating an initial point (p, q) in the first quadrant of Figure A1a through angles of $\theta = l\pi/2,\ l \in [0,3]$. Importantly, each of these four points maintains the same distance $|\boldsymbol{d}_{pq}|$ from the origin. Using the identities

$$\sum_{l=0}^{3} \sin^2(\theta_0 + l\pi/2) = \sum_{l=0}^{3} \cos^2(\theta_0 + l\pi/2) = 2, \tag{A6}$$

$$\sum_{l=0}^{3} \sin(\theta_0 + l\pi/2)\cos(\theta_0 + l\pi/2) = 0, \tag{A7}$$

yields

$$\sum_{l=0}^{3} \mathbf{A}_1 = \sum_{l=0}^{3} \begin{pmatrix} c_l^2 & c_l s_l \\ c_l s_l & s_l^2 \end{pmatrix} = \begin{pmatrix} 2 & 0 \\ 0 & 2 \end{pmatrix}, \tag{A8}$$

$$\sum_{l=0}^{3} \mathbf{A}_2 = \sum_{l=0}^{3} \begin{pmatrix} s_l^2 & -c_l s_l \\ -c_l s_l & c_l^2 \end{pmatrix} = \begin{pmatrix} 2 & 0 \\ 0 & 2 \end{pmatrix}. \tag{A9}$$

Thus, the right-hand side of Equation (A5) is proportional to the identity matrix $\mathbf{I}$.

Appendix B.2. Triangular Lattice

Performing an analogous procedure as for the square lattice, we now evaluate the sums in Equation (A5) by grouping six points at a time, selected by rotating an initial point (p,q) in the bolded sixth of Figure A1b through angles of $\theta = l\pi/3,\ l \in [0,5]$. Using the identities

$$\sum_{l=0}^{5} \sin^2(\theta_0 + l\pi/3) = \sum_{l=0}^{5} \cos^2(\theta_0 + l\pi/3) = 3, \tag{A10}$$

$$\sum_{l=0}^{5} \sin(\theta_0 + l\pi/3)\cos(\theta_0 + l\pi/3) = 0, \tag{A11}$$

yields

$$\sum_{l=0}^{5} \mathbf{A}_1 = \sum_{l=0}^{5} \begin{pmatrix} c_l^2 & c_l s_l \\ c_l s_l & s_l^2 \end{pmatrix} = \begin{pmatrix} 3 & 0 \\ 0 & 3 \end{pmatrix}, \tag{A12}$$

$$\sum_{l=0}^{5} \mathbf{A}_2 = \sum_{l=0}^{5} \begin{pmatrix} s_l^2 & -c_l s_l \\ -c_l s_l & c_l^2 \end{pmatrix} = \begin{pmatrix} 3 & 0 \\ 0 & 3 \end{pmatrix}. \tag{A13}$$

Thus, once again, the right-hand side of Equation (A5) is proportional to the identity matrix $\mathbf{I}$.

References

1. Krumhansl, J. Lattice vibrations in solids. *J. Appl. Phys.* **1962**, *33*, 307–319. [CrossRef]
2. Cochran, W. Lattice vibrations. *Rep. Prog. Phys.* **1963**, *26*, 1. [CrossRef]
3. Dean, P. Atomic vibrations in solids. *IMA J. Appl. Math.* **1967**, *3*, 98–165. [CrossRef]
4. Kittel, C. *Introduction to Solid State Physics*; Wiley: Hoboken, NJ, USA, 1996.
5. Kane, C.; Lubensky, T. Topological boundary modes in isostatic lattices. *Nat. Phys.* **2014**, *10*, 39–45. [CrossRef]
6. Lubensky, T.; Kane, C.; Mao, X.; Souslov, A.; Sun, K. Phonons and elasticity in critically coordinated lattices. *Rep. Prog. Phys.* **2015**, *78*, 073901. [CrossRef]
7. Nash, L.M.; Kleckner, D.; Read, A.; Vitelli, V.; Turner, A.M.; Irvine, W.T. Topological mechanics of gyroscopic metamaterials. *Proc. Natl. Acad. Sci. USA* **2015**, *112*, 14495–14500. [CrossRef]
8. Süsstrunk, R.; Huber, S.D. Classification of topological phonons in linear mechanical metamaterials. *Proc. Natl. Acad. Sci. USA* **2016**, *113*, E4767–E4775. [CrossRef]
9. Bragg, W.L.; Nye, J.F. A dynamical model of a crystal structure. *Proc. R. Soc. A* **1947**, *190*, 474–481.
10. Penciu, R.; Kriegs, H.; Petekidis, G.; Fytas, G.; Economou, E. Phonons in colloidal systems. *J. Chem. Phys.* **2003**, *118*, 5224–5240. [CrossRef]

11. Keim, P.; Maret, G.; Herz, U.; von Grünberg, H.H. Harmonic lattice behavior of two-dimensional colloidal crystals. *Phys. Rev. Lett.* **2004**, *92*, 215504. [CrossRef]
12. Kaya, D.; Green, N.; Maloney, C.; Islam, M. Normal modes and density of states of disordered colloidal solids. *Science* **2010**, *329*, 656–658. [CrossRef]
13. Chen, K.; Still, T.; Schoenholz, S.; Aptowicz, K.B.; Schindler, M.; Maggs, A.; Liu, A.J.; Yodh, A. Phonons in two-dimensional soft colloidal crystals. *Phys. Rev. E* **2013**, *88*, 022315. [CrossRef]
14. Ghosh, A.; Mari, R.; Chikkadi, V.; Schall, P.; Maggs, A.; Bonn, D. Low-energy modes and Debye behavior in a colloidal crystal. *Phys. A Stat. Mech. Its Appl.* **2011**, *390*, 3061–3068. [CrossRef]
15. Beatus, T.; Tlusty, T.; Bar-Ziv, R. Phonons in a one-dimensional microfluidic crystal. *Nat. Phys.* **2006**, *2*, 743–748. [CrossRef]
16. Schiller, U.D.; Fleury, J.B.; Seemann, R.; Gompper, G. Collective waves in dense and confined microfluidic droplet arrays. *Soft Matter* **2015**, *11*, 5850–5861. [CrossRef]
17. Tsang, A.C.H.; Shelley, M.J.; Kanso, E. Activity-induced instability of phonons in 1D microfluidic crystals. *Soft Matter* **2018**, *14*, 945–950. [CrossRef]
18. Ono-dit-Biot, J.C.; Soulard, P.; Barkley, S.; Weeks, E.R.; Salez, T.; Raphaël, E.; Dalnoki-Veress, K. Mechanical properties of 2D aggregates of oil droplets as model mono-crystals. *Soft Matter* **2021**, *17*, 1194–1201. [CrossRef]
19. Benjamin, T.B.; Ursell, F. The stability of the plane free surface of a liquid in vertical periodic motion. *Proc. R. Soc. A* **1954**, *225*, 505–515.
20. Miles, J.; Henderson, D. Parametrically forced surface waves. *Annu. Rev. Fluid Mech.* **1990**, *22*, 143–165. [CrossRef]
21. Walker, J. The amateur scientist. Drops of liquid can be made to float on liquid. What enables them to do so? *Sci. Am.* **1978**, *238*, 151. [CrossRef]
22. Couder, Y.; Fort, E.; Gautier, C.H.; Boudaoud, A. From bouncing to floating: Noncoalescence of drops on a fluid bath. *Phys. Rev. Lett.* **2005**, *94*, 177801. [CrossRef] [PubMed]
23. Moláček, J.; Bush, J.W.M. Drops bouncing on a vibrating bath. *J. Fluid Mech.* **2013**, *727*, 582–611. [CrossRef]
24. Couder, Y.; Protiere, S.; Fort, E.; Boudaoud, A. Dynamical phenomena: Walking and orbiting droplets. *Nature* **2005**, *437*, 208. [CrossRef] [PubMed]
25. Protière, S.; Boudaoud, A.; Couder, Y. Particle–wave association on a fluid interface. *J. Fluid Mech.* **2006**, *554*, 85–108. [CrossRef]
26. Eddi, A.; Sultan, E.; Moukhtar, J.; Fort, E.; Rossi, M.; Couder, Y. Information stored in Faraday waves: The origin of a path memory. *J. Fluid Mech.* **2011**, *674*, 433–463. [CrossRef]
27. Fort, E.; Eddi, A.; Boudaoud, A.; Moukhtar, J.; Couder, Y. Path-memory induced quantization of classical orbits. *Proc. Natl. Acad. Sci. USA* **2010**, *107*, 17515–17520. [CrossRef]
28. Bush, J.W.M. Pilot-wave hydrodynamics. *Annu. Rev. Fluid Mech.* **2015**, *47*, 269–292. [CrossRef]
29. Bush, J.W.M.; Couder, Y.; Gilet, T.; Milewski, P.A.; Nachbin, A. Introduction to focus issue on hydrodynamic quantum analogs. *Chaos Interdiscip. J. Nonlinear Sci.* **2018**, *28*, 096001. [CrossRef]
30. Bush, J.W.M.; Oza, A.U. Hydrodynamic quantum analogs. *Rep. Prog. Phys.* **2020**, *84*, 017001. [CrossRef]
31. Rahman, A.; Blackmore, D. Walking droplets through the lens of dynamical systems. *Mod. Phys. Lett. B* **2020**, *34*, 2030009. [CrossRef]
32. Couchman, M.M.P.; Turton, S.E.; Bush, J.W.M. Bouncing phase variations in pilot-wave hydrodynamics and the stability of droplet pairs. *J. Fluid Mech.* **2019**, *871*, 212–243. [CrossRef]
33. Couchman, M.M.P.; Bush, J.W.M. Free rings of bouncing droplets: Stability and dynamics. *J. Fluid Mech.* **2020**, *903*. [CrossRef]
34. Thomson, S.J.; Couchman, M.M.P.; Bush, J.W.M. Collective vibrations of confined levitating droplets. *Phys. Rev. Fluids* **2020**, *5*, 083601. [CrossRef]
35. Barnes, L.; Pucci, G.; Oza, A.U. Resonant interactions in bouncing droplet chains. *C. R. Mécanique* **2020**, *348*, 573–589. [CrossRef]
36. Protière, S.; Bohn, S.; Couder, Y. Exotic orbits of two interacting wave sources. *Phys. Rev. E* **2008**, *78*, 036204. [CrossRef]
37. Oza, A.U.; Siéfert, E.; Harris, D.M.; Moláček, J.; Bush, J.W.M. Orbiting pairs of walking droplets: Dynamics and stability. *Phys. Rev. Fluids* **2017**, *2*, 053601. [CrossRef]
38. Borghesi, C.; Moukhtar, J.; Labousse, M.; Eddi, A.; Fort, E.; Couder, Y. Interaction of two walkers: Wave-mediated energy and force. *Phys. Rev. E* **2014**, *90*, 063017. [CrossRef]
39. Arbelaiz, J.; Oza, A.U.; Bush, J.W.M. Promenading pairs of walking droplets: Dynamics and stability. *Phys. Rev. Fluids* **2018**, *3*, 013604. [CrossRef]
40. Lieber, S.I.; Hendershott, M.C.; Pattanaporkratana, A.; Maclennan, J.E. Self-organization of bouncing oil drops: Two-dimensional lattices and spinning clusters. *Phys. Rev. E* **2007**, *75*, 056308. [CrossRef]
41. Eddi, A.; Terwagne, D.; Fort, E.; Couder, Y. Wave propelled ratchets and drifting rafts. *Europhys. Lett.* **2008**, *82*, 44001. [CrossRef]
42. Eddi, A.; Decelle, A.; Fort, E.; Couder, Y. Archimedean lattices in the bound states of wave interacting particles. *Europhys. Lett.* **2009**, *87*, 56002. [CrossRef]
43. Eddi, A.; Boudaoud, A.; Couder, Y. Oscillating instability in bouncing droplet crystals. *Europhys. Lett.* **2011**, *94*, 20004. [CrossRef]
44. Thomson, S.J.; Durey, M.; Rosales, R.R. Collective vibrations of a hydrodynamic active lattice. *Proc. R. Soc. A* **2020**, *476*, 20200155. [CrossRef]
45. Oza, A.U.; Rosales, R.R.; Bush, J.W.M. A trajectory equation for walking droplets: Hydrodynamic pilot-wave theory. *J. Fluid Mech.* **2013**, *737*, 552–570. [CrossRef]

46. Thomson, S.J.; Durey, M.; Rosales, R.R. Discrete and periodic complex Ginzburg-Landau equation for a hydrodynamic active lattice. *Phys. Rev. E* **2021**, *103*, 062215. [CrossRef]
47. Moláček, J.; Bush, J.W.M. Drops walking on a vibrating bath: Towards a hydrodynamic pilot-wave theory. *J. Fluid Mech.* **2013**, *727*, 612–647. [CrossRef]
48. Turton, S.E.; Couchman, M.M.P.; Bush, J.W.M. A review of the theoretical modeling of walking droplets: Towards a generalized pilot-wave framework. *Chaos Interdiscip. J. Nonlinear Sci.* **2018**, *28*, 096111. [CrossRef]
49. Harris, D.M.; Quintela, J.; Prost, V.; Brun, P.T.; Bush, J.W.M. Visualization of hydrodynamic pilot-wave phenomena. *J. Vis.* **2017**, *20*, 13–15. [CrossRef]
50. Damiano, A.P.; Brun, P.T.; Harris, D.M.; Galeano-Rios, C.A.; Bush, J.W.M. Surface topography measurements of the bouncing droplet experiment. *Exp. Fluids* **2016**, *57*, 163. [CrossRef]
51. Tadrist, L.; Shim, J.B.; Gilet, T.; Schlagheck, P. Faraday instability and subthreshold Faraday waves: Surface waves emitted by walkers. *J. Fluid Mech.* **2018**, *848*, 906–945. [CrossRef]
52. Wind-Willassen, Ø.; Moláček, J.; Harris, D.M.; Bush, J.W.M. Exotic states of bouncing and walking droplets. *Phys. Fluids* **2013**, *25*, 082002. [CrossRef]
53. Blackman, M. Contributions to the theory of the specific heat of crystals II—On the vibrational spectrum of cubical lattices and its application to the specific heat of crystals. *Proc. R. Soc. A* **1935**, *148*, 384–406.
54. Montroll, E.W. Dynamics of a square lattice I. Frequency spectrum. *J. Chem. Phys.* **1947**, *15*, 575–591. [CrossRef]
55. Bell, R. The dynamics of disordered lattices. *Rep. Prog. Phys.* **1972**, *35*, 1315–1409. [CrossRef]
56. Kosevich, A.M. *The Crystal Lattice: Phonons, Solitons, Dislocations, Superlattices*; John Wiley & Sons: Hoboken, NJ, USA, 2006.

symmetry MDPI

Article

Lack of Plasma-like Screening Mechanism in Sedimentation of a Non-Brownian Suspension

Paweł Sznajder [1,*], Bogdan Cichocki [2] and Maria Ekiel-Jeżewska [1]

1 Institute of Fundamental Technological Research, Polish Academy of Sciences, 02-106 Warsaw, Poland; mekiel@ippt.pan.pl
2 Faculty of Physics, University of Warsaw, 02-093 Warsaw, Poland; bogdan.cichocki@fuw.edu.pl
* Correspondence: psznaj@ippt.pan.pl

Abstract: We investigate qualitatively a uniform non-Brownian sedimenting suspension in a stationary state. As a base of our analysis we take the BBGKY hierarchy derived from the Liouville equation. We then show that assumption of the plasma-like screening relations can cancel some long-range terms in the hierarchy but it does not provide integrable solutions for correlation functions. This suggests breaking the translational symmetry of the system. Therefore a non-uniform structure can develop to suppress velocity fluctuations and make the range of correlations finite.

Keywords: non-Brownian sedimentation; stability; BBGKY hierarchy; hydrodynamic screening; correlation functions; low-Reynolds-number hydrodynamics

Citation: Sznajder, P.; Cichocki, B.; Ekiel-Jeżewska, M. Lack of Plasma-like Screening Mechanism in Sedimentation of a Non-Brownian Suspension. *Symmetry* **2022**, *14*, 63. https://doi.org/10.3390/sym14010063

Academic Editor: Andrzej Herczyński and Roberto Zenit

Received: 19 November 2021
Accepted: 17 December 2021
Published: 3 January 2022

Publisher's Note: MDPI stays neutral with regard to jurisdictional claims in published maps and institutional affiliations.

1. Introduction

Sedimentation is a process of falling of particles in a fluid due to gravity (the particles are more dense then the fluid). Considerations in this paper are restricted to the limit of vanishing Reynolds number and infinite Peclet number. Reynolds number is given by

$$\mathrm{Re} = a\eta^{-1}\rho_l U_S, \tag{1}$$

where a is radius of a particle, η is dynamic viscosity coefficient of the fluid, ρ_l is density of the fluid and U_S is the Stokes velocity of a single particle falling in unbounded fluid motionless at infinity. When $\mathrm{Re} \to 0$, the fluid flow instantly adjusts to the boundary conditions. Peclet number is given by

$$\mathrm{Pe} = aD^{-1}U_S, \tag{2}$$

where D is diffusion coefficient of a single particle in a fluid. In case of $\mathrm{Pe} \to \infty$, Brownian motion is negligible compared to the motion caused by gravity and hydrodynamic interactions. In such a system there are difficulties with divergent expressions due to hydrodynamic field disturbance slowly decaying over distance (inversely proportional to distance—same as electric potential produced by an isolated charge in a vacuum). Challenges of theoretical approach to system with vanishing Reynolds number (and divergent Peclet number) are known for more than hundred years. In 1911, Smoluchowski [1,2] investigated a particle surrounded by other particles suspended in a Newtonian fluid. His observation was that, considering larger and larger systems leads to a divergent expression for the particle falling velocity, caused by the long-range velocity disturbance produced by other particles. Now this is known in literature as the Smoluchowski paradox [3]. Solution of this paradox was given by Batchelor [4] sixty years after the work of Smoluchowski and then reanalyzed by Beenakker and Mazur [5].

Batchelor's main idea was to calculate the average velocity of suspended particles $\langle U\rangle$ relative to the flow $\langle v\rangle$ of the whole suspension which resulted in $\langle U\rangle - \langle v\rangle$ where cancellation of divergent terms secures that the average relative velocity is finite. Nevertheless there is another problem within physics of such a system: calculation of the sedimentation velocity variance. This one was noticed by Caflisch and Luke [6] in 1985. They showed that fluctuations in general should grow with linear dimension L of a system which is again a result of a slowly decaying flow disturbance around a particle. One can spot that if this flow disturbance is inversely proportional to distance then integrating square of it leads to linear divergence with L. To solve that conundrum, Koch and Shaqfeh [7] proposed in 1991 that pair probability has to obey the screening relation like in plasma (as it is with the Debye–Hückel screening), namely that it has to reflect a net deficit of one particle in the vicinity of each particle. This condition corresponds to vanishing structure factor for vanishing $\boldsymbol{k}$ wave vector, $S(\boldsymbol{k} \to 0) = 0$. The idea of screening is that hydrodynamic interaction between particles results in lower probability of finding a close pair of particles. When screening occurs this effect is finely tuned in such a way that effective hydrodynamic interaction between particles within a pair has a much faster decay with the inter-particle distance—a power law is replaced by an exponential decay. Moreover, the change is such that integral of the correlation function reflects the average lack of exactly one neighboring particle around a test particle. This leads to cancellation of the leading long-range terms and results in convergent expressions for velocity fluctuations. An example of such a cancellation owing to the screening effect will be provided in Section 3.1.

Nevertheless, the idea of the screening condition being satisfied was a subject of a substantial critique [8–11] which was mainly focused on showing no evidence of screening in simulations of a sedimenting suspension. In order to cope with fluctuations many ideas were given. Some were pointing to the crucial role of walls [12], some were focusing on stratification [13] or its critique [14,15], some were considering additional shear flow [16], and others even tried to look for a solution in non-zero Re number [17]. Good scope of troubles with finding solution to fluctuation problem is given in review articles [18–20].

In this paper, the screening idea is analyzed and it is shown that the BBGKY (Bogoliubov, Born, Green, Kirkwood, Yvon) hierarchy for the correlation functions between sedimenting particles [21] in a uniform stationary state has no screening solutions at all. This is a significant problem since BGGKY hierarchy is derived straightforwardly from the Liouville equation which cannot be violated.

In the Section 2, the theoretical framework is presented. In Section 3 it is shown that even if some of long range terms in hierarchy are regularized by screening conditions, there are still other ones which cannot be screened at all. This leads to contradiction since assuming screening still leads to equations which give non-integrable solutions for the correlation functions. This suggests that the system becomes non-uniform and symmetry with respect to translations is broken. As pointed out in Section 4, this finding gives a novel, unique theoretical explanation of a non-uniform structure of non-Brownian sedimenting suspensions, previously reported in experiments and numerical simulations [13,19,22–32].

2. Methods

2.1. System and Its Theoretical Description

System consists of N identical rigid spherical particles (mono-disperse suspension) immersed in a Newtonian fluid of viscosity η. Particle radius a is small enough and viscosity of fluid is sufficiently large to ensure Reynolds number (Re) is much smaller than unity. Then one can set Re $= 0$ which reduces Navier–Stokes equations for the fluid flow to the Stokes equations [33]. Simultaneously a is large enough to keep Peclet number (Pe) much larger then 1 so Brownian motion can be neglected. Particles are uniform, spherically symmetric and their density ρ_p is larger than the density ρ_f of the surrounding fluid, which results in settling of those particles under gravity. Macroscopically system is assumed to be uniform and stationary.

The fluid flow is governed by the Stokes equations. The Green tensor for an unbounded system, with condition that fluid velocity is zero at infinity, is represented by the Oseen tensor $\mathcal{T}$ of the form [34]:

$$\mathcal{T}(\boldsymbol{r}) = \frac{1}{8\pi\eta r}(\mathbb{I} + \hat{\boldsymbol{r}}\hat{\boldsymbol{r}}). \tag{3}$$

Velocities of the fluid and particles are determined by external forces (By external forces one should understand those which are not a result of an interaction with the fluid) and boundary conditions at infinity and at the particles. It is important to solve the so-called mobility problem which is stated as below: determine velocities of the particles given their positions and total external forces acting upon them, in the absence of any external torques on the particles. The solution is represented by 3×3 Cartesian mobility matrices $\boldsymbol{\mu}_{ij}$, with the particle labels $i, j = 1, ..., N$ [35],

$$\boldsymbol{U}_i(\boldsymbol{X}) = \sum_{j=1}^{N} \boldsymbol{\mu}_{ij}(\boldsymbol{X})\boldsymbol{F}_j, \tag{4}$$

where

$$\boldsymbol{X} = \{\boldsymbol{R}_1, \ldots, \boldsymbol{R}_N\} \tag{5}$$

is the set of positions $\boldsymbol{R}_i$ of all the particle centers, $\boldsymbol{U}_i$ is the velocity of i-th particle, $\boldsymbol{F}_j$ is the external force, e.g., gravity, acting on particle j. In this paper all the forces are identical, $\boldsymbol{F}_j = \boldsymbol{F}$. In general, mobility matrices $\boldsymbol{\mu}_{ij}$ depend on positions of all the particles $\boldsymbol{X}$. Although $\boldsymbol{\mu}_{ij}$ have many-body character, they can be expressed as an infinite sum of sequences of two-particle interactions. Such a scattering expansion [36–39], analogous to the reflection method [33], can be written as,

$$\boldsymbol{\mu}_{ij} = \left(\boldsymbol{\mu}_0 + \boldsymbol{\mu}_0 \mathcal{Z}_0 \mathcal{G} \left[\sum_{k=0}^{\infty} (-\hat{\mathcal{Z}}_0 \mathcal{G})^k \right] \mathcal{Z}_0 \boldsymbol{\mu}_0 \right)_{ij}. \tag{6}$$

At the RHS of the above equation, we use the operators, defined, e.g., in [38,40,41] and briefly described below. The single particle operators: mobility $\boldsymbol{\mu}_0$, friction $\mathcal{Z}_0$ and convective friction $\hat{\mathcal{Z}}$ are diagonal in the particle labels i, and identical for each value of i because the particles are identical. Their meaning is the following,

$$\boldsymbol{\mu}_0 : \quad \text{force on a particle} \mapsto \text{velocity of that particle} \tag{7}$$

$$\mathcal{Z}_0 : \quad \text{incoming fluid velocity field at a particle} \mapsto \text{force density induced on that particle} \tag{8}$$

$$\hat{\mathcal{Z}}_0 : \quad \text{incoming fluid velocity field at a freely moving particle} \mapsto \text{force density induced on that particle} \tag{9}$$

The difference between the operators $\mathcal{Z}_0$ and

$$\hat{\mathcal{Z}}_0 = \mathcal{Z}_0 - \mathcal{Z}_0 \boldsymbol{\mu}_0 \mathcal{Z}_0 \tag{10}$$

is that in case of $\mathcal{Z}_0$, the particle translational and rotational velocities are given, while in case of $\hat{\mathcal{Z}}_0$, the particle moves freely, what by definition means that the total induced force and torque on that particle vanish [38,40,41].

The off-diagonal operator $\mathcal{G}$ contains as its elements propagators based on the Oseen tensor given by Equation (3). The elements $\mathcal{G}_{ij}$ are propagators between different particles $i \neq j$, and they represent single hydrodynamic interactions. The element $\mathcal{G}_{ij}$ depends only on the relative position $\boldsymbol{R}_{ij} = \boldsymbol{R}_i - \boldsymbol{R}_j$, and it acts in the following way,

$$\mathcal{G}_{ij}(\boldsymbol{R}_{ij}) : \text{force density on } j\text{-th particle at } \boldsymbol{R}_j \mapsto \text{generated fluid velocity at } \boldsymbol{R}_i \tag{11}$$

In general, the operators μ_0, $\mathcal{Z}_0$, $\hat{\mathcal{Z}}_0$ and $\mathcal{G}$ are complicated but one can simplify description by expressing those elements in terms of vector solid harmonics [34] as base functions given by irreducible representations of the SO(3) group. It results in multipole expansion of hydrodynamic interactions [33,42–45] which is one of among other successful applications of multipole formalism like in electromagnetism [46] or gravity [47]. The base multipole functions are characterized by three indices l, m, σ, with $l = 1, 2, \ldots, m = -l, \ldots, l$ and $\sigma = 0, 1, 2$. Using them, one can represent each operator as a large matrix with the particle labels and the multipole indices, $ilm\sigma$. In this way, superposition of the operators in Equation (6) should be understood as multiplication of the corresponding large matrices. In particular, $\mathcal{G}_{ilm\sigma,jl'm'\sigma'}$ describes the coupling between $l'm'\sigma'$ force multipole on j-th particle and $lm\sigma$ velocity multipole on i-th particle. Using the multipole representation is very convenient for asymptotic analysis since the dependence of $\mathcal{G}$ elements on distance is very simple [44],

$$\mathcal{G}_{ilm\sigma,jl'm'\sigma'} \propto R_{ij}^{-l-l'-\sigma-\sigma'+1}. \tag{12}$$

To proceed with further analysis it is needed to point out the crucial difference between the multipole elements of $\mathcal{Z}_0$ and $\hat{\mathcal{Z}}_0$. Both operators are diagonal in pairs of indices l, l', and m, m'. For $l \geq 2$ their multipole elements are identical. However, they are different for $l = 1$. In this case, the operator $\hat{\mathcal{Z}}_0$ has the only non-zero element for $(\sigma, \sigma') = (2, 2)$, determined by the relation (10).

Therefore, when $\mathcal{G}_{ij}$ is between operators $\hat{\mathcal{Z}}_0$ in a scattering sequence given by Equation (6), the slowest possible decay of the interaction is for $(l, l') = (2, 2)$ and $(\sigma, \sigma') = (0, 0)$ which, according to Equation (12), scales as $1/R_{ij}^3$. This specific term corresponds to coupling between symmetric vector dipole moments ($l = 2, \sigma = 0$) so it can be called a dipole-dipole interaction.

Using the multipole representation, it is now easier to understand the scattering expansion (6). Taking into account that the large matrices representing the single particle operators μ_0, $\mathcal{Z}_0$ and $\hat{\mathcal{Z}}_0$ are diagonal in the particle labels, and the diagonal elements are identical for each particle, from now on we will redefine the symbols μ_0, $\mathcal{Z}_0$ and $\hat{\mathcal{Z}}_0$ to denote the corresponding diagonal elements, which are matrices in the space of the multipole indices. In the new notation, the first term in the bracket at the RHS of (6) is $\mu_0 \delta_{ij}$. The second term is the single-scattering sequence $\mu_0 \mathcal{Z}_0 \mathcal{G}_{ij} \mathcal{Z}_0 \mu_0$, with multiplication of the consecutive multipole matrices. The third term is the sum of the two-scattering sequences, $-\mu_0 \mathcal{Z}_0 \mathcal{G}_{im} \hat{\mathcal{Z}}_0 \mathcal{G}_{mj} \mathcal{Z}_0 \mu_0$, with respect to $m \neq i$ and $m \neq j$, and so on.

2.2. *BBGKY Hierarchy of Equations for Correlation Functions*

To derive the BBGKY hierarchy of equations for the correlation functions of sedimenting particles one needs to consider the Liouville equation for the system. For a stationary state it reads [21]:

$$\sum_{i=1}^{N} \frac{\partial}{\partial \boldsymbol{R}_i} \cdot [\boldsymbol{U}_i(\boldsymbol{X}) \rho(\boldsymbol{X})] = 0. \tag{13}$$

where $\rho(\boldsymbol{X})$ is the probability distribution function for particle positions. To proceed further one needs to introduce m-particle microscopic density operator

$$\hat{n}_m(\boldsymbol{m}, \boldsymbol{X}) = \sum_{j_1, j_2, \ldots, j_m}{}' \prod_{i=1}^{m} \delta(\boldsymbol{r}_i - \boldsymbol{R}_{j_i}), \tag{14}$$

where $\boldsymbol{m}$ is set of positions

$$\boldsymbol{m} = \{\boldsymbol{r}_1, \ldots, \boldsymbol{r}_m\}, \tag{15}$$

also abbreviated to $\{1, \ldots, m\}$. The sum $\sum'$ is taken over all m-particle subsets and their permutations which results in sums with respect to j_i where $i = 1, \ldots, m$ and there is no duplicating indices $\forall_{i,k} j_i \neq j_k$ (see equation (3.1.20) in [48]). When averaged over positions $\boldsymbol{X}$ one obtains m-particle density functions

$$n_m(\boldsymbol{m}) = \int \mathrm{d}\boldsymbol{X}\rho(\boldsymbol{X})\hat{n}_m(m,\boldsymbol{X}) = \langle \hat{n}_m \rangle, \tag{16}$$

with

$$\langle \ldots \rangle = \int \mathrm{d}\boldsymbol{X}\rho(\boldsymbol{X})\ldots \tag{17}$$

In case of identical particles one can also recover the well-known formula,

$$n_m(\boldsymbol{m}) = \frac{N!}{(N-m)!}\int \mathrm{d}\boldsymbol{X}\rho(\boldsymbol{X})\delta(\boldsymbol{r}_1 - \boldsymbol{R}_1)\cdot\ldots\cdot\delta(\boldsymbol{r}_m - \boldsymbol{R}_m). \tag{18}$$

Averaging for $m = 1$ one gets concentration $n_1(1) = n = N/V$ where N is the number of particles and V is the volume of the system. Following [48], we introduce k-particle correlation functions $h_k(\boldsymbol{k})$, defining them by the relations,

$$n_2(1,2) = n^2(1 + h_2(1,2)) \tag{19}$$

$$n_3(1,2,3) = n^3(1 + h_2(1,2) + h_2(2,3) + h_2(1,3) + h_3(1,2,3)) \tag{20}$$

$$\begin{aligned} n_4(1,2,3,4) = n^4(&1 + h_2(1,2) + h_2(1,3) + h_2(1,4) + h_2(2,3) + h_2(2,4) + h_2(3,4) \\ &+ h_2(1,2)h_2(3,4) + h_2(1,3)h_2(2,4) + h_2(1,4)h_2(2,3) \\ &+ h_3(1,2,3) + h_3(1,2,4) + h_3(2,3,4) + h_4(1,2,3,4)) \end{aligned} \tag{21}$$

$$\vdots$$

Now multiplying the Liouville Equation (13) by density operator $\hat{n}_m$ and averaging over all positions $\boldsymbol{X}$ results in generation of the following BBGKY hierarchy of equations for m-particle density functions (It can be transformed to connected hierarchy for correlation functions [21], but for the purpose of this work it is enough to consider this hierarchy as it is.):

$$\sum_{i=1}^{m} \frac{\partial}{\partial \boldsymbol{r}_i} \cdot \langle \boldsymbol{U}_i \hat{n}_m \rangle(\boldsymbol{m}) = 0, \qquad \text{with } m = 2, 3, \ldots. \tag{22}$$

Explicitly,

$$\frac{\partial}{\partial \boldsymbol{r}_1} \cdot \langle \boldsymbol{U}_1 \hat{n}_2 \rangle(\boldsymbol{2}) + \frac{\partial}{\partial \boldsymbol{r}_2} \cdot \langle \boldsymbol{U}_2 \hat{n}_2 \rangle(\boldsymbol{2}) = 0 \tag{23}$$

$$\frac{\partial}{\partial \boldsymbol{r}_1} \cdot \langle \boldsymbol{U}_1 \hat{n}_3 \rangle(\boldsymbol{3}) + \frac{\partial}{\partial \boldsymbol{r}_2} \cdot \langle \boldsymbol{U}_2 \hat{n}_3 \rangle(\boldsymbol{3}) + \frac{\partial}{\partial \boldsymbol{r}_3} \cdot \langle \boldsymbol{U}_3 \hat{n}_3 \rangle(\boldsymbol{3}) = 0 \tag{24}$$

$$\frac{\partial}{\partial \boldsymbol{r}_1} \cdot \langle \boldsymbol{U}_1 \hat{n}_4 \rangle(\boldsymbol{4}) + \frac{\partial}{\partial \boldsymbol{r}_2} \cdot \langle \boldsymbol{U}_2 \hat{n}_4 \rangle(\boldsymbol{4}) + \frac{\partial}{\partial \boldsymbol{r}_3} \cdot \langle \boldsymbol{U}_3 \hat{n}_4 \rangle(\boldsymbol{4}) + \frac{\partial}{\partial \boldsymbol{r}_4} \cdot \langle \boldsymbol{U}_4 \hat{n}_4 \rangle(\boldsymbol{4}) = 0 \tag{25}$$

$$\vdots$$

For two identical particles, the LHS of Equation (23) is always zero due to the symmetry with respect to interchanging the particle labels, and the translational symmetry of the system. In case of two different particles, Equation (23) was solved by Batchelor [49,50], with the use of the expansion in powers of the concentration n. In case of three particles Cichocki and Sadlej [51] proposed a scheme to deal with the long range terms, but limited to mono-disperse case. The problem of long-range terms in Equations (24) and (25), and also higher order equations, will be discussed in the next section.

3. Results

3.1. Long-Range Terms and Screening

Hierarchy of the Equations (22) describes the conditions for particle current divergences which represent the law of mass conservation. By considering an increasing number of particles and performing the cluster analysis [21,39] one can obtain equations for the correlation functions. In these equations, there appear long-range terms which scale with the inter-particle distances r as $1/r$, $1/r^2$ and $1/r^3$ which makes them non-integrable in the thermodynamic limit, defined as $N, V \to \infty$ and $n = N/V = const$. Without imposing additional conditions, this would lead to non-integrable correlation functions. However, integrability of the correlation functions is necessary for stability of the system. As it is mentioned in the introduction, Koch and Shaqfeh proposed a screening mechanism for the system under consideration, similar as in a one component plasma, as condition (2.14) in [7]. This condition, formulated in terms of the pair probability g, can be written in terms of the two-particle correlation function $h_2 = g/n - 1$ as:

$$n \int \mathrm{d}\mathbf{r} h_2(\mathbf{r}) = -1, \tag{26}$$

For the higher order correlation functions this should be generalized to [52]:

$$n \int \mathrm{d}\mathbf{r}_{k+1} h_{k+1}(\mathbf{r}_1, \mathbf{r}_2, \ldots, \mathbf{r}_{k+1}) = -k h_k(\mathbf{r}_1, \ldots, \mathbf{r}_k), \qquad k = 2, 3, \ldots \tag{27}$$

Those conditions will lead to finite velocity fluctuations in the thermodynamic limit. One can try to apply the conditions (26) and (27) to eliminate (regularize) long-range terms also in the BBGKY hierarchy. However, as it will be shown below, the screening conditions eliminate many but not all long-range terms in the hierarchy. Nevertheless, the screening conditions (26) and (27) can be satisfied only if all the correlation functions have a finite range. Therefore, we will face a contradiction.

In the following, we will first provide an example how the screening condition (26) leads to cancellation of some long-range terms. Then we will show another example of long-range terms which remain unaffected by screening.

Equations (24) and (25), . . . of the BBGKY hierarchy for three, four and more particles contain long-range terms. We will focus on Equation (24) and identify such terms. To this goal, one should do substitutions using the expressions (4) for the particle velocities $\boldsymbol{U}_i$ in terms of the mobilities μ_{ij}, given as the series (6) of scattering sequences. Each sequence linking particles i and j consists of a certain number of scatterings, generated by the force $\boldsymbol{F}$ acting on particle j. The scattering sequences are next averaged, what leads to multiplication by certain correlation functions. It is useful to remember that the density operator $\hat{n}_m$ sets positions of particles $1, 2, \ldots, m$ regardless whether they are included in a particular scattering sequence for $\boldsymbol{U}_i$ or not, with $\boldsymbol{U}_i$ given by Equations (4) and (6). It means that if there are particle positions in a certain BBGKY hierarchy term $\langle \boldsymbol{U}_i \hat{n}_m \rangle(\boldsymbol{m})$ which are being integrated then they must appear in a scattering sequence of $\boldsymbol{U}_i$.

We will now provide the first example. In the expression (24) there exists a term with the single interaction $\mathcal{G}_{12}$ between particles 12, the single interaction $\mathcal{G}_{23}$ between particles 23 and the pair correlation $h_2(r_{12})$ between the particles 12. This term can be written as follows,

$$n^3 \frac{\partial}{\partial \boldsymbol{r}_1} \cdot \left[\mu_0 \mathcal{Z}_0 \mathcal{G}_{12} \hat{\mathcal{Z}}_0 \mathcal{G}_{23} \mathcal{Z}_0 \mu_0 \boldsymbol{F} h_2(\boldsymbol{r}_{12}) \right]. \tag{28}$$

The slowest decaying part of this term is proportional to r_{23}^{-2} since it is between operators $\mathcal{Z}_0$ and $\hat{\mathcal{Z}}_0$, and we know how $\mathcal{G}_{23}$ can vary with distance given (12). If $n \int \mathrm{d}r h_2(\mathbf{r}) = -1$ holds then the expression (28) can be screened by another term from Equation (24), involving in addition a fourth particle and containing integration with respect to its position r_4. This term contains the single interaction $\mathcal{G}_{12}$ between particles 12, the single interaction

$\mathcal{G}_{24}$ between particles 24, the pair correlation functions $h_2(\mathbf{r}_{12})$ and $h_2(\mathbf{r}_{34})$ between pairs of particles 12 and 34,

$$n^4 \frac{\partial}{\partial \mathbf{r}_1} \cdot \left[\int \mathrm{d}\mathbf{r}_4 \mu_0 \mathcal{Z}_0 \mathcal{G}_{12} \hat{\mathcal{Z}}_0 \mathcal{G}_{24} \mathcal{Z}_0 \mu_0 F h_2(\mathbf{r}_{12}) h_2(\mathbf{r}_{34}) \right]. \tag{29}$$

In this term $\mathcal{G}_{24}$ can be expanded in the following way,

$$\mathcal{G}_{ik} = \mathcal{G}_{ij} + \mathbf{r}_{jk} \cdot \frac{\partial}{\partial \mathbf{r}_{ij}} \mathcal{G}_{ij} + \ldots, \tag{30}$$

with $i, j, k = 2, 3, 4$. Such an expansion, when plugged into Equation (29), is justified since, according to our assumption, the correlation function $h_2(\mathbf{r}_{34})$ has short range. When taking into account the first term of (30), the expression (29) reads

$$n^3 \frac{\partial}{\partial \mathbf{r}_1} \cdot \left[\mu_0 \mathcal{Z}_0 \mathcal{G}_{12} \hat{\mathcal{Z}}_0 \mathcal{G}_{23} \mathcal{Z}_0 \mu_0 F h_2(\mathbf{r}_{12}) n \int \mathrm{d}\mathbf{r}_4 h_2(\mathbf{r}_{34}) \right]. \tag{31}$$

with the screening condition (26), the expressions (28) and (31) cancel each other. The second term of (30) gives no contribution to expression (29) because, due to symmetry of correlation function, $\int \mathbf{r} h_2(\mathbf{r}) \mathrm{d}\mathbf{r} = 0$. The correlation function has to obey general symmetry of the system, first one with respect to the rotation around vertical axis and second one with respect to parity in the direction of that axis. Higher order terms in (30) lead in Equation (29) to the hydrodynamic interaction between particles 2 and 3 which decays at least as fast as r_{23}^{-4} so it is then of a short range. Ability to construct a screening pair of terms is based on the fact that those screened interactions are the first from the right in the scattering sequence.

Such a construction is not possible otherwise which gives rise to examples contradicting the possibility of screening of all terms. An example of such term can be found in the hierarchy Equation (25) for 4 particles. This term is written below,

$$n^4 \frac{\partial}{\partial \mathbf{r}_1} \cdot \left[\mu_0 \mathcal{Z}_0 \mathcal{G}_{12} \hat{\mathcal{Z}}_0 \mathcal{G}_{23} \hat{\mathcal{Z}}_0 \mathcal{G}_{34} \mathcal{Z}_0 \mu_0 F h(\mathbf{r}_{12}) h(\mathbf{r}_{34}) \right]. \tag{32}$$

Above, the expression $\hat{\mathcal{Z}}_0 \mathcal{G}_{23} \hat{\mathcal{Z}}_0$ decays as r_{23}^{-3} and therefore corresponds to infinite range of interaction as it was explained below Equation (12). In analogous way to formation of the pair (28) and (31) the term (32) can be paired with the following one,

$$n^5 \frac{\partial}{\partial \mathbf{r}_1} \cdot \left[\int \mathrm{d}\mathbf{r}_5 \mu_0 \mathcal{Z}_0 \mathcal{G}_{12} \hat{\mathcal{Z}}_0 \mathcal{G}_{25} \hat{\mathcal{Z}}_0 \mathcal{G}_{54} \mathcal{Z}_0 \mu_0 F h(\mathbf{r}_{12}) h_3(\mathbf{r}_{34}, \mathbf{r}_{35}) \right]. \tag{33}$$

In this case, one has to expand two hydrodynamic interaction propagators, $\mathcal{G}_{25}$ and $\mathcal{G}_{51}$ using (30), with i, j, k equal to 2, 3, 5 and 5, 3, 4 respectively. However, this operation together with the screening condition (27) for $k = 2$ does not lead to cancellation of infinitely long range interactions like it was with a pair of (28) and (31).

In general, such a dipole–dipole part of the hydrodynamic interaction $\mathcal{G}_{ij}$ inside a scattering sequence, which gives rise to long range behavior of correlations, cannot be screened by the plasma-like screening conditions (27). This is due to the fact that construction of a screening pair requires also expanding preceding interaction which is on the right side of $\mathcal{G}_{ij}$. This situation results in an inability to cancel the slowest r_{ij}^{-3} term because one gets integrals which contains both correlations and interactions and they cannot be reduced by screening conditions. It means that screening can only affect first from right interaction in the scattering sequence. Therefore, only some long range terms are canceled by (27). Other ones still remain, as illustrated in the above example. This leads to a contradiction with the assumption that correlation functions are integrable.

This means that one will not obtain for correlations an integrable solution of the BBGKY hierarchy. This contradicts screening relations leading to instability of a uniform system due to divergent expressions for velocity fluctuations in the thermodynamic limit. It suggest that system has to break the translational symmetry and that it has to develop some non-uniform structure like stratification or swirls [22] which can change the velocity fluctuations magnitude [13]. However, for numerical simulations of small systems in periodic boundary conditions [9,53] fluctuations are small enough to keep the stationary state uniform. In those systems mechanism proposed by Cichocki and Sadlej [51] to regularize long range interactions may be a hint to find a solution for correlation functions.

4. Conclusions

In this article, it has been shown that in the BBGKY hierarchy [21] for the correlation function in a sedimenting uniform non-Brownian suspension there is no solution that obeys plasma-like screening relations. This was done by assuming validity of screening conditions and showing that this hypothesis cannot eliminate all long-range terms in the BBGKY hierarchy, and therefore it leads to a contradiction with our assumption. Our results suggest that there is no uniform steady state in the thermodynamic limit. The problem cannot be solved by a system invariant to translations. Our finding is that the uniformity is broken at least on a certain length scale [19,22,23]. This seems to be in an agreement with experimental and numerical investigations. They show that non-uniform structures like swirls [13,24–29] or stratification [30–32] are likely to develop in sedimenting suspensions.

Author Contributions: Conceptualization, B.C.; investigation, P.S., B.C. and M.E.-J.; formal analysis, P.S.; writing—original draft preparation, P.S.; writing—review and editing, P.S., M.E.-J. and B.C.; funding acquisition, M.E.-J. All authors have read and agreed to the published version of the manuscript.

Funding: P.S. and M.E.-J. were supported in part by the National Science Centre under grant UMO-2018/31/B/ST8/03640.

Conflicts of Interest: The authors declare no conflicts of interest.

References

1. Smoluchowski, M. Über die Wechselwirkung von Kugeln, die sich in einer zähen Flüssigkeit bewegen(Mutual interaction of spheres in a viscous medium) 28–29. *Bull. Int. Acad. Sei. Cracovie* **1911**, *2*, 182–194.
2. Smoluchowski, M. On the practical applicability of Stokes law of resistance and its modifications required in certain cases. In Proceedings of the 5th International Congress of Mathematicians, Cambridge, UK, 22–28 August 1912; pp. 195–208.
3. Beenakker, C.; Mazur, P. On the Smoluchowski Paradox in a Sedimenting Suspension. *Phys. Fluids* **1985**, *28*, 767–769. [CrossRef]
4. Batchelor, G.K. Sedimentation in a dilute dispersion of spheres. *J. Fluid Mech.* **1972**, *52*, 245. [CrossRef]
5. Beenakker, C.; Mazur, P. Is sedimentation container-shape dependent? *Phys. Fluids* **1985**, *28*, 3203. [CrossRef]
6. Caflisch, R.; Luke, J.H.C. Variance in the sedimentation speed of a suspension. *Phys. Fluids* **1985**, *28*, 759. [CrossRef]
7. Koch, D.L.; Shaqfeh, E. Screening in sedimenting suspensions. *J. Fluid Mech.* **1991**, *224*, 275. [CrossRef]
8. Ladd, A. Hydrodynamic screening in sedimenting suspensions of non-Brownian spheres. *Phys. Rev. Lett.* **1996**, *76*, 1392–1395. [CrossRef]
9. Ladd, A. Sedimentation of homogeneous suspensions of non-Brownian spheres. *Phys. Fluids* **1997**, *9*, 491–499. [CrossRef]
10. Cunha, F.R.; Sousa, A.J.; Hinch, E.J. Numerical Simulation of Velocity Fluctuations and Dispersion of Sedimentating Particles. *Chem. Eng. Commun.* **2002**, *189*, 1105–1129. [CrossRef]
11. Ladd, A.J.C. Effects of Container Walls on the Velocity Fluctuations of Sedimenting Spheres. *Phys. Rev. Lett.* **2002**, *88*, 048301. [CrossRef] [PubMed]
12. Brenner, M. Screening mechanisms in sedimentation. *Phys. Fluids* **1999**, *11*, 754. [CrossRef]
13. Mucha, P.J.; Tee, S.Y.; Weitz, D.A.; Shraiman, B.I.; Brenner, M. A model for velocity fluctuations in sedimentation. *J. Fluid Mech.* **2004**, *501*, 71–104. [CrossRef]
14. Nguyen, N.Q.; Ladd, A. Microstructure in a settling suspension of hard spheres. *Phys. Rev. E* **2004**, *69*, 050401. [CrossRef]
15. Nguyen, N.Q.; Ladd, A. Sedimentation of hard-sphere suspensions at low Reynolds number. *J. Fluid Mech.* **2005**, *525*, 73–104. [CrossRef]
16. Wang, J.; Schwarz, J.M.; Paulsen, J.D. Hyperuniformity with no fine tuning in sheared sedimenting suspensions. *Nat. Commun.* **2018**, *9*, 1–7. [CrossRef]
17. Yin, X.; Koch, D.L. Velocity fluctuations and hydrodynamic diffusion in finite-Reynolds-number sedimenting suspensions. *Phys. Fluids* **2008**, *20*, 043305. [CrossRef]

18. Ramaswamy, S. Issues in the statistical mechanics of steady sedimentation. *Adv. Phys.* **2001**, *50*, 297–341. 110050617. [CrossRef]
19. Guazzelli, E.; Hinch, J. Fluctuations and instability in sedimentation. *Annu. Rev. Fluid Mech.* **2010**, *43*, 97–116. [CrossRef]
20. Piazza, R. Settled and unsettled issues in particle settling. *Rep. Prog. Physics. Phys. Soc. (Great Britain)* **2014**, *77*, 056602. [CrossRef]
21. Cichocki, B.; Sadlej, K. Stokesian Dynamics—The BBGKY Hierarchy for Correlation Functions. *J. Stat. Phys.* **2008**, *132*, 129–151. [CrossRef]
22. Segrè, P.; Herbolzheimer, E.; Chaikin, P. Long-Range Correlations in Sedimentation. *Phys. Rev. Lett.* **1997**, *79*, 2574–2577. [CrossRef]
23. Segre, P.; Liu, F.; Umbanhowar, P.; Weltz, D. An effective gravitational temperature for sedimentation. *Nature* **2001**, *409*, 594–597. [CrossRef]
24. Bergougnoux, L.; Guazzelli, É. Dilute sedimenting suspensions of spheres at small inertia. *J. Fluid Mech.* **2021**, *914*, 1–19. [CrossRef]
25. Guazzelli, É. Evolution of particle-velocity correlations in sedimentation. *Phys. Fluids* **2001**, *13*, 1537–1540. [CrossRef]
26. Guazzelli, É. Sedimentation of small particles: How can such a simple problem be so difficult? *Comptes Rendus-Mec.* **2006**, *334*, 539–544. [CrossRef]
27. Miguel, M.C.; Pastor-Satorras, R. Velocity fluctuations and hydrodynamic diffusion in sedimentation. *Europhys. Lett.* **2001**, *54*, 45–50. [CrossRef]
28. Tee, S.Y.; Mucha, P.J.; Cipelletti, L.; Manley, S.; Brenner, M.P.; Segre, P.N.; Weitz, D.A. Nonuniversal velocity fluctuations of sedimenting particles. *Phys. Rev. Lett.* **2002**, *89*, 054501. [CrossRef] [PubMed]
29. Climent, E.; Maxey, M.R. Numerical simulations of random suspensions at finite Reynolds numbers. *Int. J. Multiph. Flow* **2003**, *29*, 579–601. [CrossRef]
30. Bergougnoux, L.; Ghicini, S.; Guazzelli, É.; Hinch, J. Spreading fronts and fluctuations in sedimentation. *Phys. Fluids* **2003**, *15*, 1875–1887. [CrossRef]
31. Gómez, D.C.; Bergougnoux, L.; Guazzelli, É.; Hinch, J. Fluctuations and stratification in sedimentation of dilute suspensions spheres. *Phys. Fluids* **2009**, *21*, 1–10. [CrossRef]
32. Gómez, D.C.; Bergougnoux, L.; Hinch, J.; Guazzelli, É. On stratification control of the velocity fluctuations in sedimentation. *Phys. Fluids* **2007**, *19*, 1–5. [CrossRef]
33. Kim, S.; Karilla, S.J. *Microhydrodynamics: Principles and Selected Applications*, Dover ed.; Dover Publications, Inc.: Mineola, NY, USA, 2005.
34. Lamb, S.H. *Hydrodynamics*, 6th ed.; Cambridge University Press: Cambridge, UK, 1932.
35. Jones, R.B.; Schmitz, R. Mobility matrix for arbitrary spherical particles in solution. *Phys. A Stat. Mech. Its Appl.* **1988**, *149*, 373–394. [CrossRef]
36. Felderhof, B.U. Hydrodynamic interaction between two spheres. *Phys. A Stat. Mech. Its Appl.* **1977**, *89*, 373–384. [CrossRef]
37. Felderhof, B.U. Many-body hydrodynamic interactions in suspensions. *Phys. A Stat. Mech. Its Appl.* **1988**, *151*, 1–16. [CrossRef]
38. Cichocki, B.; Felderhof, B.U.; Hinsen, K.; Wajnryb, E.; Bławzdziewicz, J. Friction and mobility of many spheres in Stokes flow. *J. Chem. Phys.* **1994**, *100*, 3780. [CrossRef]
39. Szymczak, P.; Cichocki, B. A diagrammatic approach to response problems in composite systems. *J. Stat. Mech.* **2008**, *2008*, P01025. [CrossRef]
40. Schmitz, R.; Felderhof, B.U. Friction matrix for two spherical particles with hydrodynamic interaction. *Phys. A Stat. Mech. Its Appl.* **1982**, *113*, 103–116. [CrossRef]
41. Cichocki, B.; Felderhof, B.U.; Schmitz, R. Hydrodynamic interactions between two spherical particles. *PCH PhysicoChemical Hydrodyn.* **1988**, *10*, 383–403.
42. Felderhof, B.U.; Jones, R.B. Hydrodynamic scattering theory of flow about a sphere. *Phys. A Stat. Mech. Its Appl.* **1986**, *136*, 77–98. [CrossRef]
43. Felderhof, B.U.; Jones, R.B. Addition theorems for spherical wave solutions of the vector Helmholtz equation. *J. Math. Phys.* **1987**, *28*, 836–839. [CrossRef]
44. Felderhof, B.U.; Jones, R.B. Displacement theorems for spherical solutions of the linear Navier–Stokes equations. *J. Math. Phys.* **1989**, *30*, 339–342. doi.org/10.1063/1.528450. [CrossRef]
45. Ekiel-Jeżewska, M.; Wajnryb, E. Precise multipole method for calculating hydrodynamic interactions between spherical particles in the Stokes flow. In *Theoretical Methods for Micro Scale Viscous Flows;* Feuillebois, F., Sellier, A., Eds.; Transworld Research Network: Kerala, India, 2009; pp. 127–172.
46. Damour, T.; Iyer, B.R. Multipole analysis for electromagnetism and linearized gravity with irreducible Cartesian tensors. *Phys. Rev. D* **1991**, *43*, 3259–3272. [CrossRef] [PubMed]
47. Thorne, K.S. Multipole expansions of gravitational radiation. *Rev. Mod. Phys.* **1980**, *52*, 299–339. [CrossRef]
48. Balescu, R. *Equilibrium and Nonequilibrium Statistical Mechanics*; A Wiley-Interscience Publication: Hoboken, NJ, USA, 1975.
49. Batchelor, G.K. Sedimentation in a dilute polydisperse system of interacting spheres. Part 1. General Theory. *J. Fluid Mech.* **1982**, *119*, 379, doi.org/10.1017/S0022112082001402. [CrossRef]
50. Batchelor, G.K.; Wen, C.S. Sedimentation in a dilute polydisperse system of interacting spheres. Part 2. Numerical results. *J. Fluid Mech.* **1982**, *124*, 379. [CrossRef]

51. Cichocki, B.; Sadlej, K. Steady-state particle distribution of a dilute sedimenting suspension. *Europhys. Lett.* **2005**, *72*, 936–942. [CrossRef]
52. Martin, P. Sum rules in charged fluids. *Rev. Mod. Phys.* **1988**, *60*, 1075–1127. [CrossRef]
53. Abade, G.C.; Cichocki, B. Sedimentation of non-Brownian suspensions. *J. Phys. Conf. Ser.* **2012**, *392*, 012002. [CrossRef]

symmetry

MDPI

Article

Symmetry and Asymmetry in the Fluid Mechanical Sewing Machine

Neil M. Ribe [1,*], Pierre-Thomas Brun [2] and Basile Audoly [3]

[1] Laboratoire FAST, Université de Paris-Saclay, CNRS, 91405 Orsay, France
[2] Department of Chemical and Biological Engineering, Princeton University, Princeton, NJ 08540, USA; pbrun@princeton.edu
[3] Laboratoire de Mécanique des Solides, CNRS, Institut Polytechnique de Paris, 91120 Palaiseau, France; audoly@lms.polytechnique.fr
* Correspondence: ribe@fast.u-psud.fr; Tel.: +33-6-33-30-83-38

Abstract: The 'fluid mechanical sewing machine' is a device in which a thin thread of viscous fluid falls onto a horizontal belt moving in its own plane, creating a rich variety of 'stitch' patterns depending on the fall height and the belt speed. This review article surveys the complex phenomenology of the patterns, their symmetries, and the mathematical models that have been used to understand them. The various patterns obey different symmetries that include (slightly imperfect) fore–aft symmetry relative to the direction of belt motion and invariance under reflection across a vertical plane containing the velocity vector of the belt, followed by a shift of one-half the wavelength. As the belt speed decreases, the first (Hopf) bifurcation is to a 'meandering' state whose frequency is equal to the frequency Ω_c of steady coiling on a motionless surface. More complex patterns can be studied using direct numerical simulation via a novel 'discrete viscous threads' algorithm that yields the Fourier spectra of the longitudinal and transverse components of the motion of the contact point of the thread with the belt. The most intriguing case is the 'alternating loops' pattern, the spectra of which are dominated by the first five multiples of $\Omega_c/3$. A reduced (three-degrees-of-freedom) model succeeds in predicting the sequence of patterns observed as the belt speed decreases for relatively low fall heights for which inertia in the thread is negligible. Patterns that appear at greater fall heights seem to owe their existence to weakly nonlinear interaction between different 'distributed pendulum' modes of the quasi-vertical 'tail' of the thread.

Keywords: viscous threads; coiling instability; fluid mechanical sewing machine

Citation: Ribe, N.M.; Brun, P.-T.; Audoly, B. Symmetry and Asymmetry in the Fluid Mechanical Sewing Machine. *Symmetry* **2022**, *14*, 772. https://doi.org/10.3390/sym14040772

Academic Editors: Andrzej Herczyński and Roberto Zenit

Received: 11 February 2022
Accepted: 2 April 2022
Published: 8 April 2022

1. Introduction

Everyone is familiar with the beautiful 'liquid thread coiling' instability (also known as 'liquid rope coiling') that occurs when a thin stream of honey falls onto a piece of toast (Figure 1a). Liquid coiling is a classic example of a symmetry-breaking bifurcation. In this case, the basic state is an axisymmetric stagnation flow (Figure 1b). When the fall height H exceeds a critical value (for small heights) or the volumetric flow rate Q becomes less than a critical value (for large heights), the stagnation flow becomes unstable to small lateral perturbations (incipient bending) of the thread's axis. The instability eventually saturates to form a coil with a finite radius R that rotates about the vertical with an angular frequency Ω. The resulting structure comprises three distinct parts in general: an upper quasi-vertical 'tail', the 'coil' in which the thread is strongly bent, and an underlying 'pile' of fluid previously laid down by the coiling. The essential difference between the tail and the coil is that deformation of the former is dominated by stretching, whereas the deformation of the coil is dominated by bending. Since the equations that describe bending are of higher order in the arcwise derivatives than those describing stretching, the coil can be thought of as a boundary layer that ensures the satisfaction of all the relevant boundary conditions at the contact point with the pile.

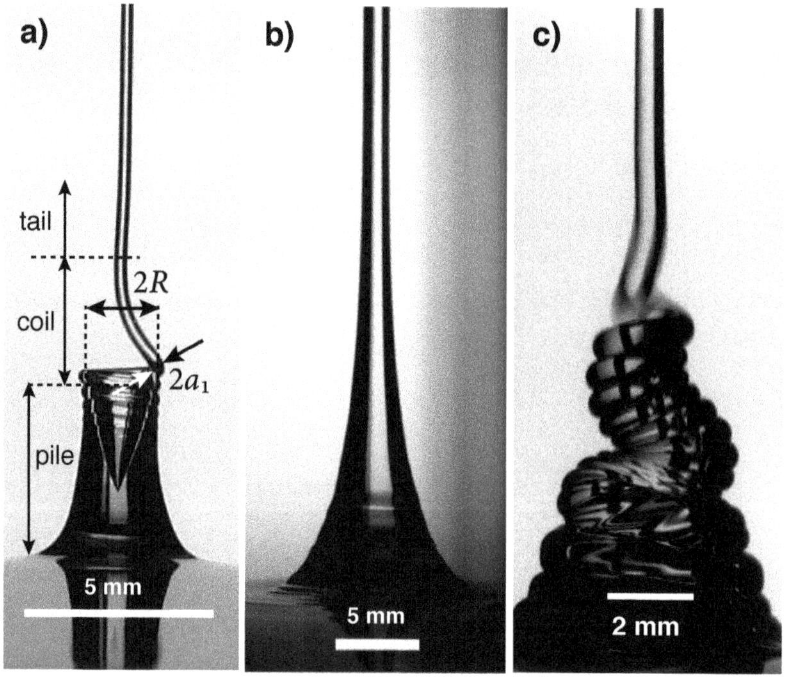

Figure 1. Behaviors of a viscous thread falling onto a motionless surface. (**a**) Steady coiling for $\nu = 6150$ cS, $Q = 0.032$ mL s^{-1}, and $H = 7$ cm. The definitions of the coil radius R and the radius a_1 of the thread at the contact point are shown. The structure comprises a long and nearly vertical tail, the coil, and a pile of fluid previously laid down by the coiling. (**b**) Axisymmetric stagnation flow for $\nu = 3450$ cS, $Q = 3.55$ mL s^{-1}, and $H = 23.5$ cm. (**c**) Coiling with an unstable pile for $\nu = 3450$ cS, $Q = 0.68$ mL s^{-1}, and $H = 25$ cm.

In laboratory experiments, the formation of a pile on the surface is unavoidable. Under some conditions, this pile is a steady-state feature (Figure 1a), while under others it is itself unstable to small perturbations, leading to periodic collapse or quasi-steady secondary buckling (Figure 1c). Especially in these latter cases, the time-dependence of the pile corresponds to a rather 'dirty' unsteady boundary condition at the bottom of the freely coiling thread. It is thus natural to ask: what would happen if we could continuously remove all the fluid laid down by the coiling, thereby doing away with the pile? An easy way to do this comes to mind: it is to let the fluid fall, not onto a motionless surface, but onto a surface moving with a constant horizontal speed V. The result is the "fluid mechanical sewing machine" (FMSM), a name first coined by Chiu-Webster and Lister [1].

Figure 2 shows a simple laboratory apparatus that realizes the FMSM. A fluid thread ejected at a constant volumetric rate Q from a nozzle falls onto a belt in the form of a closed loop that is kept in motion by two rollers. When V is sufficiently high, the thread is stretched in the downstream direction and remains confined to a vertical plane, as shown in Figure 2. However, when V is less than a critical value, the thread becomes unstable to out-of-plane oscillations. This oscillatory motion leaves complex traces of fluid on the belt that resemble stitch patterns when viewed from above, hence the name FMSM.

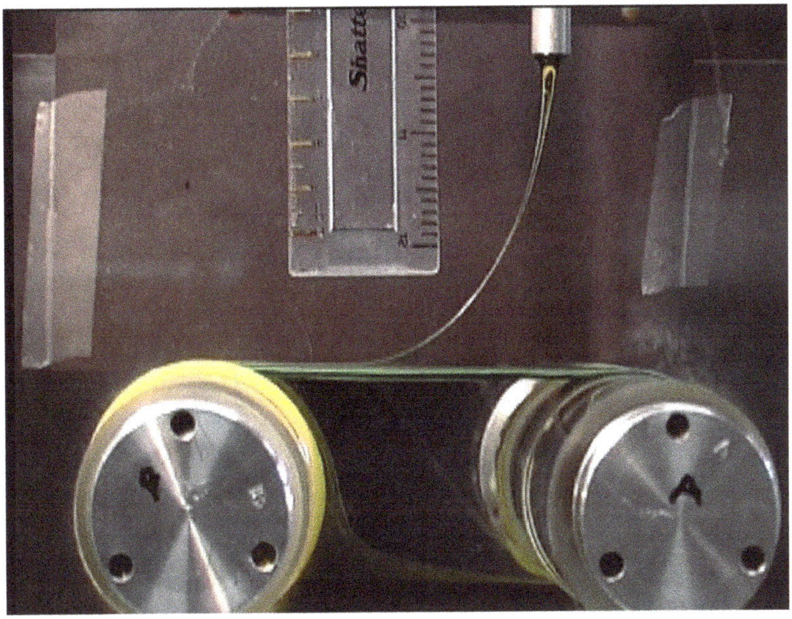

Figure 2. A laboratory apparatus for studying the FMSM. The units on the ruler are centimeters (**left**) and inches (**right**). Photograph courtesy of J. R. Lister.

The first investigations of the dynamics of threads of viscous fluid falling onto surfaces were the experimental studies of liquid thread coiling by G. Barnes and collaborators [2,3]. Since then, liquid thread coiling has been studied in depth using laboratory experiments [4–14] linear stability analysis [11,15,16], scaling analysis [8], asymptotic analysis [17], and numerical analysis based on slender-thread theory [9–14,18]. Interest in the more complicated case of a moving surface—the FMSM—began with the pioneering study of Chiu-Webster and Lister [1], who performed laboratory experiments and proposed a theoretical model in which the thread deforms by stretching alone, without bending. In the same year, Ribe et al. [19] proposed a theoretical model that included bending. They performed a linear stability analysis of the steady dragged state of the thread, and found that the predicted critical belt speed and frequency for the onset of out-of-plane oscillations (meandering) agreed closely with the experimental measurements of [1]. Extensive laboratory experiments with improved apparatus were performed by Morris et al. [20], who determined a detailed phase diagram of the stitch patterns as a function of fall height and belt speed. They showed that the onset of meandering is a Hopf bifurcation, and applied equivariant bifurcation theory together with symmetry constraints to determine generic amplitude equations for interacting modes of the thread's motion. Blount and Lister [21] used matched asymptotic expansions to determine the structure and stability of a dragged viscous thread in the limit of extreme slenderness. Brun et al. [22] performed numerical simulations of the FMSM patterns using an algorithm based on a discrete formulation of the slender-thread equations, and classified the different patterns according to their Fourier spectra. Finally, Brun et al. [23] proposed a reduced (three-degrees-of-freedom) model for non-inertial FMSM patterns, and showed that the model equations accurately predict the sequence of bifurcations that occur as the belt speed changes.

This review article begins in Section 2 by surveying the complex phenomenology of the stitch patterns and classifying them according to their symmetries. Section 3 reviews liquid thread coiling on a motionless surface, the physics of which underlies the FMSM. Section 4 examines the initial bifurcation from a steady dragged thread to meandering via a linear stability analysis based on the theory of slender viscous threads. Section 5 presents direct numerical simulations of the stitch patterns using a 'discrete viscous threads' (DVT) numerical model. Section 6 presents a spectral analysis of selected stitch patterns based on DVT simulations. Section 7 discusses a reduced model for inertia-free stitch patterns that explains several curious features of the more complicated DVT simulations. Finally, Section 8 discusses the (slightly imperfect) fore–aft symmetry of many of the patterns, and closes with some suggestions for further research.

2. Stitch Patterns

Figure 3 shows an essentially complete catalog of the stitch patterns observed experimentally in the FMSM. In this preliminary presentation, the patterns are shown 'all in a jumble', with no attempt to indicate the relationships between them. Apart from the disordered pattern and the steady dragged thread, all the other patterns are periodic in the coordinate x parallel to the belt's motion, with a well-defined primary wavelength λ.

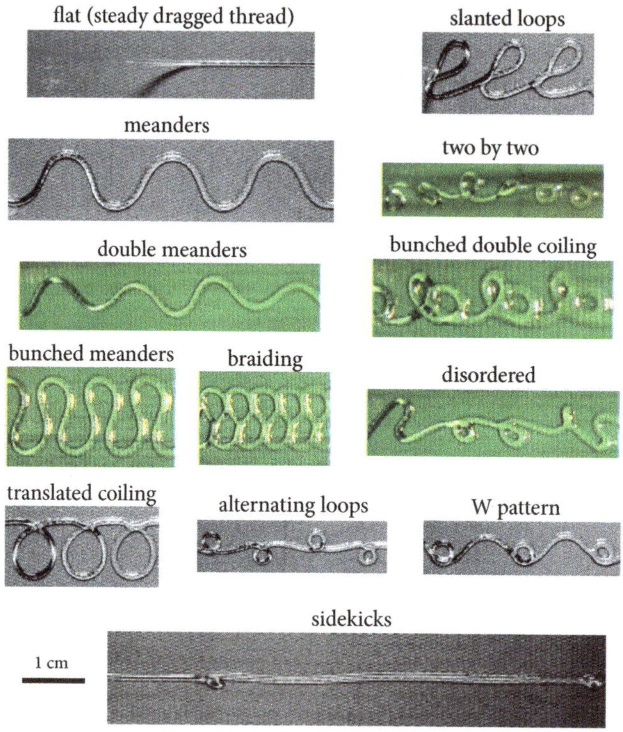

Figure 3. Stitch patterns observed in the FMSM. Photographs courtesy of J.R. Lister.

It is instructive to classify the patterns by their different symmetries. Let y be the horizontal coordinate normal to the direction of the belt's motion, with origin on the line below the center of the nozzle. Because not all patterns are graphs $y(x)$, we describe each pattern by the pair of functions $x(s)$, $y(s)$, where s is the arclength along the thread's axis. Let λ_s be the 'arcwise wavelength', i.e., the total arclength of the thread contained within one wavelength λ in the x-direction. Table 1 shows the symmetries of the patterns. The most highly symmetric pattern is of course the steady dragged thread with $y(s) = 0$, which exhibits (trivial) mirror symmetry across the vertical plane $y = 0$. Among the periodic patterns, three (meanders, alternating loops, and two-by-two) have the symmetry $y(s + \lambda_s/2) = -y(s)$. This corresponds to reflecting the pattern across the line $y = 0$ and then shifting it longitudinally by an amount $\lambda_s/2$. Next comes the symmetry

$y(s_0 - s) = y(s)$ for some s_0, which corresponds to fore–aft symmetry with respect to the direction of belt motion. In addition to the patterns already mentioned, translated coiling and the W-pattern exhibit this symmetry. Finally, the lowest symmetry is simple periodicity such that $y(s + \lambda_s) = y(s)$. All the patterns except the disordered one are periodic.

Table 1. Symmetries of stitch patterns observed experimentally. An X indicates a pattern having the symmetry shown at the top of each column.

Pattern	$y(s) = 0$	$y(s + \lambda_s/2) = -y(s)$	$y(s_0 - s) = y(s)$	$y(s + \lambda_s) = y(s)$
Flat (dragged thread)	X	X	X	X
Meanders	–	X	X	X
Alternating loops	–	X	X	X
Two-by-two	–	X	X	X
Translated coiling	–	–	X	X
W pattern	–	–	X	X
Bunched meanders	–	–	–	X
Braiding	–	–	–	X
Bunched double coiling	–	–	–	X
Slanted loops	–	–	–	X
Double meanders	–	–	–	X
Sidekicks	–	–	–	X
Disordered	–	–	–	–

To understand better the relationships among the patterns, it is helpful to examine a phase diagram showing where each pattern is observed in the space of belt speed and fall height. Figure 4 shows such a diagram for the values of the viscosity ν and the flow rate Q given in the caption. For fall heights $H \leq 5.5$ cm, the patterns appear in the order flat - meanders - alternating loops - translated coiling as the belt speed decreases. For greater fall heights, alternating loops disappear, only to reappear again around $H = 7.8$ cm. The phase diagram is particularly rich for $H \geq 7.2$ cm, where additional patterns such as the W-pattern, double coiling, biperiodic meanders, and slanted loops are seen. Much of the region $H \geq 7.2$ cm is given over to disordered patterns, which occur in 'patches' in the H-V space bounded by regions of periodic patterns.

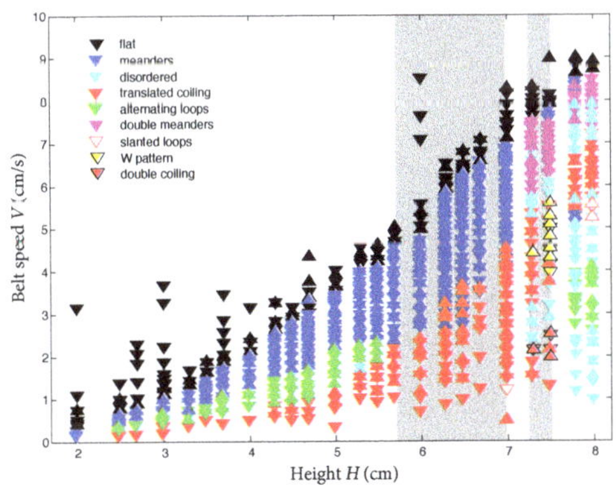

Figure 4. Stitch patterns of the FMSM as a function of fall height H and belt speed V, for silicone oil with $\nu = 277$ S and $Q = 0.027$ cm^3 s^{-1}. Upright and inverted triangles indicate observations made while increasing and decreasing the belt speed, respectively. Grey shaded regions indicate ranges of H for which the frequency of liquid thread coiling, calculated for the parameters of the experiment using the method of Ribe [18], is multivalued. Figure adapted from figure 3 of [20].

3. Basics of Liquid Thread Coiling

As the dynamics of liquid thread coiling underly those of the FMSM, it is important to start by developing a basic understanding of the former. Liquid thread coiling can best be understood by considering how its angular frequency Ω depends on the fall height H, which is the main control parameter. Figure 5 shows a typical variation of Ω as a function of H. As H increases, the relative importance of the viscous, gravitational, and inertial forces acting on the coiled portion of the thread changes, leading to four distinct regimes of liquid thread coiling. At very low fall heights, both gravity and inertia are negligible compared to viscous forces. The coiling in this 'viscous' (V) regime is purely kinematic, driven by the forced extrusion of the fluid from a nozzle, and the coiling frequency scales as [18]

$$\Omega_V \sim \frac{U_1}{H} \tag{1}$$

where $U_1 = Q/(\pi a_1^2)$ is the typical axial velocity of the fluid and a_1 is (to recall) the thread radius at the top of the pile. At somewhat larger fall heights, the viscous forces are balanced by gravity while inertia remains negligible. In this 'gravitational' (G) regime, the coiling frequency scales as [17]

$$\Omega_G \sim \frac{U_1}{\delta}\left(\ln\frac{H}{\delta}\right)^{-1/2} \tag{2}$$

where $\delta \sim (\nu Q/g)^{1/4}$ is the length scale over which gravity balances viscous forces in the coil and ν is the kinematic viscosity. The logarithmic term appears because the tail behaves like a catenary, which is deflected by an amount $\propto F|\ln F|$ by a horizontal force F associated with bending in the coil. At very large fall heights, the dominant forces in the coil are viscous forces and inertia. This is the 'inertial' (I) regime, in which Ω scales as [8]

$$\Omega_I \sim \left(\frac{Q^4}{\nu a_1^{10}}\right)^{\frac{1}{3}} \tag{3}$$

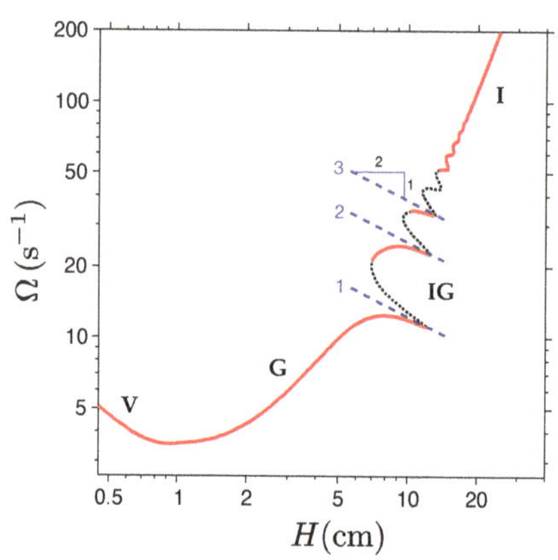

Figure 5. Coiling frequency Ω as a function of the fall height H for $\rho = 0.97$ g cm^{-3}, $\nu = 10^5$ cS, $\gamma = 21.5$ dyne cm^{-1}, $d = 0.068$ cm and $Q = 0.00215$ mL s^{-1}. The red curve was calculated numerically using the method of Ribe [18]. Portions of the curve corresponding to the different steady coiling regimes are indicated: V (viscous), G (gravitational), IG (inertiogravitational), and I (inertial). Dotted portions of the curve represent coiling that is unstable to small perturbations. Dashed lines with slope $-1/2$ are the first three pendulum frequencies of the tail of the thread, with the order of each mode indicated by the number to the left.

At fall heights intermediate between the G and I regimes, there appears an 'inertio-gravitational' (IG) regime in which Ω is a multivalued function of H. In this regime the tail of the thread behaves as a distributed pendulum that enters into resonance with the coil whenever the frequency Ω_G fixed by the coil is close to one of the natural oscillation frequencies of the tail. These frequencies are [10]

$$\Omega^{(n)} = C_n \left(\frac{g}{H}\right)^{1/2}, \quad n = 1, 2, ... \tag{4}$$

where C_n are constants of proportionality. The coiling frequency $\Omega_{IG}(H)$ in this regime defines a set of resonance peaks centered on lines with slope $-1/2$ on a log–log plot of Ω vs. H (Figure 5).

Finally, the relation between a_1 and H depends on the dimensionless fall height $\hat{H} \equiv H(g/\nu^2)^{1/3}$. For $\hat{H} < 2$, the weight of the fluid in the tail is balanced primarily by the viscous resistance to stretching, and $a_1 \sim (\nu Q/gH^2)^{1/2}$. For $\hat{H} > 10$, the weight is balanced primarily by the vertical momentum flux, and $a_1 \sim (Q^2/gH)^{1/4}$.

4. Onset of Meandering

We now examine more closely the first bifurcation in the FMSM from a steady dragged thread to meandering. This bifurcation was studied experimentally by Chiu-Webster and Lister [1] and Morris et al. [20], and theoretically by Ribe et al. [19], Morris et al. [20], and Blount and Lister [21]. Figure 6a compares experimental measurements (circles) of the critical belt speed V_c [1] with the prediction (solid line) of a numerical linear stability analysis of the equations governing the motion of a slender viscous thread with inertia deforming by stretching, bending, and twisting [19]. The parameters of the experiment are given in the figure caption. The experiments and the numerics agree remarkably well.

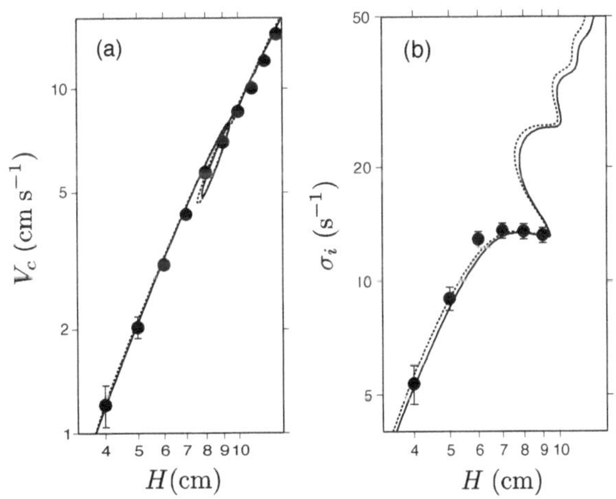

Figure 6. Onset of meandering for golden syrup (ρ = 1.438 g cm^{-3}, ν = 350 S), Q = 0.044 cm^3 s^{-1}, and d = 1.0 cm. (**a**) Critical belt speed V_c as a function of fall height. Circles: experimental measurements. Solid line: prediction of the linear stability analysis described in the text. Dotted line (nearly indistinguishable from the solid line): axial velocity $U_1 = Q/\pi a_1^2$ at the bottom of a thread coiling on a motionless surface for the same experimental parameters. (**b**) Angular frequency of oscillation σ_i as a function of fall height. Circles: experimental measurements. Solid line: prediction of the linear stability analysis. Dotted line: angular frequency Ω of steady liquid thread coiling on a motionless surface for the same experimental parameters. Figure adapted from figures 5 and 7 of [19].

To understand the origin of the critical belt speed, it is revealing to compare the quantity V_c with the vertical speed $U_1 \equiv Q/(\pi a_1^2)$ of a coiling viscous thread that has fallen through a distance H onto a stationary surface. The function $U_1(H)$, calculated numerically using the method of Ribe [18] for the parameters of the FMSM experiment in question, is shown by the dashed line in Figure 6a. The solid and dashed lines are nearly indistinguishable, indicating that $V_c = U_1$ with negligible error. This result has a simple kinematical interpretation: meandering begins when the belt speed becomes too slow to advect away in a straight line the fluid falling onto it.

Further insight can be obtained by comparing the angular frequency σ_i of incipient meandering (i.e., the imaginary part of the growth rate predicted by linear stability analysis) with the angular frequency Ω of liquid thread coiling with the same parameters. These two quantities are shown as functions of H by solid and dashed lines in Figure 6b, respectively. The two curves track each other closely apart from a small systematic offset. This means that for a given fall height the frequency of incipient meandering is nearly equal to the frequency of liquid thread coiling on a motionless surface.

Deeper physical insight into the mechanism of incipient meandering is provided by the asymptotic stability analysis of Blount and Lister [21]. These authors focus on the 'gravitational heel' structure of the lowermost part of the thread when $V \approx U_1$. The shape of this heel (Figure 7) results from a steady balance of the weight of the fluid and the viscous forces that resist bending of the thread. The asymptotic analysis shows that the thread is stable to meandering if bending forces in the heel pull the tail in the same direction as the belt motion, and unstable if the tail is pushed against the belt motion. Blount and Lister [21] liken the latter situation to the heel 'losing its balance'.

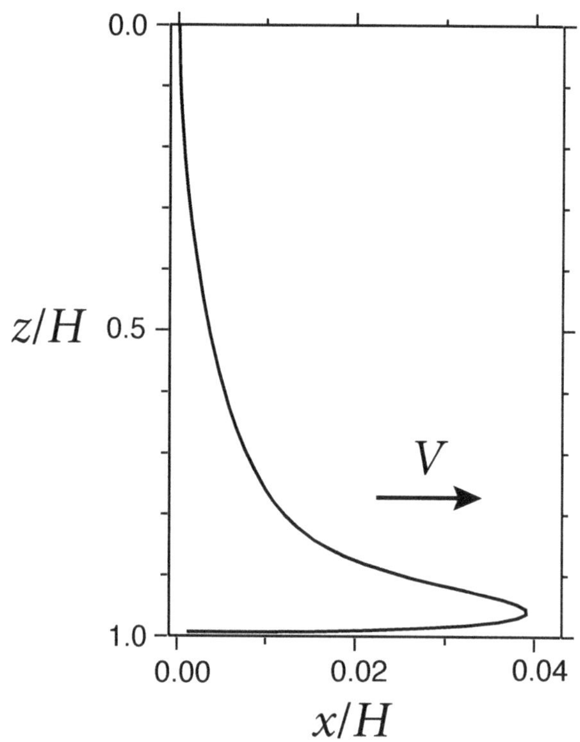

Figure 7. 'Gravitational heel' shape of a dragged viscous thread at the onset of meandering, for the parameters of the experiment of Figure 6. The shape was calculated numerically using the method of Ribe et al. [19]. The horizontal scale is exaggerated by a factor of 15 relative to the vertical scale. Figure adapted from figure 6 of [19].

The linear stability analyses discussed above are limited to meandering of infinitesimal amplitude. Finite-amplitude meandering was studied by Morris et al. [20] using a phenomenological model in which the amplitude A of the meandering is governed by a Landau equation

$$\tau\dot{A} = \epsilon A - \mu A|A|^2 + \text{h.o.t.} \tag{5}$$

where $\epsilon = (V_c - V)/V_c$ is the degree of supercriticality, τ is a time scale related to the linear growth rate of the instability, and h.o.t. denotes higher-order terms. For small amplitudes $A \ll 1$, we can write down perturbation expansions for the position $(x(t), y(t))$ of the contact point. Now the reflectional symmetry of the catenary implies that the out-of-plane variable $y(t)$ must be an odd function of A, while the in-plane variable $x(t)$ must be an even function of A. Thus the position of the contact point must have the form

$$y(t) = A\sin\omega t + O(A^3), \tag{6}$$

$$x(t) = x_0 + B\cos(2\omega t + \phi) + O(A^4) \tag{7}$$

where $B = O(A^2)$, $\omega = \omega_c + O(A^2)$, ω_c is the frequency of the onset of meandering, ϕ is a constant phase, and x_0 is the unperturbed displacement to within an $O(A^2)$ correction. The absolute speed $W(t)$ of the contact point is

$$W^2 = (\dot{x} - V)^2 + \dot{y}^2 = V^2 + 4B\omega V\sin(2\omega t + \phi) + \frac{A^2\omega^2}{2}(1 + \cos 2\omega t) + O(A^4). \tag{8}$$

We now assume that W is constant and equal to the free-fall speed U_1, which is in turn equal to the critical belt speed V_c. Setting $W = V_c$ in (8), we find

$$V_c^2 = V^2 + \frac{A^2\omega^2}{2} + O(A^4), \tag{9}$$

$$B = \frac{A^2\omega}{8V} + O(A^4), \quad \phi = \frac{\pi}{2} + O(A^2). \tag{10}$$

Rearranging (9), we obtain

$$A^2 = \frac{2(V_c^2 - V^2)}{\omega^2} \approx \frac{4V_c^2}{\omega_c^2}\frac{V_c - V}{V_c}, \tag{11}$$

which when compared with the steady form of (5) gives

$$\mu = (\omega_c/2V_c)^2. \tag{12}$$

This generic argument yields the amplitude of the meanders close to threshold. Morris et al. [20] showed in their Figure 8 that the kinematic estimate (12) agreed closely with experimental measurements for fall heights $H > 5.5$ cm.

5. Direct Numerical Simulation

Brun et al. [22] performed a direct numerical simulation of the FMSM using a 'discrete viscous threads' (DVT) algorithm that is most fully discussed in Audoly et al. [24]. Such simulations allow one rapidly to construct a phase diagram of the FMSM patterns by varying adiabatically the fall height H and the belt speed V. Brun et al. [22] presented their phase diagrams in the space of dimensionless fall height $\hat{H}$ and belt speed $\hat{V}$ defined by

$$\hat{H} = H\left(\frac{g}{\nu^2}\right)^{1/3}, \quad \hat{V} = V(\nu g)^{-1/3}. \tag{13}$$

A given phase diagram is uniquely identified by the values of the following three dimensionless groups that are independent of both H and V:

$$\Pi_1 = \left(\frac{\nu^5}{gQ^3}\right)^{1/5}, \quad \Pi_2 = \left(\frac{\nu Q}{gd^4}\right)^{1/4}, \quad \Pi_3 = \frac{\gamma d^2}{\rho \nu Q}. \tag{14}$$

Figure 8 shows the phase diagram predicted using DVT for $\Pi_1 = 670$, $\Pi_2 = 0.37$, and $\Pi_3 = 1.84$ (vertical lines), together with experimental data from Morris et al. [20] for the same parameters (circles). The fall heights $\hat{H} \leq 0.8$ here are 'low' values in the sense that inertia is negligible throughout the thread. The agreement between the numerics and the experiments is quite good, although a tendency for the predicted boundaries to be slightly higher than the experimental ones is noticeable.

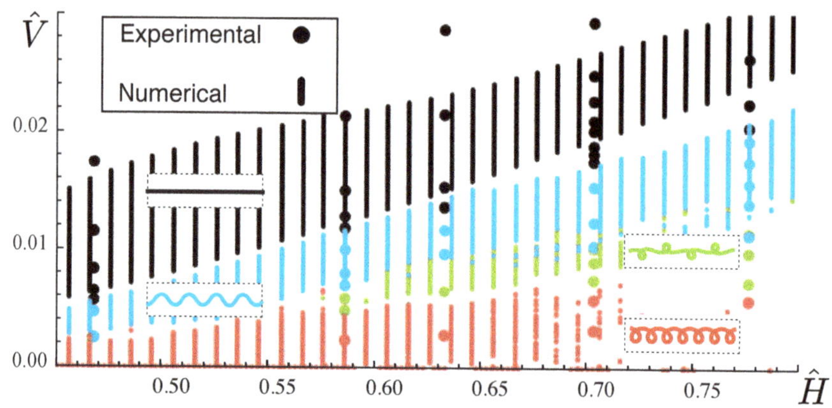

Figure 8. Phase diagram for the patterns at low fall heights $\hat{H} \leq 0.8$. Numerically predicted and experimentally observed patterns as functions of $\hat{H}$ and $\hat{V}$ are indicated by vertical lines and circles, respectively. The patterns corresponding to each color are shown as insets. White spaces between vertical bars of different colors indicate ranges of belt speeds for which the automatic pattern recognition gave ambiguous results. Figure adapted from figure 6 of [22].

For fall heights $\hat{H} > 0.8$, Brun et al. [22] encountered numerical difficulties if surface tension was present ($\Pi_3 > 0$). Accordingly, they assumed $\Pi_3 = 0$ in order to be able to construct a phase diagram for $\hat{H}$ up to 1.5. Figure 9 compares this phase diagram (main portion) with the experimentally determined phase diagram for the same values of Π_1 and Π_2, but $\Pi_3 = 1.84$ (inset). The topology of the two diagrams is broadly similar. Nevertheless, it should be noted that the maximum values of $\hat{H}$ in the two diagrams are different, such that a given pattern appears at larger values of $\hat{H}$ in the laboratory (where surface tension is unavoidable) than in the numerics with $\Pi_3 = 0$. There are also a number of significant differences in detail. In particuliar, the W, slanted loop, and alternating loop patterns are observed in the experiments for $\hat{H} > 1.7$, but do not occur in the high-$\hat{H}$ portion of the numerical phase diagram.

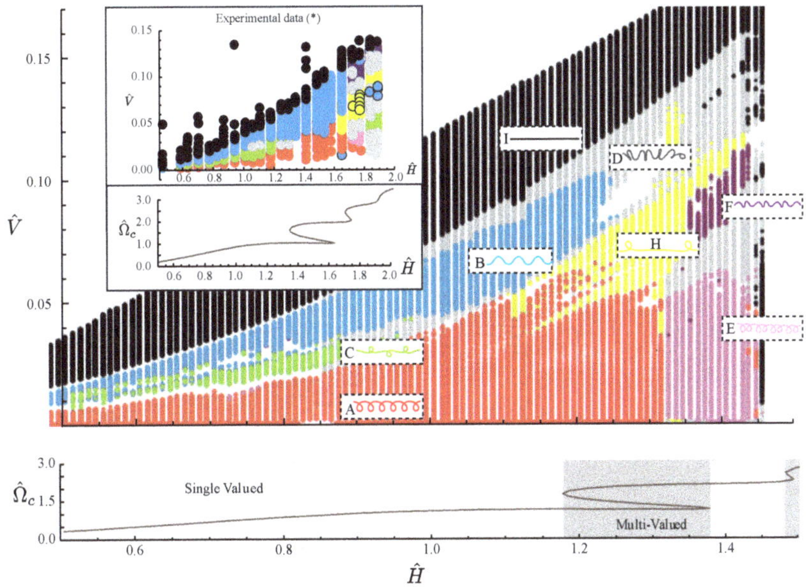

Figure 9. Main portion: phase diagram of the FMSM constructed using DVT for $\Pi_1 = 670, \Pi_2 = 0.37$, and no surface tension ($\Pi_3 = 0$). The patterns shown as insets include translated coiling (A, red), meanders (B, blue), alternating loops (C, green), disordered patterns (D, grey), double coiling (E, pink), double meanders (F, purple), and stretched coiling (H, yellow). The dimensionless frequency $\hat{\Omega}_c(\hat{H})$ of liquid thread coiling is shown beneath the phase diagram, with multivalued intervals highlighted. Inset: phase diagram determined experimentally [20] with $\Pi_1 = 670$, $\Pi_2 = 0.37$ and $\Pi_3 = 1.84$ (**top**). Additional patterns include the W pattern (yellow circled in black) and slanted loops (blue circled in black). The corresponding coiling frequency $\hat{\Omega}_c(\hat{H})$ is also shown (**bottom**). Figure adapted from figure 7 of [22].

6. Spectral Analysis of the Patterns

A revealing way to characterize the different FMSM patterns is to calculate the Fourier spectra of the time-varying longitudinal coordinate $x(t)$ and transverse coordinate $y(t)$ of the contact point of the thread with the belt. Figure 10 illustrates the spectral characteristics of three FMSM patterns: translated coiling (A), alternating loops (B), and double coiling (C). In each of the three panels, part (a) shows the trace on the belt calculated using DVT (left) and the trajectory of the contact point in the frame of the nozzle (right). Part (b), 'Spectral analysis', shows the Fourier transforms of the longitudinal (red) and transverse (blue) motion of the contact point (top), together with the corresponding steady coiling frequency $\hat{\Omega}(\hat{H})$ (bottom). Translated coiling, the simplest of the three patterns, is characterized by a quasi-circular movement of the contact point around the center of the nozzle. Both the longitudinal and transverse motions are dominated by the same single frequency, which is close to the steady coiling frequency for the height in question ($\hat{H} \approx 1.3$). Turning to the more complicated alternating loops pattern, we see that the longitudinal and transverse spectra are now dominated by different frequencies, with five distinct peaks in total. The largest peak occurs for longitudinal motion, with a frequency locked to the value $2\Omega_c/3$. There is also a secondary peak at $4\Omega_c/3$. The transverse motion on the other hand has dominant frequencies $\Omega_c/3$, Ω_c, and $5\Omega_c/3$. Finally, for the double coiling pattern both the longitudinal and transverse motions are dominated by the same two frequencies in the ratio 2:1. These frequencies are very close to the two steady coiling frequencies that coexist at the height in question ($\hat{H} \approx 1.42$).

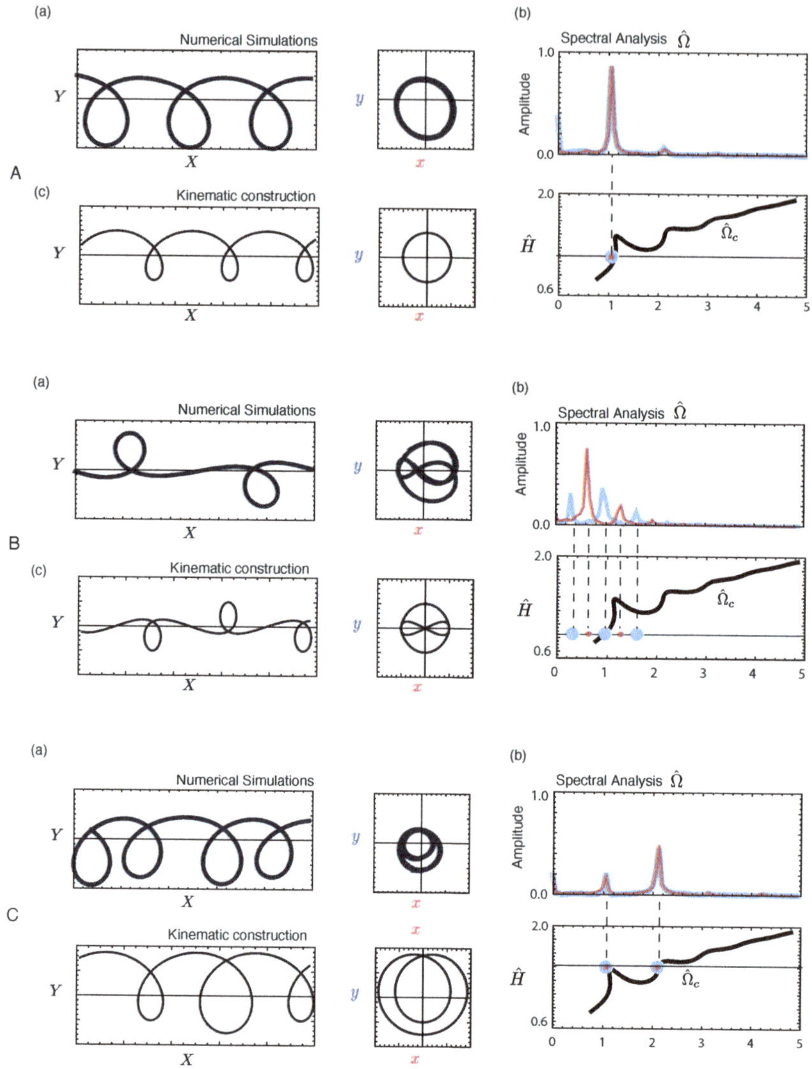

Figure 10. Spectral characteristics of three FMSM patterns: translated coiling (**A**), alternating loops (**B**), and double coiling (**C**). See the main text for a detailed discussion. Figure adapted from figure 10 of [22].

As a final step in our analysis, we reconstruct the patterns by summing the first few terms in a Fourier expansion of the form

$$x(t) + iy(t) = \sum_{j=1}^{N_x} \alpha_j \cos(\omega_j^x t + \phi_j^x) + i \sum_{j=1}^{N_y} \beta_j \cos(\omega_j^y t + \phi_j^y). \quad (15)$$

where $i = \sqrt{-1}$. In (15), ω_j^x, ϕ_j^x, and α_j are the angular frequency, phase, and amplitude of mode j of the longitudinal motion of the contact point, while ω_j^y, ϕ_j^y, and β_j are the analogous quantities for the transverse motion. It turns out that all the patterns can be reconstructed accurately by retaining at most two frequencies in each direction, i.e.,

$N_x \leq 2$ and $N_y \leq 2$. Table 2 presents the values of the relative frequencies, phases, and approximate amplitudes for seven selected FMSM patterns, including the three discussed in the previous paragraph. A frequency of unity is assigned to the largest peak in either the longitudinal or the transverse spectrum. Returning now to Figure 10, we focus on part (c) of each panel, which shows the reconstructed trace (left) and contact point trajectory (right) predicted in (15) with the parameters given in Table 2. No attempt has been made to match the wavelength of the patterns in parts (a). In all three cases, the reconstructed traces and contact-point trajectories are idealized versions of the (slightly) less regular traces/trajectories predicted by the full DVT simulations. The kinematic reconstructions of these three patterns exhibit perfect fore–aft symmetry, as do those of the other patterns considered by Brun et al. [22] (meanders, double meanders, stretched coiling, and the W pattern).

Table 2. Spectral characteristics of several FMSM patterns. The parameters shown are those that appear in the kinematic model (15). The frequency 1 is assigned to the peak in the spectrum with the largest amplitude. A star indicates a frequency that is locked to the steady coiling frequency $\hat{\Omega}_c$.

Pattern	ω_1^x	ω_2^x	ω_1^y	ω_2^y	ϕ_1^x	ϕ_2^x	ϕ_1^y	ϕ_2^y	α_1	α_2	β_1	β_2
translated coiling	1 *	–	1 *	–	0	–	0	–	1.0	–	1.0	–
meanders	2	–	1 *	–	0	–	$\pi/4$	–	0.2	–	1.0	–
alternating loops	1	–	1/2	3/2 *	$\pi/2$	–	0	0	1.0	–	0.5	0.5
double coiling	1/2	1 *	1/2	1 *	$\pi/2$	$\pi/2$	0	0	0.5	1.5	0.1	1.5
double meanders	1/2	–	–	1 *	$\pi/4$	–	0	–	1.0	0.0	0.0	1.5
stretched coiling	1	2 *	1	2 *	$\pi/2$	$\pi/2$	0	0	1.0	0.1	0.5	0.1
W pattern	1	2 *	1	2 *	$\pi/2$	$\pi/2$	0	0	1.0	0.2	0.2	0.5

7. A Reduced Model for Non-Inertial Patterns

At this point, we have a self-consistent and fairly complete description of the FMSM patterns based on direct numerical simulation. However, the very high spatial order (= 21) of the system of differential equations governing unsteady viscous threads leads us to ask: is it possible to formulate a reduced model for the FMSM with only a few degrees of freedom that can predict the patterns? The answer turns out to be yes, at least for the patterns observed at low fall heights that do not involve inertia. Figure 4 shows that for $H \leq 5.5$ cm, the sequence of patterns that occur as the belt speed decreases is flat $\rightarrow$ meanders $\rightarrow$ alternating loops $\rightarrow$ translated coiling. Moreover, DVT simulations show additionally that the W-pattern occurs in a narrow range of belt speeds between meanders and alternating loops, but only if V is increasing and not if it is decreasing. The experiments and the DVT simulations together therefore suggest that the complete order of the stitch patterns is translated coiling $\rightarrow$ alternating loops $\rightarrow$ W pattern $\rightarrow$ meanders $\rightarrow$ flat if V is increasing, but flat $\rightarrow$ meanders $\rightarrow$ alternating loops $\rightarrow$ translated coiling if V is decreasing.

To understand these transitions, Brun et al. [23] proposed a reduced model with three degrees of freedom. As previously, we denote by U_1 the speed at which the falling thread impinges on the belt. Let s be the arclength along the trace of the thread on the belt, such that $s = 0$ is the point that was laid down on the belt at time $t = 0$ and $s = U_1 t$ is the current point of contact $\boldsymbol{r}$. Then the position on the belt $\boldsymbol{q}(s,t)$ of the point s at time t is

$$\boldsymbol{q}(s,t) = \boldsymbol{r}(s) + V(t - s/U_1)\boldsymbol{e}_x \tag{16}$$

where the second term represents the change of position due to advection by the belt velocity $V\boldsymbol{e}_x$. The unit tangent vector at the point of contact is

$$\boldsymbol{t} \equiv \partial_s \boldsymbol{q}|_{s=U_1 t} = \boldsymbol{r}'(s) - (V/U_1)\boldsymbol{e}_x. \tag{17}$$

Now let $\{r(s), \psi(s)\}$ be the polar coordinates of the contact point $\boldsymbol{r}(s)$, and let $\theta(s)$ be the angle between $\boldsymbol{e}_x$ and $\boldsymbol{t}$. Referring to Figure 11, we resolve $\boldsymbol{t}$, $\boldsymbol{r}'$, and $\boldsymbol{e}_x$ into the polar basis $(\boldsymbol{e}_r, \boldsymbol{e}_\psi)$. The two components of (17) then take the form

$$r' = \cos(\theta - \psi) + \frac{V}{U_1}\cos\psi, \tag{18}$$

$$r\psi' = \sin(\theta - \psi) - \frac{V}{U_1}\sin\psi. \tag{19}$$

We need one more equation to close the system. To find it, we note that the DVT model predicts that the curvature θ' of the trace at the contact point depends only on r and $\theta - \psi$ to a good approximation:

$$\theta' = \kappa(r, \theta - \psi). \tag{20}$$

The function κ is determined by fitting to the predictions of the DVT model [23].

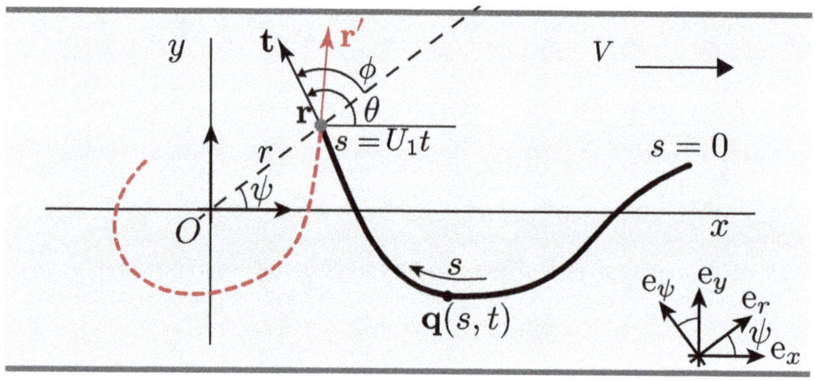

Figure 11. Geometry of a falling viscous thread in the vicinity of its contact point with a moving belt. Figure adapted from figure 3 of [23].

Equation (7) are three coupled nonlinear autonomous first-order differential equations for $r(s)$, $\psi(s)$ and $\theta(s)$. They involve only a single parameter, the dimensionless belt speed V/U_1. The equations remain unchanged when r and s are nondimensionalized by, e.g., the radius R_c of steady coiling on a motionless surface. We call Equation (7) the 'geometric model' (GM). The GM equations were solved using the NDSolve function of Mathematica [25]. The parameter V/U_1 was varied quasi-statically during the simulations, over a time scale large compared with the dominant period of the patterns. Additional solutions were obtained using the continuation and bifurcation analysis software package AUTO-07p [26] to determine the type of each bifurcation encountered as V/U_1 was varied. The convergence of the solutions was verified by showing that they remained essentially identical when the tolerance parameters were increased by an order of magnitude.

Figure 12 presents a detailed comparison of the GM (green) and DVT (brown) predictions. Part (a) shows the orbits $\boldsymbol{r}(s)$ for the four periodic patterns in the frame of the nozzle as predicted by the GM. Part (b) shows the corresponding patterns $\boldsymbol{q}(s,t)$ predicted by both DVT and the GM for four different values of $V/U_1 = 0.28$ (meanders), 0.52 (alternating loops), 0.66 (W pattern) and 0.83 (meanders). The GM captures all the patterns originally predicted by the DVT simulations, in the proper order as V/U_1 varies. The horizontal lines in part (c) show the stability ranges of the patterns. In the original DVT simulations, bifurcations between the patterns were found to be hysteretic: the value of V/U_1 at which each transition occurs depends on whether V/U_1 is increasing or decreasing. This feature of the solutions is reproduced by the GM, which in addition allows one to characterize the type of bifurcation involved in each case. The bifurcations are fold points, except for two: a torus bifurcation from translated coiling to alternating loops as V/U_1 increases, and a period-doubling bifurcation from the W pattern to alternating loops as V/U_1 decreases. Finally, the GM explains why the W pattern was observed in DVT simulations with an increasing belt velocity, but not with a decreasing one.

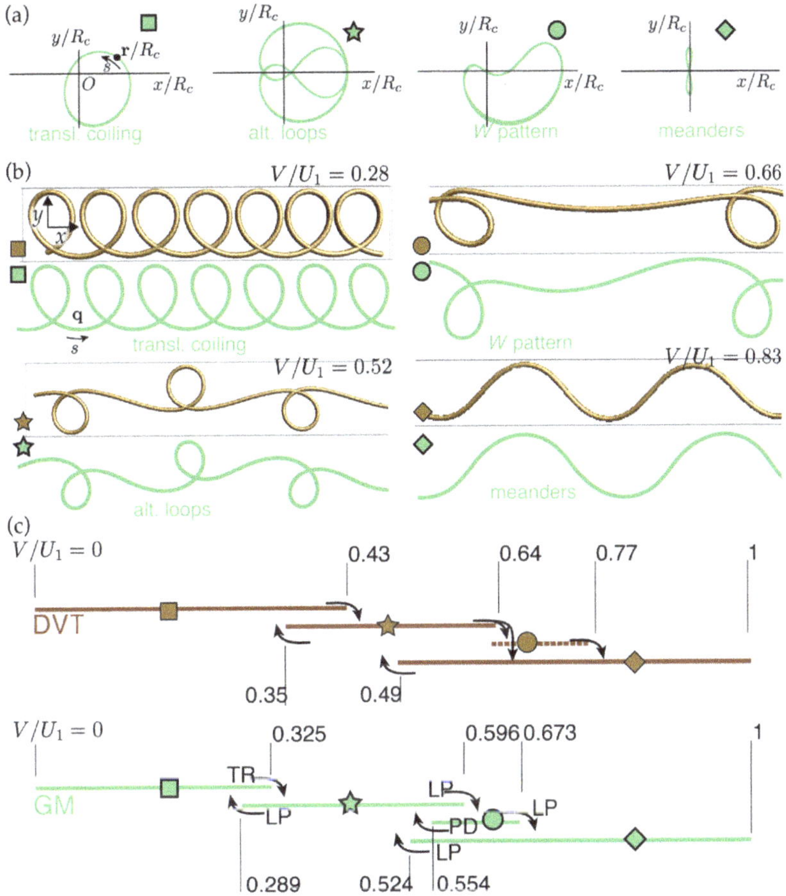

Figure 12. Comparison of the predictions of the GM (green) and full DVT simulations (brown). See the main text for a detailed discussion. Figure adapted from figure 5 of [23].

8. Discussion

A striking feature of many of the FMSM patterns is their fore–aft symmetry with respect to the direction of motion of the belt. Experimentally observed patterns having this symmetry include meanders, alternating loops, two-by-two, translated coiling, and the W pattern. Moreover, kinematic reconstructions reveal that other patterns belong to this group, including double coiling, double meanders, and stretched coiling. Viewing such patterns from above provides no indication of which way the belt is moving. The situation for patterns lacking fore–aft symmetry is quite different: once any of these patterns have been associated with a direction of belt motion by means of one observation, subsequent observations will always be able to indicate which way the belt is moving.

Figure 3 and Table 1 shows that several patterns lack fore–aft symmetry: bunched meanders, braiding, bunched double coiling, slanted loops, double meanders, and sidekicks. Different factors seem to be responsible for this. First, in some of these patterns (braiding, bunched double coiling, slanted loops), the free part of the falling thread makes contact with an 'earlier' portion of the thread that was previously laid down on the belt. Such self-interaction does not occur in the DVT simulations, where the fluid laid down on the belt is continuously 'whisked away' to make the boundary condition clean. In laboratory experiments, however, self-interaction in the case of certain patterns cannot be avoided, and breaks the fore–aft symmetry. A second symmetry-breaking factor is a large amplitude of the pattern relative to its wavelength. An example is bunched meanders, which weakly break the fore–aft symmetry of their smaller-amplitude counterparts (i.e., normal meanders). A third potential symmetry-breaking factor may be the presence of two dominant frequencies, as in double meanders.

However, a closer look shows that fore–aft symmetry is always imperfect even for patterns that look symmetric to the naked eye. This is clear from the DVT spectra of the longitudinal motion (red lines) shown in the right-hand colum of Figure 10. These all have noticeable power at zero frequency, indicating a constant offset of the contact point trajectory in the x-direction relative to the center of the nozzle. This offset is visible in the contact point trajectories shown at upper right in each panel of the left-hand column of Figure 10. Physically, the offset reflects the fact that the gravitational heel structure of the thread near the belt (Figure 7) does not itself have fore–aft symmetry, because the thread is dragged in only one direction by the belt. In this context it is worth noting that kinematic reconstructions of the DVT contact point trajectories that neglect the power at zero frequency do exhibit fore–aft symmetry (left-hand column of Figure 10, lower-right portion of each panel).

While existing studies provide significant physical insight, much remains to be done before we fully understand the FMSM. On the experimental side, observations obtained using much higher values of the dimensionless parameter $\Pi_1 = (\nu^5/gQ^3)^{1/5}$ would be desirable. The reason is that the multivaluedness of the steady coiling frequency $\Omega(H)$ increases with Π_1 [10]. Values of $\Pi_1 \approx 10^5$ are easy to achieve in the laboratory using high-viscosity silicone oil and very low flow rates [11]. A greater degree of multivaluedness indicates a greater number of possible nonlinear interactions among the inertio-gravitational pendulum modes. Such interactions should lead to an even richer phase diagram of FMSM patterns than the one we have studied here. From the theoretical point of view, the FMSM can be regarded as a weakly forced nonlinear oscillator. An ambitious but worthy goal would be to develop an asymptotic nonlinear oscillator theory for the FMSM patterns by considering the mode interactions mentioned above.

Author Contributions: Writing—original draft preparation, N.M.R.; writing—review and editing, P.-T.B. and B.A. All authors have read and agreed to the published version of the manuscript.

Funding: This research received no external funding.

Informed Consent Statement: Not applicable.

Data Availability Statement: Not applicable.

Acknowledgments: We should like to thank the many collaborators without whose contributions the present review could not have been written: M. Bergou, S. Chiu-Webster, N. Clauvelin, J. H. P. Dawes, T. S. Eaves, E. Grinspun, J. R. Lister, S. W. Morris, and M. Wardetzky. We also thank four anonymous referees for constructive suggestions that helped to improve the original manuscript.

Conflicts of Interest: The authors declare no conflict of interest.

Abbreviations

The following abbreviations are used in this manuscript:

DVT	Discrete Viscous Threads
FMSM	Fluid Mechanical Sewing Machine
G	Gravitational coiling regime

GM Geometric model
I Inertial coiling regime
IG Inertiogravitational coiling regime
V Viscous coiling regime

References

1. Chiu-Webster, S.; Lister, J.R. The fall of a viscous thread onto a moving surface: A 'fluid mechanical sewing machine'. *J. Fluid Mech.* **2006**, *569*, 89–111. [CrossRef]
2. Barnes, G.; Woodcock, R. Liquid rope-coil effect. *Am. J. Phys.* **1958**, *26*, 205–209. [CrossRef]
3. Barnes, G.; MacKenzie, R. Height of fall versus frequency in liquid rope-coil effect. *Am. J. Phys.* **1959**, *27*, 112–115. [CrossRef]
4. Cruickshank, J.O. Viscous Fluid Buckling: A Theoretical and Experimental Analysis with Extensions to General Fluid Stability. Ph.D. Thesis, Iowa State University, Ames, IA, USA, 1980.
5. Cruickshank, J.O.; Munson, B.R. Viscous fluid buckling of plane and axisymmetric jets. *J. Fluid Mech.* **1981**, *113*, 221–239. [CrossRef]
6. Huppert, H.E. The intrusion of fluid mechanics into geology. *J. Fluid Mech.* **1986**, *173*, 557–594. [CrossRef]
7. Griffiths, R.W.; Turner, J.S. Folding of viscous plumes impinging on a density or viscosity interface. *Geophys. J.* **1988**, *95*, 397–419. [CrossRef]
8. Mahadevan, L.; Ryu, W.S.; Samuel, A.D.T. Fluid 'rope trick' investigated. *Nature* **1998**, *392*, 140; Erratum in *Nature* **2000**, *403*, 502. [CrossRef]
9. Maleki, M.; Habibi, M.; Golestanian, R.; Ribe, N.M.; Bonn, D. Liquid rope coiling on a solid surface. *Phys. Rev. Lett.* **2004**, *93*, 214502. [CrossRef]
10. Ribe, N.M.; Huppert, H.E.; Hallworth, M.A.; Habibi, M.; Bonn, D. Multiple coexisting states of liquid rope coiling. *J. Fluid Mech.* **2006**, *555*, 275–297. [CrossRef]
11. Ribe, N.M.; Habibi, M.; Bonn, D. Stability of liquid rope coiling. *Phys. Fluids* **2006**, *18*, 084102. [CrossRef]
12. Habibi, M.; Maleki, M.; Golestanian, R.; Ribe, N.M.; Bonn, D. Dynamics of liquid rope coiling. *Phys. Rev. E* **2006**, *74*, 066306. [CrossRef] [PubMed]
13. Habibi, M.; Rahmani, Y.; Bonn, D.; Ribe, N.M. Buckling of liquid columns. *Phys. Rev. Lett.* **2010**, *104*, 074301. [CrossRef] [PubMed]
14. Habibi, M.; Hosseini, S.H.; Khatami, M.H.; Ribe, N.M. Liquid supercoiling. *Phys. Fluids* **2014**, *26*, 024101. [CrossRef]
15. Cruickshank, J.O. Low-Reynolds-number instabilities in stagnating jet flows. *J. Fluid Mech.* **1988**, *193*, 111–127. [CrossRef]
16. Tchavdarov, B.; Yarin, A.L.; Radev, S. Buckling of thin liquid jets. *J. Fluid Mech.* **1993**, *253*, 593–615. [CrossRef]
17. Blount, M.J. Bending and Buckling of a Falling Viscous Thread. Ph.D. Thesis, University of Cambridge, Cambridge, UK, 2010.
18. Ribe, N.M. Coiling of viscous jets. *Proc. R. Soc. London Ser. A Math. Phys. Eng. Sci.* **2004**, *460*, 3223–3239. [CrossRef]
19. Ribe, N.M.; Lister, J.R.; Chiu-Webster, S. Stability of a dragged viscous thread: Onset of 'stitching' in a fluid mechanical 'sewing machine'. *Phys. Fluids* **2006**, *18*, 124105. [CrossRef]
20. Morris, S.; Dawes, J.H.P.; Ribe, N.M.; Lister, J.R. Meandering instability of a viscous thread. *Phys. Rev. E* **2008**, *77*, 066218. [CrossRef]
21. Blount, M.J.; Lister, J.R. The asymptotic structure of a slender dragged viscous thread. *J. Fluid Mech.* **2011**, *674*, 489–521. [CrossRef]
22. Brun, P.T.; Ribe, N.M.; Audoly, B. A numerical investigation of the fluid mechanical sewing machine. *Phys. Fluids* **2012**, *24*, 043102. [CrossRef]
23. Brun, P.T.; Audoly, B.; Ribe, N.M.; Eaves, T.S.; Lister, J.R. Liquid ropes: A geometrical model for thin viscous jet instabilities. *Phys. Rev. Lett.* **2015**, *114*, 174501. [CrossRef] [PubMed]
24. Audoly, B.; Clauvelin, N.; Brun, P.T.; Bergou, M.; Grinspun, E.; Wardetzky, M. A discrete geometric approach for simulating the dynamics of thin viscous threads. *J. Comp. Phys.* **2013**, *253*, 18–49. [CrossRef]
25. Wolfram Research, Inc. *Mathematica*, Version 13.0.0; Wolfram Research, Inc.: Champaign, IL, USA, 2021. Available online: https://www.wolfram.com/mathematica (accessed on 1 February 2022).
26. Doedel, E.J.; Oldeman, B.E. *AUTO-07P: Continuation and Bifurcation Software for Ordinary Differential Equations*; Technical Report; Concordia University: Montréal, QC, Canada, 2007.

symmetry MDPI

Article

Viscous Thread Falling on a Spinning Surface

Maciej Lisicki [1], Łukasz Adamowicz [1], Andrzej Herczyński [2,*] and Henry Keith Moffatt [3]

[1] Institute of Theoretical Physics, Faculty of Physics, University of Warsaw, 02-093 Warsaw, Poland; mklis@fuw.edu.pl (M.L.); l.adamowicz3@student.uw.edu.pl (Ł.A.)
[2] Department of Physics, Boston College, Chestnut Hill, MA 02467, USA
[3] Department of Applied Mathematics and Theoretical Physics, University of Cambridge, Cambridge CB3 0WA, UK; hkm2@cam.ac.uk
* Correspondence: andrzej@bc.edu

Abstract: A rotational version of the fluid-mechanical sewing machine (FMSM) is investigated experimentally. A thin thread of silicon oil was dispensed at a constant flow rate Q from a height H and fell on a table rotating at an angular speed ω, at a distance R from the axis. In all experimental runs, the values of Q and H were kept constant while the radius R was changed manually after each full rotation. Preliminary results show that the usual stitching patterns ensue as the local linear speed $V = \omega R$ approaches the critical transition speeds seen in the FMSM scenario but with subtle asymmetries introduced by rotational (centrifugal) effects. In some instances, arcs and loops of the traces were noticeably more pronounced when directed outward compared to those pointing toward the axis of rotation. Moreover, we observed stitching patterns not reported before. Overall, the symmetry-breaking features, while clearly visible, were rather subtle. Their morphological characteristics, such as differences in local curvature of traces relative to those in FMSM, are estimated to be below 10% in most cases.

Keywords: viscous flows; coiling instability; free-surface flows; rotating fluids

Citation: Lisicki, M.; Adamowicz, Ł.; Herczyński, A.; Moffatt, H.K. Viscous Thread Falling on a Spinning Surface. *Symmetry* **2022**, *14*, 1550. https://doi.org/10.3390/sym14081550

Academic Editor: Mikhail Sheremet

Received: 28 June 2022
Accepted: 24 July 2022
Published: 28 July 2022

1. Introduction

In 2006, Chiu-Webster and Lister [1] discovered and investigated the fascinating behavior of a viscous thread falling onto a translating surface. The phenomenon has attracted sustained attention since then and because a variety of intricate "stitching" patterns could be obtained—depending on the flow rate, fall height, the speed of the belt, and the material properties of the liquid—it was dubbed the fluid-mechanical sewing machine (FMSM).

The FMSM can be regarded as a generalization of the more familiar coiling effect, first described in detail by Barnes and Woodcock [2]. The behavior of a viscous stream falling on a planar surface is itself a highly complex phenomenon which has only recently been understood to comprise four distinct modes, with the corresponding four scaling laws: viscous, gravitational, inertial-gravitational, and inertial [3–6]. Indeed, at belt speed zero, the FMSM "simply" becomes coiling instability. At increased belt speed, a phase space of FMSM patterns emerges that has been elucidated in detail experimentally [7]. In order to rationalize experimental observations, numerous theoretical works explored the onset of stability of the falling thread [8], also in the asymptotic limit of an extremely slender thread [9], providing insights into the nature of the instability. Further numerical studies [10–12] elucidated the sequence of bifurcations seen when the belt speed changes. A comprehensive review [13] is included in this issue.

An aspect which has received scant attention thus far are the phenomena arising when, instead of falling from a stationary source onto a moving surface, the viscous thread's source is translating and the stream falls onto a stationary substrate. The two scenarios are not equivalent. In the former case, the disturbance of the thread originates at the low end of the stream and propagates upwards; in the latter, the disturbance is imparted at the upper end and propagates downwards. The second scenario is, in fact, commonplace (think of maple syrup poured on toast) and has been identified in artistic explorations of abstract expressionist painter Jackson Pollock [14], who poured viscous pigment while moving the source in the air above a horizontally stretched canvas—effectively painting in the air.

It should be emphasized that in all investigations of FMSM to date, the focus has been on the steady-state motion at a constant belt speed. In experimental studies, transitions from one fixed speed to another have clearly been observed and documented in numerous videos, but theoretical and numerical analyses have yet to address the case when the belt is accelerating.

The present contribution is a preliminary experimental investigation of a viscous thread falling on a rotating, rather than translating, platform. In this case, the stream is dragged by a moving surface, which is rotating about a fixed axis and is thus locally subject to centripetal acceleration. The situation has an interesting practical application. In Chinese culinary tradition, Shangdong pancake is a giant crispy crepe, 30–40 cm in diameter, which is made on a spinning wheel; during the process, sweet (and very viscous) syrup is sometimes poured, in a thin stream, over the pancake. The real motivation for this study, however, came from a suggestion, inspired by the FMSM, that Pollock's painting technique could lead to novel artistic effects when inverted, so that a stream of viscous pigment issuing from a fixed source would be falling on a moving canvas. In such a case, a spinning circular canvas would be the most practical solution and the resulting painting would presumably be a tondo [15].

The remainder of this paper is organized as follows. In Section 2, the details of the experimental procedure are given. In Section 3, selected recorded traces are presented for various groups of dimensionless parameters. In Section 4, experimental results are compared to those obtained in previous FMSM experiments. Brief concluding remarks follow in Section 5.

2. Experimental Setup

A photo of the experimental apparatus is shown in Figure 1. The main component is a stainless steel rotating table of diameter 55 cm. A thin (0.5 cm thick) Plexiglas plate of slightly smaller diameter is attached on the top to provide a smooth substrate for paper with circular and radial gridlines, onto which the viscous thread falls. The ruled papers served three functions: they could be quickly changed, obviating the need to scrape and clean the rotating table between experimental runs; they were used to measure the rotational speed of the table from the videos; and they were used to measure radial positions of the recorded traces. For more accurate measurements of the angular speed, a simple tachometer was employed, which clocked the passing of bright markers affixed to the rim of the table. The rotating table was powered by a rubber belt, which was connected to an electric step motor (FPN Warszawa, Poland, model 182, 40 W power). The range of stable angular speeds obtainable with this apparatus was ω = 0.09–0.25 Hz. At a given setting of the motor, the system maintained the angular speed with the accuracy of about 5%, as measured using the tachometer. A ruled piece of paper was placed on the rotating table to record the motion of the stream as it fell on the surface. A new sheet was placed on the table after each experimental run.

The flow of viscous fluid was ensured by a syringe equipped with the adjustable automatic dispenser (New Era Instruments NE-1000 syringe pump). The syringe was connected through a tube to a copper nozzle outlet of diameter $d = 4 \times 10^{-3}$ m and could be moved radially over the table along a metal rail. In all experimental runs, the vertical position of the nozzle and the flow rate were kept constant, with the volume flux Q of either 2 or 3 mL/min. The nozzle height above the rotating surface was in the range of H = 4–10 $\times 10^{-2}$ m. During experiments, the distance of the nozzle from the rotation axis, R, was adjusted manually.

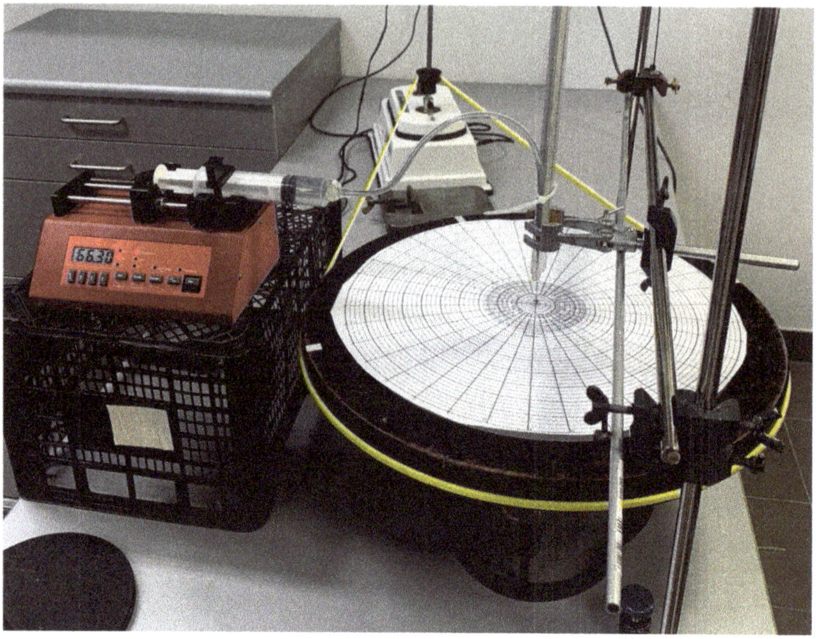

Figure 1. Photograph of the experimental system. The rotating table is connected to a motor by a yellow rubber band. Constant flux of silicon oil was maintained by the syringe pump on the left. The metal rail over the table allowed the nozzle to be moved radially.

Following Morris et al. [7], silicon oil (manufactured by Tornado) was used as the working liquid. Experiments were conducted at room temperature, which was about 21 °C with variations below one degree. Properties of the silicon oil were not measured but were assumed to be those reported by the manufacturer: density $\rho = 996\ \mathrm{kg/m^3}$, kinematic viscosity $\nu = 0.03\ \mathrm{m^2/s}$, and surface tension $\sigma = 21.5 \times 10^{-3}\ \mathrm{N/m}$.

Since there are six fixed physical parameters $(d, Q, \rho, \nu, \sigma, g)$, and three base SI units are needed (kg, m, s), the setup can be characterized by three dimensionless parameters, as guaranteed by the Buckingham Π theorem. Following [7], the three parameters and their approximate values (for $Q = 3\ \mathrm{mL/min} = 5 \times 10^{-8}\ \mathrm{m^3/s}$) are given as

$$\Pi_1 = \left(\frac{\nu^5}{gQ^3}\right)^{1/5} \approx 456, \qquad \Pi_2 = \left(\frac{\nu Q}{gd^4}\right)^{1/4} \approx 0.879, \qquad \Pi_3 = \frac{\sigma d^2}{\rho \nu Q} \approx 0.23. \tag{1}$$

The first parameter, Π_1, is smaller but within 9% of the value reported in [7]. The other two dimensionless parameters differ by a factor of 2.4 and 0.13, respectively, primarily due to the different nozzle diameter d, which was twice as large in [7] as here.

In any particular experimental run, while the syringe pump was dispensing liquid at a constant rate, the two variable flow parameters, H and ω, were kept constant, while the distance to the axis of rotation, R, was varied. In each case, once a constant angular speed ω was ensured, it was necessary to reach the steady-state behavior of the viscous thread, which required waiting for a few seconds from the moment the stream was released onto the rotating table. It was also necessary to prevent the doubling-up of the viscous thread on itself (i.e., its falling on the trace laid-down earlier) by changing the radius R after each full rotation. The two requirements severely limited useful observation time, particularly for smaller radii R and at faster rotation rates.

In what follows, the four variable parameters are reported in dimensionless forms, defined as

$$\hat{H} = \frac{H}{L'}, \qquad \hat{R} = \frac{R}{L'}, \qquad \hat{\omega} = \frac{\omega}{\omega'}, \qquad \hat{Q} = \frac{Q}{Q'}. \tag{2}$$

with the characteristic length and angular frequency for our system

$$L' = \left(\frac{\nu^2}{g}\right)^{1/3} = 4.51 \text{ cm}, \qquad \omega' = \left(\frac{g^2}{\nu}\right)^{1/3} = 0.067 \text{ Hz}. \tag{3}$$

Furthermore, there are two ways of introducing the normalizing flow rate Q', based on dimensionless Π_1 or Π_2. We chose the latter, as we expect that the natural length scale for the flow rate is the diameter of the nozzle d. Thus, the characteristic flow rate is

$$Q' = \frac{gd^4}{\nu} = 5.02 \text{ mL/min}, \tag{4}$$

and it is based on the length scale independent of L'. The approximate corresponding ranges for the experiments reported here were

$$4 < H < 8 \text{ cm}, \quad \text{or} \quad 0.9 < \hat{H} < 1.8, \tag{5}$$

$$0.5 < R < 14 \text{ cm}, \quad \text{or} \quad 0.1 < \hat{R} < 3.1, \tag{6}$$

$$0.09 < \omega < 0.22 \text{ Hz}, \quad \text{or} \quad 1.3 < \hat{\omega} < 3.2, \tag{7}$$

$$Q \in \{2, 3\} \text{ mL/min}, \quad \text{or} \quad \hat{Q} \in \{0.40, 0.60\}. \tag{8}$$

3. Experimental Results

Experiments were carried out for two flow rates, $\hat{Q} = 0.40$ and $\hat{Q} = 0.60$, four heights, $\hat{H} = 0.89$, 1.11, 1.40, and 1.73, and eight angular speeds, $\hat{\omega} = 1.33$, 1.43, 1.82, 1.88, 1.92, 2.20, 2.56, and 3.22. It was not possible to set an a priori selected angular speed because the simple electrical motor used did not have a fine speed adjustment or a speed (or voltage) indicator. After setting the speed to the approximate value required, it was measured repeatedly using a tachometer until the system stabilized and twenty consecutive measurements of ω differed by the minimal fluctuation obtainable, which also provided a measure of uncertainty; only then did experimental runs commence. This method accounts for a relatively large and variable experimental uncertainty in ω, as indicated by the gray vertical strips in Figures 2 and 3.

All experimental runs were filmed from above and the videos were used to delineate the transitions between different "stitching" patterns. Figure 4, extracted from a film of a single experimental run at $\hat{H} = 1.7$, $\hat{\omega} = 2.20 \pm 0.05$, and R decreasing from approximately 8 to 2 cm, shows a typical sequence of patterns, from the flat trace (no instabilities), to sidekicks, meanders or alternating meanders, and translated coiling.

Disordered patterns, as shown in Figure 4b, appeared between meanders or alternating meanders and translated coiling. Such chaotic behavior is transitory and occurs between more stable patterns, usually after the system is disturbed, for example, by dragging the thread radially inward or outward. The phase–space plots for this case are shown, for the two values of flow rate $\hat{Q}$, in Figure 5.

Figure 2 displays results as a function of angular speed for $\hat{Q} = 0.40$ at two slightly different heights, $\hat{H} = 0.9$ and $\hat{H} = 1.1$. At these two smaller heights, a similar sequence of stitching patterns was generated as the radius was decreased, transitioning from flat trace to meanders, alternating loops, and translated coiling. Distinct sidekicks were not observed, while sustained chaotic patterns appeared only at the larger height and lower rotational rates, consistent with the trends encapsulated in Figure 5.

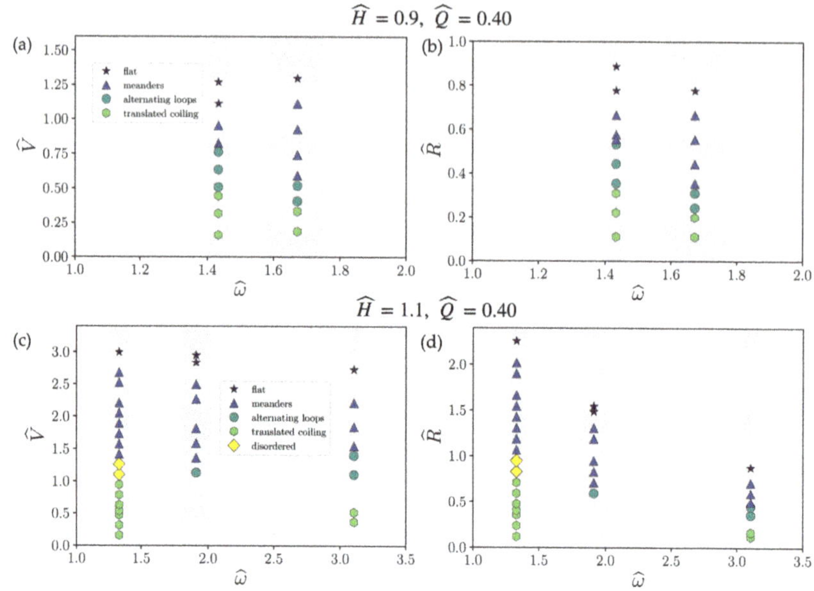

Figure 2. Comparison of patterns obtained at different radial distances (**a**,**c**) and thus different velocities $\widehat{V} = \widehat{\omega}\,\widehat{R}$ (**b**,**d**) for a selection of angular speeds $\widehat{\omega}$ for two similar heights of $\widehat{H} = 0.9$ (**a**,**b**) and $\widehat{H} = 1.1$ (**c**,**d**) for a constant flow rate $\widehat{Q} = 0.40$. Shaded areas show the standard deviation of the measured angular velocity.

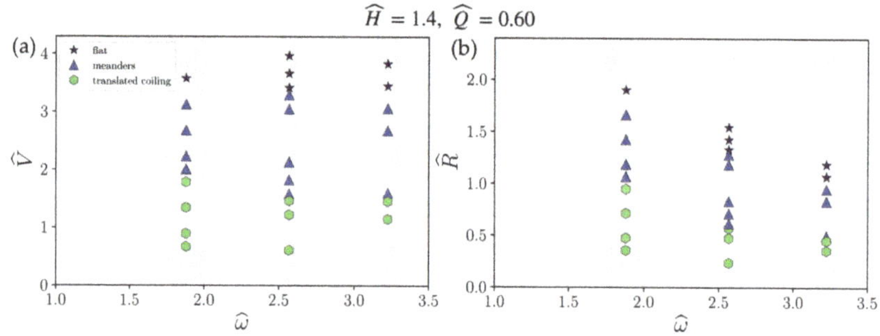

Figure 3. Phase diagram of patterns obtained at different radial distances (**a**) and different velocities (**b**) for a selection of angular speeds $\widehat{\omega}$ for the height $\widehat{H} = 1.4$ at the higher flow rate $\widehat{Q} = 0.60$. Shaded areas show the standard deviation of the measured angular velocity.

Some of the data corresponding to translated coiling in Figure 2, for $\widehat{Q} = 0.40$, $\widehat{H} = 1.1$, and $\widehat{\omega} = 1.33$, are illustrated in the photographs of Figure 6. Panel (a) shows the traces created at $R = 0.7$ cm ($\widehat{R} = 0.16$); panel (b) at $R = 1.5$ cm ($\widehat{R} = 0.33$); and panel (c) at $R = 3$ cm ($\widehat{R} = 0.67$). A striking feature of this sequence is the reversal of the direction of coiling from panel (b) to panel (c), and the corresponding shift from outward (radially away from the axis) to inward (radially toward the axis) loops. Whether this transition was forced by the circular geometry itself, the impact of pulling the stream radially out, or a combination of the two effects remains to be investigated. The falling stream is certainly prone to disturbances which are amplified and determine the transient patterns until fluctuations are damped and a stable pattern emerges.

Behavior at the larger of two flow rates, $\widehat{Q} = 0.60$ and $\widehat{H} = 1.4$, is summarized in Figure 3. Only two sustained stitching patterns, meanders and translated coiling, were observed in this case; double meanders, sidekicks, and alternating loops were not observed. It is possible that even at this larger height, as compared to $\widehat{H} = 0.9$ and $\widehat{H} = 1.1$ in the experiments of Figure 2, the larger flow rate damped-out these particular stitching patterns.

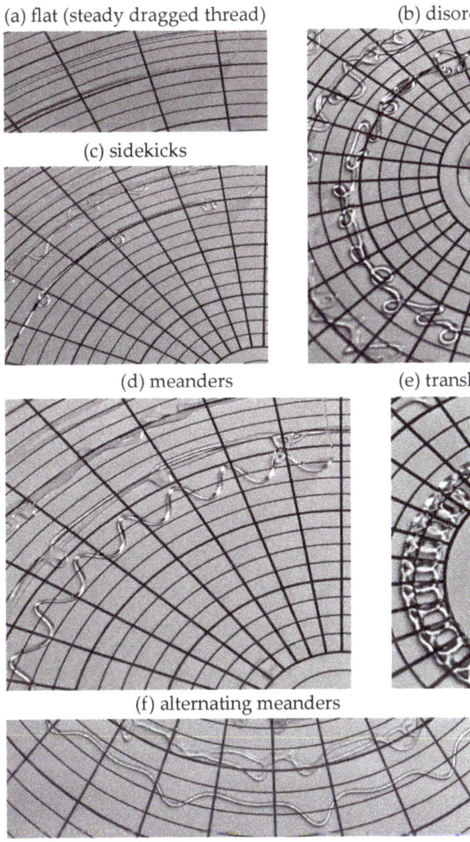

Figure 4. Some examples of stitch patterns of the spinning FMSM observed in a single experimental run with $\widehat{H} = 1.7$, $\widehat{\omega} = 2.20 \pm 0.05$, and $\widehat{Q} = 0.40$. Radial spokes are drawn every 10°, while concentric circles are drawn at a radial distance of 0.5 cm from each other. See Figure 5 for details of radii $\hat{R}$ and velocities $\hat{V}$ at which these patterns were obtained.

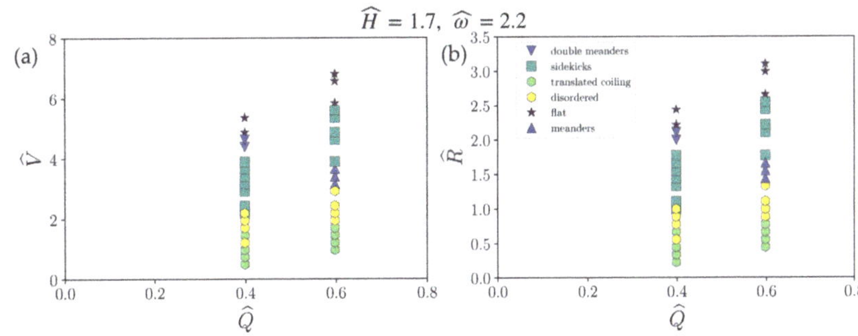

Figure 5. Comparison of patterns obtained at different radial distances (**a**) or velocities $\widehat{V}$ (**b**) for a nozzle positioned at $\widehat{H} = 1.7$ at a fixed angular velocity $\hat{\omega} = 2.2$ and two different flow rates.

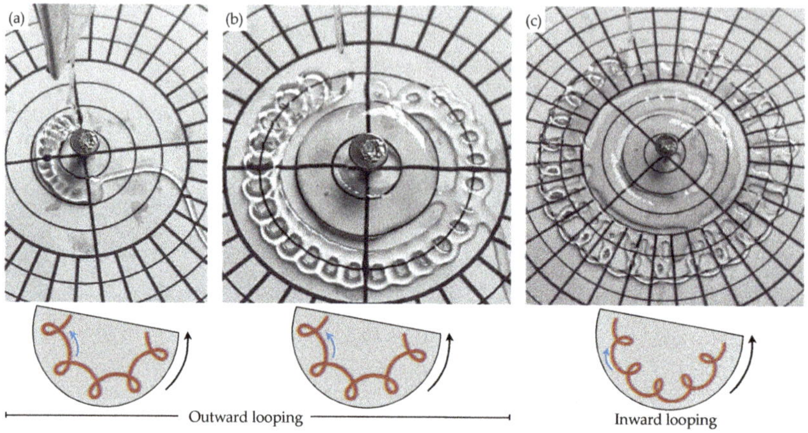

Figure 6. Geometric details of alternating loops patterns within a single experiment with $\widehat{H} = 1.1$, $\widehat{\omega} = 1.33 \pm 0.08$, and $\widehat{Q} = 0.40$. (**a**,**b**) At closer distances from the axis, "outward looping" coils are seen, which means that the angular velocity of the coiling stream (indicated by blue arrows) is of the same sign as the angular velocity of the spinning table (black arrow). (**c**) Further away from the center, the direction of local rotation of the coiling stream reverses.

4. Comparison with FMSM

The phase diagrams of Figures 2 and 3 indicate that the clearly observed patterns, while fewer in number, generally followed a similar order for decreasing $\widehat{V}$, as in classical (linear) FMSM [1,7,10,13], with meanders, sidekicks, and translated coils as the most common forms. The "missing" forms, such as slanted loops, two-by-two, bunched double coiling, bunched meanders, braiding, and W pattern (in the terminology of Chiu-Webster and Lister [1]) were not reliably captured in the experimental runs reported here. Conceivably, these forms might have been missed or suppressed within the limited parameter space explored.

Several stitching patterns were observed only briefly, after the experimental conditions were adjusted and the system was transitioning towards a new pattern. This happened in particular when the radius was changed while other parameters were kept constant. Indeed, the stitching patterns were extremely sensitive to the initial conditions and also subject to hysteresis. A significant limitation of the apparatus was the absence of an oil scraping mechanism. This meant that reliable observations for any given conditions could only be carried out over one full rotation of the table. For larger angular speeds, and smaller radii, this constituted a severe constraint and is probably the reason why some "classical" patterns were not seen.

Quantitative comparisons with the classical FMSM can be only approximate since the experiments reported here did not reproduce the exact conditions of the earlier investigations. The principal difference lies in the initial diameter of the stream, which was $d = 4$ mm in the present investigation and $d = 8$ mm in [7]. The initial speeds of the stream exiting the nozzle were also different, approximately 0.3 cm/s (or 0.4 cm/s for the larger flow rate) in the rotating system and 0.05 cm/s in the linear case, respectively. Finally, flow rates in dimensional terms differed by about 20% (30%).

However, a rough comparison of the thresholds between different stitching patterns can be attempted, relying on the transition from the flat trace to meanders, the most robust indicator. Choosing Figure 2c, wherein highly consistent data for three different values of angular speed are provided at a smaller value of Q, at $H = 5$ cm, this first transition occurred at the local translational speed of the turntable $V \sim 0.8$ cm/s. For the same height, Figure 4 in [13] (adopted from Figure 3 in [7]) indicates the belt speed of $V \sim 2$ cm/s. We conjecture that at smaller flow rates and doubled diameter of the oil dispensing nozzle, the stream in [7] must have been much thinner at contact with the surface than in our experiments. For a conclusive comparison, the same fluid dispensing mechanism should ideally be used in both linear and rotational experiments.

In the regular, translational FMSM apparatus, the circulation of the viscous thread, as in the Magnus effect, breaks the transverse symmetry of the flow. This raises the question of the additional breaking of symmetry due to the local centripetal acceleration of the rotating surface. There are, conceivably, three separate classes of symmetry-breaking manifestations. The first would be a deformation of the shapes of the arcs with respect to the "classical" FMSM traces; the second, additional breaking of the symmetry with respect to the center line of the trace; the third, evidence that either outward or inward arcs or loops become more likely.

For the range of parameters accessible with the present apparatus, the effect of circular geometry on the form of the traces, and their transverse (inward–outward) asymmetry, was subtle and, in most cases, difficult to discern. All the same, some of the transitory patterns observed have apparently not been seen before. The two most conspicuous novel forms are shown in Figure 7. The first, in panel (a), is a pattern with pairs of inward loops separated by an outward loop, which we call two-by-one in analogy to the two-by-two pattern seen earlier (Figure 3 in [13]). Whether a two-by-one pattern can be obtained in a translational FMSM remains to be verified. The second, which we call asymmetric alternating loops, corresponds to the alternating loops seen before, but with an evident symmetry-breaking feature. Here, the S-shaped links connecting subsequent loops display a pronounced centrifugal bulge in the outward-pointing arcs preceding inward loops.

(a) two by one (b) asymmetric alternating loops

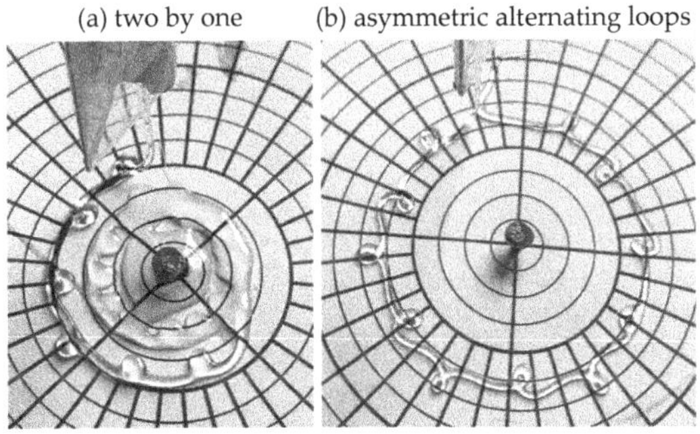

Figure 7. Novel FMSM stitch patterns observed in spinning table experiments. We call the pattern (**a**) two-by-one, since each two inward loops are succeeded by an outward loop. It was observed for $\widehat{H} = 0.89$, $\widehat{\omega} = 1.42 \pm 0.03$, and $\widehat{Q} = 0.40$. The pattern (**b**), seen here for $\widehat{H} = 1.1$, $\widehat{\omega} = 1.92 \pm 0.10$, and $\widehat{Q} = 0.40$, is the analogue of alternating loops in the linear case but its details are affected by centrifugal effects.

It remains uncertain whether in the rotating system outward pointing loops are more likely to occur than inward facing ones (or vice versa). Some of the recorded runs, such as the one illustrated in Figure 6, appear to indicate a tendency to form outward coils at smaller radii, when centrifugal effects at a given V would be expected to be the strongest. Further systematic experiments will be needed to elucidate these effects, if they are indeed present.

5. Concluding Remarks

The present study is a preliminary experimental exploration of a rotational version of the fluid-mechanical sewing machine. Although limited in scope, it is the first investigation of a viscous thread falling on a locally accelerating surface. The photographs of the traces shown in Figures 4, 6 and 7, taken by a fixed camera above the turntable, therefore depict a non-inertial reference frame.

As in previous FMSM studies, a single working liquid was used, silicon oil, and the diameter of the nozzle d was kept the same in all experiments, while the flow rate Q was limited to two values. The remaining flow parameters, ω, R, and H, were varied over a small range of discrete values. However, even within this limited dataset, a number of new results can be extracted.

Since the observational time was limited to one period of revolution ($2\pi/\omega$) for any fixed radius R in any given experimental run, some of the stitching patterns documented in earlier experiments were missed. Nevertheless, the sequence of patterns observed at decreasing $V = \omega R$ qualitatively followed the usual order for translational FMSM: flat line, meanders, alternating loops (when present), and translated coiling.

The rotational effects in the form of the stitching patterns proved, in general, subtle. Symmetry-breaking features, such as differences in morphology relative to the equivalent traces in FMSM, for example, local curvature, are estimated to be below 10% in most cases. One notable exception was an asymmetric variant of the alternating loop pattern observed at $\hat{H} = 1.1$, $\omega = 1.92$, $\hat{R} = 0.5$, and $\hat{Q} = 0.4$, Figure 7b, wherein centrifugal deformations are clearly discernible—the arcs connecting loops directed away from the axis are more pronounced than those toward the axis, while the outward loops are more elongated than the inward loops. Again, additional experiments will be needed to elucidate this new form and search for other symmetry-breaking features, expected to be more evident at larger values of R and ω, and for a given V, at decreasing R.

The distinction between inward and outward loops is enabled by the circular geometry of the setup. Similar distinction arises when a solid, elastic rope falls on a rotating platform and undergoes buckling instability. Amnuanpol explored this scenario [16] and found two new modes of coiling: circles with outward loops (hypotrochoids) and circles with inward loops (epitrochoids). The former dominate for large fall height, feeding speed, and angular speed, whereas the latter is transitional between outward loops and circles. A similar transition for a liquid rope is documented in Figure 6, but we do not have sufficient data to ascertain whether the two phenomena are close analogues.

Detailed quantitative comparisons with the "classical" FMSM will require additional systematic experiments, with three dimensionless parameters delineating the dynamics, Π_1, Π_2, and Π_3, matching those in [7] more closely, and also a similar mechanism for dispensing the liquid. Adding a scraping mechanism would allow much longer observation times under any given conditions so that the system could reliably reach steady-state equilibrium. This would be particularly useful at large heights, H, when the thread is very thin and thus extremely susceptible to external perturbations.

We conclude by noting that a painting technique based on the scenario described here could perhaps open new artistic opportunities within the abstract idiom. Imagine a few paint dispensers suspended and movable over a spinning circular canvas. By carefully varying their radial positions, an intricate network of meandering and looping skeins could be obtained. A subtle interplay of symmetric and asymmetric patterns would then naturally emerge.

Author Contributions: Conceptualization and methodology, H.K.M., M.L. and A.H.; experiments, M.L. and Ł.A.; writing—original draft preparation, A.H. and M.L.; writing—review and editing, M.L., A.H. and H.K.M. All authors have read and agreed to the published version of the manuscript.

Funding: This research received no external funding.

Institutional Review Board Statement: Not applicable.

Informed Consent Statement: Not applicable.

Data Availability Statement: Data are available from the Authors upon request.

Acknowledgments: We wish to thank Krzysztof Karpierz for providing access to experimental equipment and Brian Zhou for helping to identify the Shangdong pancake as a culinary analogue of our setup. We also gratefully acknowledge help from Neil Ribe, who read the manuscript and offered many helpful suggestions.

Conflicts of Interest: The authors declare no conflict of interest.

Abbreviations

The following abbreviations are used in this manuscript:

FMSM	Fluid-mechanical sewing machine

References and Note

1. Chiu-Webster, S.; Lister, J.R. The fall of a viscous thread onto a moving surface: A 'fluid-mechanical sewing machine'. *J. Fluid Mech.* **2006**, *569*, 89–111. [CrossRef]
2. Barnes, G.; Woodcock, R. Liquid Rope-Coil Effect. *Am. J. Phys.* **1958**, *26*, 205–209. [CrossRef]
3. Mahadevan, L.; Ryu, W.S.; Samuel, A.D. Fluid 'rope trick' investigated. *Nature* **1998**, *392*, 140–140. [CrossRef]
4. Ribe, N.M.; Huppert, H.E.; Hallworth, M.A.; Habibi, M.; Bonn, D. Multiple coexisting states of liquid rope coiling. *J. Fluid Mech.* **2006**, *555*, 275–297. [CrossRef]
5. Ribe, N.M.; Habibi, M.; Bonn, D. Liquid Rope Coiling. *Annu. Rev. Fluid Mech.* **2012**, *44*, 249–266. [CrossRef]
6. Ribe, N.M. Liquid rope coiling: A synoptic view. *J. Fluid Mech.* **2017**, *812*, R2. [CrossRef]
7. Morris, S.W.; Dawes, J.H.P.; Ribe, N.M.; Lister, J.R. Meandering instability of a viscous thread. *Phys. Rev. E* **2008**, *77*, 066218. [CrossRef] [PubMed]
8. Ribe, N.M.; Lister, J.R.; Chiu-Webster, S. Stability of a dragged viscous thread: Onset of "stitching" in a fluid-mechanical "sewing machine". *Phys. Fluids* **2006**, *18*, 124105. [CrossRef]
9. Blount, M.J.; Lister, J.R. The asymptotic structure of a slender dragged viscous thread. *J. Fluid Mech.* **2011**, *674*, 489–521. [CrossRef]
10. Habibi, M.; Najafi, J.; Ribe, N.M. Pattern formation in a thread falling onto a moving belt: An "elastic sewing machine". *Phys. Rev. E* **2011**, *84*, 016219. [CrossRef]
11. Brun, P.T.; Ribe, N.M.; Audoly, B. A numerical investigation of the fluid mechanical sewing machine. *Phys. Fluids* **2012**, *24*, 043102. [CrossRef]
12. Brun, P.T.; Audoly, B.; Ribe, N.M.; Eaves, T.S.; Lister, J.R. Liquid Ropes: A Geometrical Model for Thin Viscous Jet Instabilities. *Phys. Rev. Lett.* **2015**, *114*, 174501. [CrossRef] [PubMed]
13. Ribe, N.M.; Brun, P.T.; Audoly, B. Symmetry and Asymmetry in the Fluid Mechanical Sewing Machine. *Symmetry* **2022**, *14*, 772. [CrossRef]
14. Herczyński, A.; Chernuschi, C.; Mahadevan, L. Painting with drops, jets and films. *Phys. Today* **2011**, *64*, 31–36. [CrossRef]
15. Pollock did create one circular painting, *Tondo* (1948, private collection, diameter 58.7 cm), but, as far as is known, it was executed in his "usual" manner, by moving the source of pigment above the stationary, horizontal canvas.
16. Amnuanpol, S. Buckling instability on a rotating plane. *EPL Europhys. Lett.* **2021**, *135*, 50003. [CrossRef]

symmetry

MDPI

Article

Possible Role of Non-Stationarity of Magnetohydrodynamic Turbulence in Understanding of Geomagnetic Excursions

Krzysztof Andrzej Mizerski

Department of Magnetism, Institute of Geophysics, Polish Academy of Sciences, ul. Ksiecia Janusza 64, 01-452 Warsaw, Poland; kamiz@igf.edu.pl; Tel.: +48-22-6915-904

Abstract: The natural simplifying assumptions often put forward in the theoretical investigations of the magnetohydrodynamic turbulence are that the turbulent flow is statistically isotropic, homogeneous and stationary. Of course, the natural turbulence in the planetary interiors, such as the liquid core of the Earth is neither, which has important consequences for the dynamics of the planetary magnetic fields generated via the hydromagnetic dynamo mechanism operating in the interiors of the planets. Here we concentrate on the relaxation of the assumption of statistical stationarity of the turbulent flow and study the effect of turbulent wave fields in the Earth's core, which induces non-stationarity, on the turbulent resistivity in the non-reflectionally symmetric flow and the geodynamo effect. It is shown that the electromotive force, including the so-called α-effect and the turbulent magnetic diffusivity $\bar{\eta}$, induced by non-stationary turbulence, evolves slowly in time. However, the turbulent $\bar{\alpha}$ coefficient, responsible for the dynamo action and $\bar{\eta}$ evolve differently in time, thus creating periods of enhanced and suppressed turbulent diffusion and dynamo action somewhat independently. In particular, periods of enhanced $\bar{\alpha}$ may coincide with periods of suppressed diffusion, leading to a stable and strong field period. On the other hand, it is shown that when enhanced diffusion occurs simultaneously with suppression of the α-effect, this leads to a sharp drop in the intensity of the large-scale field, corresponding to a geomagnetic excursion.

Keywords: magnetohydrodynamics; non-stationary turbulence; dynamo theory; geomagnetic excursions; geomagnetic reversals

Citation: Mizerski, K.A. Possible Role of Non-Stationarity of Magnetohydrodynamic Turbulence in Understanding of Geomagnetic Excursions. *Symmetry* **2021**, *13*, 1881. https://doi.org/10.3390/sym13101881

Academic Editors: Andrzej Herczyński and Roberto Zenit

Received: 7 September 2021
Accepted: 1 October 2021
Published: 5 October 2021

1. Introduction

The terrestrial magnetic field is generated by the hydromagnetic dynamo action in the Earth's liquid core driven thermally and compositionally [1–3]. The turbulent mechanism of dynamo action, which is typically invoked as responsible for the generation of large-scale fields, is the so-called α-effect, based on nonlinear interactions of the small-scale fluctuating components of the non-reflectionally symmetric turbulent state, which generate the large-scale electromotive force (EMF). For dynamo action to occur, the wave field must exhibit chirality, i.e., lack of reflectional symmetry [4–6]. The turbulent flow in the core of the Earth contains a rich wave field composed of nonlinearly interacting magnetohydrodynamic waves such as, e.g., the inertial waves [7], the so-called MAC waves [8,9] or magnetic Rossby waves [10]; for a more complete review see [11] on the dynamics of the core. These interactions could have a profound effect on the dynamics of the core, in particular the dynamo process as they practically rule out statistical stationarity of the core turbulence—a feature often invoked in theoretical investigations of magnetohydrodynamic turbulence, as it greatly simplifies the mathematical approach.

In fact, non-stationarity has been recently shown by [12–15], as an effective mechanism of generation of the large-scale electromotive force through interactions of waves with distinct but close frequencies of oscillations (the beating effect). Such an EMF slowly evolves on timescales comparable or larger than the typical time scales of variation of the large-scale field. It is shown here that statistically non-stationary, helical turbulence induces a slowly varying in time α-effect and turbulent magnetic diffusion, and their variations are out of phase.

Some well-known features of the evolution of the geomagnetic field are the so-called geomagnetic excursions and polarity reversals, the former being the periods of a significant drop in the large-scale field's intensity. In fact, from the point of view of the possible impact that such a phenomenon could have on the high-tech human civilization, it is the decrease in the intensity of the geomagnetic dipole, which is crucial, since it is the dipolar field that greatly prevents the solar wind from entering the atmosphere. Whether or not the actual polarity reversal takes place is of much less importance and hence in here, for short, under the term 'excursion', we will contain both the actual excursions and reversals. The aim of this paper is to comment on the phenomenon of geomagnetic excursions in light of the non-stationarity of the core turbulence induced by wave-interactions and the slow time evolution of the α-effect and turbulent diffusion. This evolution is shown to create periods of enhanced diffusion, which may be correlated with suppression of the α-effect, thus creating favourable conditions for the excursions.

Numerical simulations do not reveal any significant large-scale alterations in the flow of the conducting liquid during a reversal [16–18], although it must be said that the currently available computer power does not allow to reach the strongly constrained Earth-like parameter regime. The first stage of a polarity reversal in simulations by [18] is associated with local intensification of turbulence, where the vigorous flow twists and bends the field lines to locally reverse the magnetic field direction. The enhanced turbulence intensifies the reversed field, which spreads into surrounding regions until the polarity in the core becomes mixed and the dipole moment is weakened. This suggests that the magnetic excursions and polarity reversals are manifestations of a chaotic turbulent behaviour of the liquid core, which is in qualitative agreement with the temporal variations of turbulent transport coefficients conjectured here.

2. Mathematical Formulation of the Mean Field Dynamo Problem

Evolution of the large-scale magnetic fields induced by the complex flow of an incompressible conducting fluid is governed by the following dynamical equations

$$\frac{\partial \mathbf{U}}{\partial t} + (\mathbf{U}\cdot\nabla)\mathbf{U} = \mathbf{F} - \nabla\Pi + \frac{1}{\mu_0\rho}(\mathbf{B}\cdot\nabla)\mathbf{B} + \nu\nabla^2\mathbf{U}, \tag{1a}$$

$$\frac{\partial \mathbf{B}}{\partial t} + (\mathbf{U}\cdot\nabla)\mathbf{B} = (\mathbf{B}\cdot\nabla)\mathbf{U} + \eta\nabla^2\mathbf{B}, \tag{1b}$$

$$\nabla\cdot\mathbf{U} = 0 \qquad \nabla\cdot\mathbf{B} = 0, \tag{1c}$$

where the velocity field of the fluid flow is denoted by $\mathbf{U}(t,\mathbf{x})$, the magnetic field by $\mathbf{B}(t,\mathbf{x})$ and the total pressure

$$\Pi = \frac{p}{\rho} + \frac{B^2}{2\mu_0\rho}. \tag{2}$$

Without loss of generality, we may assume solenoidal forcing

$$\nabla\cdot\mathbf{F} = 0, \tag{3}$$

and for the purpose of simplicity, we rescale the magnetic field in the following way

$$\frac{\langle\mathbf{B}\rangle}{\sqrt{\mu_0\rho}} \to \langle\mathbf{B}\rangle,$$

so that the prefactor $1/\mu_0\rho$ in the Lorentz-force term in the Navier–Stokes (1a) equation is lost.

Next, denoting by angular brackets the ensemble mean,

$$\langle\cdot\rangle - \text{ensemble mean}$$

let us introduce the standard Reynolds decomposition

$$\mathbf{U} = \langle \mathbf{U} \rangle + \mathbf{u}, \quad \mathbf{B} = \langle \mathbf{B} \rangle + \mathbf{b}, \quad p = \langle p \rangle + p', \tag{4}$$

and write down separately the equations for the mean fields $\langle \mathbf{U} \rangle$ and $\langle \mathbf{B} \rangle$ and the turbulent fluctuations $\mathbf{u}$ and $\mathbf{b}$; this yields

$$\begin{aligned} \frac{\partial \langle \mathbf{U} \rangle}{\partial t} + (\langle \mathbf{U} \rangle \cdot \nabla)\langle \mathbf{U} \rangle =& \langle \mathbf{F} \rangle - \nabla \langle \Pi \rangle + (\langle \mathbf{B} \rangle \cdot \nabla)\langle \mathbf{B} \rangle + \nu \nabla^2 \langle \mathbf{U} \rangle \\ & - \nabla \cdot (\langle \mathbf{uu} \rangle - \langle \mathbf{bb} \rangle), \end{aligned} \tag{5a}$$

$$\frac{\partial \langle \mathbf{B} \rangle}{\partial t} = \nabla \times (\langle \mathbf{U} \rangle \times \langle \mathbf{B} \rangle) + \nabla \times \langle \mathbf{u} \times \mathbf{b} \rangle + \eta \nabla^2 \langle \mathbf{B} \rangle, \tag{5b}$$

$$\nabla \cdot \langle \mathbf{B} \rangle = 0, \quad \nabla \cdot \langle \mathbf{U} \rangle = 0. \tag{5c}$$

where

$$\mathcal{E} = \langle \mathbf{u} \times \mathbf{b} \rangle \tag{6}$$

is the large-scale electromotive force (EMF) and

$$\begin{aligned} \frac{\partial \mathbf{u}}{\partial t} - \nu \nabla^2 \mathbf{u} + (\langle \mathbf{U} \rangle \cdot \nabla)\mathbf{u} + (\mathbf{u} \cdot \nabla)\langle \mathbf{U} \rangle - (\langle \mathbf{B} \rangle \cdot \nabla)\mathbf{b} - (\mathbf{b} \cdot \nabla)\langle \mathbf{B} \rangle + \nabla \Pi' \\ = \mathbf{f} - \nabla \cdot (\mathbf{uu} - \mathbf{bb}) + \nabla \cdot (\langle \mathbf{uu} \rangle - \langle \mathbf{bb} \rangle), \end{aligned} \tag{7a}$$

$$\begin{aligned} \frac{\partial \mathbf{b}}{\partial t} - \eta \nabla^2 \mathbf{b} + (\langle \mathbf{U} \rangle \cdot \nabla)\mathbf{b} - (\langle \mathbf{B} \rangle \cdot \nabla)\mathbf{u} + (\mathbf{u} \cdot \nabla)\langle \mathbf{B} \rangle - (\mathbf{b} \cdot \nabla)\langle \mathbf{U} \rangle \\ = \nabla \times (\mathbf{u} \times \mathbf{b} - \langle \mathbf{u} \times \mathbf{b} \rangle), \end{aligned} \tag{7b}$$

$$\nabla \cdot \mathbf{b} = 0, \quad \nabla \cdot \mathbf{u} = 0. \tag{7c}$$

Furthermore, we adopt the "first-order smoothing approximation" [6] in which squares and products of fluctuating quantities are ignored

$$\frac{\partial \mathbf{u}}{\partial t} - \nu \nabla^2 \mathbf{u} + (\langle \mathbf{U} \rangle \cdot \nabla)\mathbf{u} + (\mathbf{u} \cdot \nabla)\langle \mathbf{U} \rangle - (\langle \mathbf{B} \rangle \cdot \nabla)\mathbf{b} - (\mathbf{b} \cdot \nabla)\langle \mathbf{B} \rangle + \nabla \Pi' = \mathbf{f}, \tag{8a}$$

$$\frac{\partial \mathbf{b}}{\partial t} - \eta \nabla^2 \mathbf{b} + (\langle \mathbf{U} \rangle \cdot \nabla)\mathbf{b} - (\langle \mathbf{B} \rangle \cdot \nabla)\mathbf{u} + (\mathbf{u} \cdot \nabla)\langle \mathbf{B} \rangle - (\mathbf{b} \cdot \nabla)\langle \mathbf{U} \rangle = 0, \tag{8b}$$

$$\nabla \cdot \mathbf{b} = 0, \quad \nabla \cdot \mathbf{u} = 0. \tag{8c}$$

Introducing

$$G_{ij} = \frac{\partial \langle U \rangle_i}{\partial x_j}, \qquad \overline{G}_{ij} = \frac{\partial \langle B \rangle_i}{\partial x_j}, \tag{9}$$

taking the Fourier transforms of (8a)–(8c) and eliminating the pressure from the Fourier transformed velocity equation with the use of the projection operator

$$P_{ij}(\mathbf{k}) = \delta_{ij} - \frac{k_i k_j}{k^2}, \tag{10}$$

one obtains

$$\left[-\mathrm{i}(\omega - \mathbf{k} \cdot \langle \mathbf{U} \rangle) + \nu k^2\right] \hat{\mathbf{u}} + \mathbf{P} \cdot \mathbf{G} \cdot \hat{\mathbf{u}} - \mathrm{i}\mathbf{k} \cdot \langle \mathbf{B} \rangle \hat{\mathbf{b}} - \mathbf{P} \cdot \overline{\mathbf{G}} \cdot \hat{\mathbf{b}} = \hat{\mathbf{f}} \tag{11a}$$

$$\left[-\mathrm{i}(\omega - \mathbf{k} \cdot \langle \mathbf{U} \rangle) + \eta k^2\right] \hat{\mathbf{b}} = \mathrm{i}\mathbf{k} \cdot \langle \mathbf{B} \rangle \hat{\mathbf{u}} - \overline{\mathbf{G}} \cdot \hat{\mathbf{u}} + \mathbf{G} \cdot \hat{\mathbf{b}}, \tag{11b}$$

$$\mathbf{k} \cdot \hat{\mathbf{b}} = 0, \quad \mathbf{k} \cdot \hat{\mathbf{u}} = 0, \tag{11c}$$

where we have already used (11c) in projecting the velocity equation on the plane perpendicular to the wave vector **k**. Next we rearrange the terms to put all the terms involving gradients of means on the right hand side

$$\hat{\mathbf{u}} = \frac{1}{\gamma_u}\hat{\mathbf{f}} - \frac{1}{\gamma_u}\mathbf{P}\cdot\mathbf{G}\cdot\hat{\mathbf{u}} - \frac{(\mathbf{k}\cdot\langle\mathbf{B}\rangle)^2}{\gamma_u\gamma_\eta^2}\mathbf{G}\cdot\hat{\mathbf{u}} - \frac{\mathrm{i}\mathbf{k}\cdot\langle\mathbf{B}\rangle}{\gamma_u\gamma_\eta}\overline{\mathbf{G}}\cdot\hat{\mathbf{u}} + \mathrm{i}\frac{\mathbf{k}\cdot\langle\mathbf{B}\rangle}{\gamma_u\gamma_\eta}\mathbf{P}\cdot\overline{\mathbf{G}}\cdot\hat{\mathbf{u}}, \tag{12a}$$

$$\hat{\mathbf{b}} = \mathrm{i}\frac{\mathbf{k}\cdot\langle\mathbf{B}\rangle}{\gamma_\eta}\hat{\mathbf{u}} - \frac{1}{\gamma_\eta}\overline{\mathbf{G}}\cdot\hat{\mathbf{u}} + \frac{1}{\gamma_\eta}\mathbf{G}\cdot\hat{\mathbf{b}}, \tag{12b}$$

where

$$\gamma_\nu = -\mathrm{i}(\omega - \mathbf{k}\cdot\langle\mathbf{U}\rangle) + \nu k^2, \qquad \gamma_\eta = -\mathrm{i}(\omega - \mathbf{k}\cdot\langle\mathbf{U}\rangle) + \eta k^2, \tag{13a}$$

$$\gamma_u = \gamma_\nu + \frac{(\mathbf{k}\cdot\langle\mathbf{B}\rangle)^2}{\gamma_\eta}. \tag{13b}$$

and assume scale separation between the means and the fluctuations, so that the gradients of means can be assumed small and treated in a perturbational manner. The large scale EMF, by the use of (6) and (12b), can be expressed in the following way

$$\begin{aligned}\epsilon_{ijk}\Big\langle \hat{u}_j(\omega,\mathbf{k})\hat{b}_k(\omega',\mathbf{k}')\Big\rangle = {}& \mathrm{i}\frac{k'_n\langle B\rangle_n}{\gamma'_\eta}\epsilon_{ijk}\langle\hat{u}_j\hat{u}'_k\rangle \\ & - \epsilon_{ijk}\frac{\partial\langle B\rangle_k}{\partial x_p}\frac{1}{\gamma'_\eta}\Big\langle\hat{u}_j\hat{u}'_p\Big\rangle + \epsilon_{ijk}\frac{\partial\langle U\rangle_k}{\partial x_p}\frac{1}{\gamma'_\eta}\Big\langle\hat{u}_j\hat{b}'_p\Big\rangle,\end{aligned} \tag{14}$$

where we have used a short notation $\hat{u}'_j = \hat{u}_j(\omega',\mathbf{k}')$. On substituting once again for $\hat{b}'_p$ from (12b) we obtain at the leading order

$$\begin{aligned}\epsilon_{ijk}\Big\langle \hat{u}_j(\omega,\mathbf{k})\hat{b}_k(\omega',\mathbf{k}')\Big\rangle = {}& \mathrm{i}\frac{k'_n\langle B\rangle_n}{\gamma'_\eta}\epsilon_{ijk}\langle\hat{u}_j\hat{u}'_k\rangle \\ & - \epsilon_{ijk}\frac{\partial\langle B\rangle_k}{\partial x_p}\frac{1}{\gamma'_\eta}\Big\langle\hat{u}_j\hat{u}'_p\Big\rangle + \mathrm{i}\epsilon_{ijk}\langle B\rangle_n\frac{\partial\langle U\rangle_k}{\partial x_p}\frac{k'_n}{\gamma'^2_\eta}\Big\langle\hat{u}_j\hat{u}'_p\Big\rangle,\end{aligned} \tag{15}$$

where higher-order terms in the gradients **G** and $\overline{\mathbf{G}}$ have been neglected.

3. Simple Kinematic Theory for Homogeneous Turbulence

It is a common practice to assume that the background small scale turbulence has statistical properties that are given beforehand, and thus the situation becomes kinematic. For example, we may assume that the turbulence is homogeneous and isotropic and thus the turbulence correlation tensor takes the general form

$$\langle\hat{u}_i(\mathbf{k},\omega)\hat{u}_j(\mathbf{k}',\omega')\rangle = \Big[E(\omega,\omega',k)P_{ij}(\mathbf{k}) + \mathrm{i}H(\omega,\omega',k)\epsilon_{ijk}k_k\Big]\delta(\mathbf{k}+\mathbf{k}'), \tag{16}$$

where the $H(\omega,\omega',k)$ is responsible for the lack of reflectional symmetry required for the large-scale dynamo process [6]. Since the mean velocity $\langle\mathbf{U}\rangle$ only creates a shift of the frequency $\omega \to \omega - \mathbf{k}\cdot\langle\mathbf{U}\rangle$, we simply absorb it into the frequency. The electromotive force takes the form

$$
\begin{aligned}
\epsilon_{ljk} \int \mathrm{d}^4 q \mathrm{e}^{\mathrm{i}(\mathbf{k}\cdot\mathbf{x}-\omega t)} \int \mathrm{d}^4 q' \mathrm{e}^{\mathrm{i}(\mathbf{k}'\cdot\mathbf{x}-\omega' t)} \Big\langle \hat{u}_j(\omega,\mathbf{k}) \hat{b}_k(\omega',\mathbf{k}') \Big\rangle \\
= \frac{8\pi}{3} \langle B \rangle_l \int \mathrm{d}k \int \mathrm{d}\omega \int \mathrm{d}\omega' \mathrm{e}^{-\mathrm{i}(\omega+\omega')t} \frac{k^4}{\gamma'_\eta} H(\omega,\omega',k) \\
- \frac{8\pi}{3} [\nabla \times \langle \mathbf{B} \rangle]_l \int \mathrm{d}k \int \mathrm{d}\omega \int \mathrm{d}\omega' \mathrm{e}^{-\mathrm{i}(\omega+\omega')t} \frac{k^2}{\gamma'_\eta} E(\omega,\omega',k) \\
- \frac{4\pi}{3} \Big[(\nabla \langle \mathbf{U} \rangle)^T \cdot \langle \mathbf{B} \rangle \Big]_l \int \mathrm{d}k \int \mathrm{d}\omega \int \mathrm{d}\omega' \mathrm{e}^{-\mathrm{i}(\omega+\omega')t} \frac{k^4}{\gamma'^2_\eta} H(\omega,\omega',k),
\end{aligned} \tag{17}
$$

where we have used (15), (16), (A8a) and (A8b) in Appendix B and we have also introduced a short notation

$$
\mathbf{q} = (\omega,\, \mathbf{k}), \qquad \mathrm{d}^4 q = \mathrm{d}^3 k \, \mathrm{d}\omega. \tag{18}
$$

Next we consider the case of non-stationary turbulence and assume the following

$$
E(\omega,\omega',k) = \mathtt{e}\Big(\omega^2,k\Big) \Delta(\omega,\omega';\tilde{\omega},\Gamma), \qquad H(\omega,\omega',k) = \mathtt{h}\Big(\omega^2,k\Big) \Delta(\omega,\omega';\tilde{\omega},\Gamma), \tag{19}
$$

where the 'non-stationarity' function

$$
\Delta(\omega,\omega';\tilde{\omega},\Gamma) = \delta(\omega+\omega') + \Gamma \frac{\delta(\omega+\omega'+\tilde{\omega}) - \delta(\omega+\omega'-\tilde{\omega})}{2\mathrm{i}}, \tag{20}
$$

with $\Gamma = \text{const}$, is chosen in such a way so that the correlations in the real space have a simple sinusoidal time dependence

$$
\big\langle u_i(\mathbf{x},t) u_i(\mathbf{x}',t) \big\rangle \sim 1 + \Gamma \sin(\tilde{\omega} t). \tag{21}
$$

Let us set for simplicity $\langle \mathbf{U} \rangle = 0$, though it is straightforward to include it, and concentrate on the α-effect only. Consequently, for the EMF in the real space, we obtain

$$
\begin{aligned}
\epsilon_{ljk} \int \mathrm{d}^4 q \mathrm{e}^{\mathrm{i}(\mathbf{k}\cdot\mathbf{x}-\omega t)} \int \mathrm{d}^4 q' \mathrm{e}^{\mathrm{i}(\mathbf{k}'\cdot\mathbf{x}-\omega' t)} \Big\langle \hat{u}_j(\omega,\mathbf{k}) \hat{b}_k(\omega',\mathbf{k}') \Big\rangle \\
= \frac{8\pi}{3} \eta \langle B \rangle_l \int \mathrm{d}k \int \mathrm{d}\omega \frac{k^6 \mathtt{h}(\omega^2,k)}{|\gamma_\eta|^2} \\
- \mathrm{i} \frac{4\pi\Gamma}{3} \langle B \rangle_l \int \mathrm{d}k \int \mathrm{d}\omega k^4 \mathtt{h}\Big(\omega^2,k\Big) \left[\frac{\mathrm{e}^{\mathrm{i}\tilde{\omega} t}}{\mathrm{i}(\omega+\tilde{\omega}) + \eta k^2} - \frac{\mathrm{e}^{-\mathrm{i}\tilde{\omega} t}}{\mathrm{i}(\omega-\tilde{\omega}) + \eta k^2} \right] \\
- \frac{8\pi}{3} \eta [\nabla \times \langle \mathbf{B} \rangle]_l \int \mathrm{d}k \int \mathrm{d}\omega \frac{k^4 \mathtt{e}(\omega^2,k)}{|\gamma_\eta|^2} \\
+ \mathrm{i} \frac{4\pi\Gamma}{3} [\nabla \times \langle \mathbf{B} \rangle]_l \int \mathrm{d}k \int \mathrm{d}\omega k^2 \mathtt{e}\Big(\omega^2,k\Big) \left[\frac{\mathrm{e}^{\mathrm{i}\tilde{\omega} t}}{\mathrm{i}(\omega+\tilde{\omega}) + \eta k^2} - \frac{\mathrm{e}^{-\mathrm{i}\tilde{\omega} t}}{\mathrm{i}(\omega-\tilde{\omega}) + \eta k^2} \right] \\
= \Big\{ \eta \Big[I_1^{(\alpha)} + \Gamma I_2^{(\alpha)} \sin(\tilde{\omega} t) \Big] + \Gamma \tilde{\omega} I_3^{(\alpha)} \cos(\tilde{\omega} t) \Big\} \langle B \rangle_l \\
- \Big\{ \eta \Big[I_1^{(\eta)} + \Gamma I_2^{(\eta)} \sin(\tilde{\omega} t) \Big] + \Gamma \tilde{\omega} I_3^{(\eta)} \cos(\tilde{\omega} t) \Big\} [\nabla \times \langle \mathbf{B} \rangle]_l
\end{aligned} \tag{22}
$$

where we have made use of the fact that for any even function of ω, say $f_e(\omega)$, the integral $\int_{-\infty}^{\infty} \omega f_e(\omega) \mathrm{d}\omega = 0$ vanishes and the integrals $I_j^{(\alpha)}$, $I_j^{(\eta)}$, for $j = 1, 2, 3$ are given in the Appendix A. In other words, the full EMF, including the molecular diffusion, takes the following form

$$
\boldsymbol{\mathcal{E}}_{\text{tot}} = \bar{\alpha} \langle \mathbf{B} \rangle - \bar{\eta} \nabla \times \langle \mathbf{B} \rangle, \tag{23}
$$

where

$$
\bar{\alpha} = \eta I_1^{(\alpha)} + \frac{\eta \Gamma I_2^{(\alpha)}}{\cos \phi_\alpha} \sin(\tilde{\omega} t + \phi_\alpha), \tag{24a}
$$

$$\bar{\eta} = \eta + \eta I_1^{(\eta)} + \frac{\eta \Gamma I_2^{(\eta)}}{\cos\phi_\eta} \sin(\tilde{\omega}t + \phi_\eta), \tag{24b}$$

$$\tan\phi_\alpha = \frac{\tilde{\omega} I_3^{(\alpha)}}{\eta I_2^{(\alpha)}}, \qquad \tan\phi_\eta = \frac{\tilde{\omega} I_3^{(\eta)}}{\eta I_2^{(\eta)}}. \tag{24c}$$

Clearly, there is a significant phase shift between the $\bar{\alpha}$ coefficient and the turbulent diffusion $\bar{\eta}$.

The excursions and in particular polarity reversals have been observed in numerical simulations, although none of the numerical models were able to reach the very demanding, extreme parameter regime of the core. It was reported, however, that increasing the vigour of convection tends to increase the frequency of polarity reversals [16,17]. In simulations of [18], the first stage of a polarity reversal is characterised by intensification of turbulence in some region of the core where the vigorous flow twists and bends the field lines to locally reverse the magnetic field direction. Since the magnitude of the turbulent flow in that region is particularly vigorous, so is the local magnetic Reynolds number, thus favouring conditions for amplification of the reversed field, which spreads into surrounding regions until the polarity in the core becomes mixed. This implies a sharp decay of the dipole moment. Consequently, we may expect that the magnetic excursions and polarity reversals are manifestations of a chaotic turbulent behaviour of the liquid core and no significant large-scale phenomena seem to take place during excursions.

The real core turbulence is obviously bound to be non-stationary, including propagation of well-known waves such as the inertial waves [7], MAC-waves [8,9] or the magnetic Rossby waves [10], likely forming wave packets. These wave fields interact nonlinearly to form the turbulent electromotive force (EMF), and interactions of waves with distinct but close frequencies were shown to be effective in creation of the EMF by [12–14]. For simplicity, we will neglect higher-order interactions and include two-wave interactions only, so that in the simplest case when only two waves with distinct frequencies but the same wave vector are present (homogeneity but non-stationarity), the turbulent point correlations at time t can be written as

$$\begin{aligned}
&\langle \mathbf{u}(\mathbf{x},t)\mathbf{u}(\mathbf{x},t)\rangle \\
&\quad = \left\langle \left[\hat{\mathbf{u}}_1 e^{i(\omega_1 t + \mathbf{k}\cdot\mathbf{x})} + \hat{\mathbf{u}}_2 e^{i(\omega_2 t + \mathbf{k}\cdot\mathbf{x})} + c.c.\right]\left[\hat{\mathbf{u}}_1^* e^{-i(\omega_1 t + \mathbf{k}\cdot\mathbf{x})} + \hat{\mathbf{u}}_2^* e^{-i(\omega_2 t + \mathbf{k}\cdot\mathbf{x})} + c.c.\right]\right\rangle \\
&\quad = \left\langle \hat{\mathbf{u}}_1\hat{\mathbf{u}}_2^* e^{i\tilde{\omega}t}\right\rangle + \left\langle \hat{\mathbf{u}}_2\hat{\mathbf{u}}_1^* e^{-i\tilde{\omega}t}\right\rangle + \langle \hat{\mathbf{u}}_1\hat{\mathbf{u}}_1^*\rangle + \langle \hat{\mathbf{u}}_2\hat{\mathbf{u}}_2^*\rangle,
\end{aligned} \tag{25}$$

where

$$\tilde{\omega} = \Delta\omega = \omega_1 - \omega_2, \tag{26}$$

c.c. means the complex conjugate, and we have assumed that the means of quantities oscillating with frequencies ω_1, ω_2 and $\omega_1 + \omega_2$, which correspond to mean field variations on the timescales from days to decades can be assumed to vanish (without loss of generality, we may assume $\omega_1 > 0$, $\omega_2 > 0$). However, since the core perturbations often possess close frequencies of oscillation, the frequency difference $\tilde{\omega}$ can be small and even smaller than the typical frequencies of oscillations of the large-scale field; in such a case, the first two terms in the second line of (25) cannot be neglected and lead to long time variations of the turbulence correlation tensor. This is in keeping with the chosen form of the 'non-stationarity' function (20), which leads to correlations in the form (21), corresponding to the form based on two-wave interactions.

In reality even such two-wave interactions lead to more than one (say N) $\tilde{\omega}$-mode, because the turbulent wave field consists of a number of distinct waves so that more generally

$$\langle u_i(\mathbf{x},t)u_i(\mathbf{x}',t)\rangle \sim 1 + \sum_{n=1}^{N} \Gamma_n \sin(\tilde{\omega}_n t + \psi_n), \tag{27}$$

where the phase shift ψ_n accounts for both the sine-type and cosine-type time dependence of the correlation tensor.

4. Dynamic Theory

Instead of assuming some given form of the turbulence correlation tensor, we can use Equation (12a) to express it in the following form

$$\begin{aligned}\langle \hat{u}_i(\mathbf{q})\hat{u}_j(\mathbf{q}')\rangle =&\frac{1}{\gamma_u\gamma_u'}\left\langle \hat{f}_i\hat{f}_j'\right\rangle \\ &-\left[\frac{1}{\gamma_u\gamma_u'^2}P_{js}'G_{sp}+\frac{(\mathbf{k}'\cdot\langle\mathbf{B}\rangle)^2}{\gamma_u\gamma_u'^2\gamma_\eta'^2}G_{jp}+\mathrm{i}\frac{\mathbf{k}'\cdot\langle\mathbf{B}\rangle}{\gamma_u\gamma_u'^2\gamma_\eta'}\frac{k_j'k_s'}{k'^2}\overline{G}_{sp}\right]\left\langle \hat{f}_i\hat{f}_p'\right\rangle \\ &-\left[\frac{1}{\gamma_u^2\gamma_u'}P_{is}G_{sp}+\frac{(\mathbf{k}\cdot\langle\mathbf{B}\rangle)^2}{\gamma_u^2\gamma_\eta^2\gamma_u'}G_{ip}+\mathrm{i}\frac{\mathbf{k}\cdot\langle\mathbf{B}\rangle}{\gamma_u^2\gamma_\eta\gamma_u'}\frac{k_ik_s}{k^2}\overline{G}_{sp}\right]\left\langle \hat{f}_p\hat{f}_j'\right\rangle\end{aligned} \tag{28}$$

where higher-order terms in the gradients $\mathbf{G}$ and $\overline{\mathbf{G}}$ have been neglected, and we have used the short notation $\mathbf{q}=(\omega,\,\mathbf{k})$. The EMF can then be calculated with the use of (15) and (28) and on assuming the force correlations in a statistically homogeneous form

$$\left\langle \hat{f}_i(\mathbf{k},\omega)\hat{f}_j(\mathbf{k}',\omega')\right\rangle =\left[F_0(\omega,\omega',k)P_{ij}(\mathbf{k})+\mathrm{i}F_1(\omega,\omega',k)\epsilon_{ijk}k_k\right]\delta(\mathbf{k}+\mathbf{k}'), \tag{29}$$

we have derived the formula for the mean EMF in Appendix B. This involves the mean velocity gradients, which within the considered model correspond to effects such as the cross-helicity dynamo or the shear-current effect [19–21]. Although it is straightforward to include them in the calculations, for clarity, we drop all the mean velocity terms (we set $\langle\mathbf{U}\rangle=0$) and concentrate only on the α-effect and the effects of turbulent diffusion. For simplicity, we will also consider the case when $\nabla\langle B\rangle^2\times\langle\mathbf{B}\rangle$ is negligible, hence the mean EMF is reduced to (cf. Appendix B)

$$\begin{aligned}\mathcal{E}_l =&\epsilon_{ljk}\int \mathrm{d}^4q\mathrm{e}^{\mathrm{i}(\mathbf{k}\cdot\mathbf{x}-\omega t)}\int \mathrm{d}^4q'\mathrm{e}^{\mathrm{i}(\mathbf{k}'\cdot\mathbf{x}-\omega' t)}\left\langle \hat{u}_j(\omega,\mathbf{k})\hat{b}_k(\omega',\mathbf{k}')\right\rangle \\ =&\mathcal{I}_1\left(\nu,\eta,\langle B\rangle^2\right)\langle B\rangle_l-\left[\mathcal{I}_2\left(\nu,\eta,\langle B\rangle^2\right)+\langle B\rangle^2\mathcal{I}_3\left(\nu,\eta,\langle B\rangle^2\right)\right][\nabla\times\langle\mathbf{B}\rangle]_l,\end{aligned} \tag{30}$$

where the integrals $\mathcal{I}_1,\mathcal{I}_2$ and $\mathcal{I}_3$ are provided in (A10a)–(A10c) in Appendix B.

Note that lack of reflectional symmetry is introduced here by the term proportional to $F_1(\omega,\omega',k)$ in the force corrections (29), hence it is crucial for calculation of the α-effect, which is not generated by reflectionally symmetric flows [6]. In natural systems, the reflectional symmetry is broken by the presence of background rotation, which introduces the Coriolis force into the dynamics. However, rapidly rotating flows are naturally anisotropic; thus for simplicity, we have chosen an isotropic model.

4.1. Stationary Turbulence

First let us explicitly demonstrate the simplest result obtained for stationary turbulence, when

$$F_0(\omega,\omega',k)=\mathcal{F}_0\left(\omega^2,k\right)\delta(\omega+\omega'),\quad F_1(\omega,\omega',k)=\mathcal{F}_1\left(\omega^2,k\right)\delta(\omega+\omega'), \tag{31}$$

i.e., the force correlations are given by

$$\left\langle \hat{f}_i(\mathbf{k},\omega)\hat{f}_j(\mathbf{k}',\omega')\right\rangle =\left[\mathcal{F}_0\left(\omega^2,k\right)P_{ij}(\mathbf{k})+\mathrm{i}\mathcal{F}_1\left(\omega^2,k\right)\epsilon_{ijk}k_k\right]\delta(\mathbf{k}+\mathbf{k}')\delta(\omega+\omega'). \tag{32}$$

In such a case, the integrals in (30) can be calculated to yield

$$\mathcal{I}_1^{(st)}\left(\nu,\eta,\langle B\rangle^2\right)=4\pi\eta\int \mathrm{d}k\int_{-1}^{1}\mathrm{d}X\int \mathrm{d}\omega\frac{k^6X^2\mathcal{F}_1(\omega^2,k)}{|\gamma_u|^2|\gamma_\eta|^2}, \tag{33a}$$

$$\mathcal{I}_2^{(st)}\left(\nu,\eta,\langle B\rangle^2\right)=\pi\eta\int \mathrm{d}k\int_{-1}^{1}\mathrm{d}X\int \mathrm{d}\omega\frac{k^4(1+X^2)\mathcal{F}_0(\omega^2,k)}{|\gamma_u|^2|\gamma_\eta|^2}, \tag{33b}$$

$$\begin{aligned}&\mathcal{I}_3^{(st)}\left(\nu,\eta,\langle B\rangle^2\right)\\&=\pi\eta\int \mathrm{d}k\int_{-1}^{1}\mathrm{d}X\int \mathrm{d}\omega\frac{k^6\mathcal{F}_0(\omega^2,k)}{|\gamma_u|^4|\gamma_\eta|^4}\left(\omega^2-\nu\eta k^4-(\mathbf{k}\cdot\langle\mathbf{B}\rangle)^2\right)\left(13X^4-10X^2+1\right),\end{aligned} \tag{33c}$$

where $X=\cos\theta$ and θ is the polar angle in the spherical coordinates (k,θ,ϕ) in the wave-vector space, hence the induction equation takes the following form

$$\frac{\partial\langle\mathbf{B}\rangle}{\partial t}=\nabla\times(\bar{\alpha}_{st}\langle\mathbf{B}\rangle)-\nabla\times(\bar{\eta}_{st}\nabla\times\langle\mathbf{B}\rangle), \tag{34}$$

with

$$\bar{\alpha}_{st}=4\pi\eta\int \mathrm{d}k\int_{-1}^{1}\mathrm{d}X\int \mathrm{d}\omega\frac{k^6X^2\mathcal{F}_1(\omega^2,k)}{|\gamma_u|^2|\gamma_\eta|^2}, \tag{35a}$$

$$\begin{aligned}\bar{\eta}_{st}=\eta\Bigg\{1+\pi\int \mathrm{d}k\int_{-1}^{1}\mathrm{d}X\int \mathrm{d}\omega\frac{k^4\mathcal{F}_0(\omega^2,k)}{|\gamma_u|^2|\gamma_\eta|^2}\Bigg[1+X^2\\+\langle B\rangle^2\frac{k^2\left(\omega^2-\nu\eta k^4-k^2\langle B\rangle^2X^2\right)(13X^4-10X^2+1)}{|\gamma_u|^2|\gamma_\eta|^2}\Bigg]\Bigg\},\end{aligned} \tag{35b}$$

where we have made use of the fact that for any even function of ω, say $f_e(\omega)$ the integral $\int_{-\infty}^{\infty}\omega f_e(\omega)\mathrm{d}\omega=0$ vanishes.

4.2. Non-Stationary Turbulence

The dynamics of the Earth's core is strongly influenced by the wave field, composed of the fundamental modes such as the inertial waves, the MAC waves and the magnetic Rossby waves. These waves are described by different dispersion relations, which imply the existence of a large number of waves with distinct frequencies of oscillation. As argued at the end of Section 3, their interactions lead to a time-dependent turbulent correlation tensor. We consider the simplest case of only two-wave interactions and introduce a simple model by postulating that the turbulence is forced by the following isotropic, homogeneous but non stationary forcing (cf. (20))

$$F_0(\omega,\omega',k)=\mathcal{F}_0\left(\omega^2,k\right)\Delta(\omega,\omega';\tilde{\omega},\Gamma),\quad F_1(\omega,\omega',k)=\mathcal{F}_1\left(\omega^2,k\right)\Delta(\omega,\omega';\tilde{\omega},\Gamma), \tag{36}$$

i.e.,

$$\begin{aligned}&\left\langle\hat{f}_i(\mathbf{k},\omega)\hat{f}_j(\mathbf{k}',\omega')\right\rangle\\&=\left[\mathcal{F}_0\left(\omega^2,k\right)P_{ij}(\mathbf{k})+\mathrm{i}\mathcal{F}_1\left(\omega^2,k\right)\epsilon_{ijk}k_k\right]\delta(\mathbf{k}+\mathbf{k}')\Delta(\omega,\omega';\tilde{\omega},\Gamma).\end{aligned} \tag{37}$$

In the above, the function $\Delta(\omega,\omega';\tilde{\omega},\Gamma)$ is the 'non-stationarity' function given in (20), so that indeed the two-wave interaction form

$$\langle f_i(\mathbf{x},t)f_i(\mathbf{x}',t)\rangle\sim 1+\Gamma\sin(\tilde{\omega}t) \tag{38}$$

is preserved. Of course, equally well, one could choose this to be a cosine-type time dependence. The parameter $\tilde{\omega}$ has the physical interpretation of frequency shift between distinct modes, i.e., MAC or magnetic Rossby waves. The calculation of the mean EMF in the non-stationary turbulence is postponed until Appendix C. Although the limit is not geophysically relevant because of very small viscosity of the core, to make analytical progress, we have considered

$$\tilde{\omega} \ll \nu k^2,\ \eta k^2, \tag{39}$$

hence

$$\frac{\tilde{\omega}}{2\Omega E_\ell} \ll 1, \qquad E_\ell = \frac{\nu}{2\Omega \ell^2}, \tag{40}$$

where E_ℓ is the Ekman number based on the length scale ℓ corresponding to most energetic turbulent eddies in the core. The induction equation then reads

$$\frac{\partial \langle \mathbf{B} \rangle}{\partial t} = \nabla \times (\bar{\alpha}_{ns} \langle \mathbf{B} \rangle) - \nabla \times (\bar{\eta}_{ns} \nabla \times \langle \mathbf{B} \rangle), \tag{41}$$

where, again, $\bar{\eta}_{ns}$ is the full effective magnetic diffusivity, including molecular effects and

$$\bar{\alpha}_{ns} = \bar{\alpha}_{st} + \bar{\alpha}_{st}\Gamma \sin(\tilde{\omega}t) - \bar{\alpha}_{\Delta\omega}\Gamma \cos(\tilde{\omega}t), \tag{42a}$$

$$\bar{\eta}_{ns} = \eta + \bar{\eta}_{st} + \bar{\eta}_{st}\Gamma \sin(\tilde{\omega}t) - \bar{\eta}_{\Delta\omega}\Gamma \cos(\tilde{\omega}t), \tag{42b}$$

where the coefficients $\bar{\alpha}_{st}$ and $\bar{\eta}_{st}$ are given in (35a) and (35b), respectively, whereas formulae for $\bar{\alpha}_{\Delta\omega}$ and $\bar{\eta}_{\Delta\omega}$ can be found in (A24), (A26), (A27a) and (A27b) in Appendix C. The latter expressions can be rearranged to yield

$$\bar{\alpha}_{ns} = \bar{\alpha}_{st} + \frac{\bar{\alpha}_{st}\Gamma}{\cos\phi_\alpha} \sin(\tilde{\omega}t - \phi_\alpha), \tag{43a}$$

$$\bar{\eta}_{ns} = \eta + \bar{\eta}_{st} + \frac{\bar{\eta}_{st}\Gamma}{\cos\phi_\eta} \sin(\tilde{\omega}t - \phi_\eta), \tag{43b}$$

where

$$\tan\phi_\alpha = \frac{\bar{\alpha}_{\Delta\omega}}{\bar{\alpha}_{st}}, \qquad \tan\phi_\eta = \frac{\bar{\eta}_{\Delta\omega}}{\bar{\eta}_{st}}. \tag{44}$$

5. The Mean EMF in Non-Stationary, Low-*Pm* Turbulence

Let us introduce the Hartmann number based on characteristic fluctuational length scale $\mathcal{M}$ and a new variable ϖ,

$$\mathcal{M}^2(k) = \frac{\langle B \rangle^2}{\nu\eta k^2}, \qquad \varpi = \frac{\omega}{\sqrt{\nu\eta}k^2}. \tag{45}$$

We can evaluate the integrals expressing the turbulent diffusivity and the α-effect in non-stationary turbulence (cf. Appendix C) in the geophysically relevant limit of a strong magnetic field

$$\mathcal{M}^2 = \frac{\langle B \rangle^2}{\nu\eta k^2} \gg 1, \tag{46}$$

however, on top of this assumption we will also need

$$\forall_k \ Pm.\mathcal{M}^2(k) \ll 1, \tag{47}$$

in order to make analytical progress. This means we assume large Hartmann numbers (strong field), but low magnetic Prandtl numbers $Pm = \nu/\eta \ll 1$. Since in the Earth's core the Hartmann number

$$M = \frac{\langle B \rangle L}{\sqrt{\nu\eta}}, \tag{48}$$

based on the core depth L is of the order 10^7, and $Pm \approx 5 \times 10^{-7}$, the latter assumption $Pm.\mathcal{M}^2(k) \ll 1$ for the Hartmann number $\mathcal{M}(k)$ based on fluctuational wavelengths is not necessarily satisfied. Still, it is reasonable at least in a large part of the short-wavelength spectrum, and thus, it is put forward in order to simplify the calculations. Note that since we consider the *strong field* limit, it is not applicable to a linear dynamo regime, i.e., quenching of the electromotive force is significant. Let us take the following simple statistical model (cf. e.g., [22,23], etc.)

$$\mathcal{F}_0\left(\omega^2, k\right) = \frac{D_0}{k^3}, \qquad \mathcal{F}_1\left(\omega^2, k\right) = \frac{D_1}{k^5}, \tag{49}$$

with $D_0 = \text{const} > 0$, $D_1 = \text{const} \leq k_{\min} D_0$ [6], and introduce explicitly the scale separation, i.e., we introduce the scale of the largest/most energetic turbulent eddies $\ell = 2\pi/k_{\min}$, where $k_{\min}$ is the smallest fluctuational wave number. This allows to calculate the relations between coefficients $\bar{\alpha}_{st}$, $\bar{\eta}_{st}$ and $\bar{\alpha}_{\Delta\omega}$, $\bar{\eta}_{\Delta\omega}$ (see Appendix C for general formulae), which yields

$$\bar{\alpha}_{\Delta\omega} \approx \frac{\pi\tilde{\omega}}{12 k_{\min}\sqrt{Pm}\langle B\rangle}\bar{\alpha}_{st}, \qquad \bar{\eta}_{\Delta\omega} \approx \frac{15\tilde{\omega}}{28 k_{\min}^2 \nu}\bar{\eta}_{st}, \tag{50}$$

and we recall here the assumptions that led to the final form of the transport coefficients

$$\frac{\sqrt{\eta\nu}k_{\min}}{\langle B\rangle} = \frac{1}{\mathcal{M}(k_{\min})} \ll 1, \qquad \frac{\tilde{\omega}}{\nu k_{\min}^2} \ll 1, \qquad Pm.\mathcal{M}^2(k_{\min}) \ll 1. \tag{51}$$

Note, however, that these assumptions were only necessary to clearly demonstrate how the turbulent diffusivity and the α coefficient can be calculated in non-stationary turbulence, and in particular, we have shown that the significant phase shift in the slow time dependence between the turbulent diffusivity and the α coefficient is a rather general feature of non-stationary turbulence. Under the current assumptions (51) the phase shifts (44) can be calculated to yield

$$\tan\phi_\alpha = \frac{\bar{\alpha}_{\Delta\omega}}{\bar{\alpha}_{st}} \approx \frac{\tilde{\omega}\ell}{24\sqrt{Pm}\langle B\rangle} = \frac{1}{24}\frac{\tilde{\omega}\ell^2}{\nu}\frac{\sqrt{\nu\eta}}{\langle B\rangle\ell} = \frac{1}{24}\frac{\tilde{\omega}\ell^2}{\nu}\frac{1}{M_\ell}, \tag{52a}$$

$$\tan\phi_\eta = \frac{\bar{\eta}_{\Delta\omega}}{\bar{\eta}_{st}} \approx \frac{15}{112\pi^2}\frac{\tilde{\omega}\ell^2}{\nu}, \tag{52b}$$

where

$$M_\ell = \frac{\langle B\rangle\ell}{\sqrt{\nu\eta}} \gg 1, \tag{53}$$

is the Hartmann number based on the maximal fluctuational lengthscale, which in the Earth's core can be expected to be at the order of 10^4 or even 10^5. It follows that $\phi_\alpha \approx 0$ is negligibly small and the phase shift between the α-effect and the turbulent magnetic diffusivity is entirely determined by ϕ_η, which is at least a few orders of magnitude larger

$$\phi_\eta \gg \phi_\alpha. \tag{54}$$

We emphasise that ϕ_η is proportional to the frequency shift $\tilde{\omega}$ between the slowly evolving MAC/magnetic Rossby waves, which is not uniquely determined, and in fact, more realistically, the non-stationary turbulence should be modelled with a superposition of at least a few (say N) distinct values $\tilde{\omega}_n$, $n = 1, \dots, N$, each generating a different phase shift $\phi_{\eta\, n}$ between the α-effect and the turbulent diffusivity.

Toy Model-Energy of a Force-Free Mode at $\langle \mathbf{U}\rangle = 0$

The aim of the simple calculation shown in this section is to demonstrate how a phase shift between the field-amplifying α-effect and the resistive decay can lead to magnetic field excursions, through an explicit numerical solution obtained for a force-free mode defined by

$$\nabla \times \langle \mathbf{B}\rangle = \kappa(\mathbf{x})\langle \mathbf{B}\rangle, \tag{55}$$

where $\kappa(\mathbf{x})$ is a scalar function of position; this ensures $\langle \mathbf{B} \rangle \times (\nabla \times \langle \mathbf{B} \rangle) = 0$. Such force-free states are known to exist and have been intensively investigated, e.g., in seminal works of [24,25] (cf. also more recent works of [26,27]). On defining the magnetic energy

$$E_m = \frac{1}{2}\langle B \rangle^2, \tag{56}$$

the induction equation in the absence of large-scale flows, $\langle \mathbf{U} \rangle = 0$, yields for the energy

$$\frac{\partial E_m}{\partial t} = 2\tilde{\alpha}\kappa(\mathbf{x})E_m - 2\tilde{\eta}\kappa(\mathbf{x})^2 E_m. \tag{57}$$

Based on the previous analysis we may propose the following

$$\tilde{\alpha} = \frac{\tilde{\alpha}_0}{1+\bar{A}E_m^2}[1+\Gamma\cos(\tilde{\omega}t - \phi_\alpha)], \qquad \tilde{\eta} = \eta + \frac{\tilde{\eta}_0}{1+\bar{A}E_m^2}\left[1+\Gamma\cos(\tilde{\omega}t - \phi_\eta)\right], \tag{58}$$

where instead of the magnetic field dependence obtained in the strong field limit with $Pm \ll 1$ considered in previous sections, we added a standard quenching factor, which models action of the Lorentz force in the simplest way (cf. [28,29]), with $\bar{A} = \text{const}$. Taking

$$\phi_\alpha = 0, \qquad \phi_\eta = \pi, \tag{59}$$

gives a well-suited case study, since it implies that the maximal enhancement of the field takes place when the effect of the resistive decay is the weakest and vice versa—the strongest resistive decay of the mean field is associated with the weakest amplification by the α-effect; the latter situation leads to possible excursions (field decay), whereas the former one to 'stable' periods with a strong magnetic field. In such a case, we obtain

$$\begin{aligned}\frac{\partial E_m}{\partial t} =& 2\tilde{\alpha}_0\kappa(1+\cos\tilde{\omega}t)\frac{E_m}{1+\bar{A}E_m^2}, \\ &- 2\kappa^2\left[\eta\left(1+\bar{A}E_m^2\right) + \tilde{\eta}_0 - \tilde{\eta}_0\cos(\tilde{\omega}t)\right]\frac{E_m}{1+\bar{A}E_m^2},\end{aligned} \tag{60}$$

where we have inserted $\Gamma = 1$. A selected solution of the latter equation is shown on Figure 1, where indeed regular short-lived excursions are manifested in the evolution of the magnetic energy, i.e., short periods with a suppressed magnetic field.

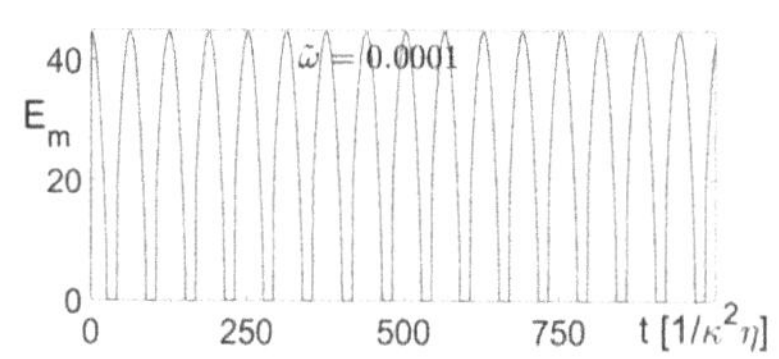

Figure 1. Evolution of the magnetic energy in non-stationary turbulence according to Equation (60), when non-stationary force correlations $\langle f_i(\mathbf{x},t)f_i(\mathbf{x}',t)\rangle \sim 1+\sin(\tilde{\omega}t)$ are defined by a single-mode time dependence with the frequency $\tilde{\omega} = 0.0001$; the remaining parameters are $\eta = 0.001$, $\tilde{\eta} = 0.1$, $\tilde{\alpha} = 1$, $\bar{A} = 1$, $\kappa = 1$. The plot is shown in diffusive time units $1/\kappa^2\eta = 10^3$.

More generally, accounting for the possibility of a more complex time dependence of the force correlations with more than one slow $\tilde{\omega}$-mode, the magnetic energy evolution equation can be rewritten in the form

$$\frac{\partial E_m}{\partial t} = 2\tilde{\alpha}_0 \kappa \left[1 + \sum_n \Gamma_n \cos(\tilde{\omega}_n t + \psi_n) \right] \frac{E_m}{1 + \bar{A} E_m^2} - 2\kappa^2 \left[\eta \left(1 + \bar{A} E_m^2 \right) + \tilde{\eta}_0 + \tilde{\eta}_0 \sum_n \Gamma_n \cos(\tilde{\omega}_n t + \phi_n) \right] \frac{E_m}{1 + \bar{A} E_m^2}. \tag{61}$$

Selected solutions are shown in Figure 2 for the cases of two and three $\tilde{\omega}$-modes. Especially in the latter case, it is evident that longer periods of relative stability of the magnetic energy are separated by much shorter excursions.

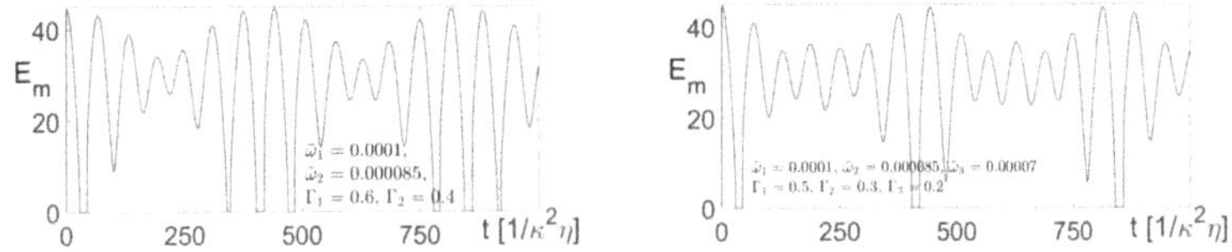

(a) (b)

Figure 2. Evolution of the magnetic energy in non-stationary turbulence according to Equation (61) with $\psi_n = 0$, $\phi_n = \pi$, when non-stationary force correlations $\langle f_i(\mathbf{x}, t) f_i(\mathbf{x}', t) \rangle$ are defined by a two-mode time dependence with the frequencies $\tilde{\omega}_1 = 0.0001$ and $\tilde{\omega}_2 = 0.000085$ (**a**) and a three-mode time dependence with $\tilde{\omega}_1 = 0.0001$, $\tilde{\omega}_2 = 0.000085$ and $\tilde{\omega}_3 = 0.00007$ (**b**); the remaining parameters are $\eta = 0.001$, $\tilde{\eta} = 0.1$, $\tilde{\alpha} = 1$, $\bar{A} = 1$, $\kappa = 1$. The plot is shown in diffusive time units $1/\kappa^2\eta = 10^3$.

6. Discussion of Relevance of the Results to the Problem of Geomagnetic Excursions

The presented toy model is obviously a great simplification. Firstly, it considers force-free modes, but it also relies on derivations obtained under restrictive assumptions of statistic homogeneity and isotropy. The toy model, however, is only used to visualise the studied effect of non-stationarity of the turbulent coefficients $\tilde{\alpha}$ and $\tilde{\eta}$, which is generic and independent of the simplifying assumptions.

It needs to be stressed that in the presented analysis, the large-scale EMF is calculated with the use of fluctuational equations, but then we concentrate on the evolution of the large-scale field, leaving the dynamics of the small-scale component of the magnetic field unexplored. It follows that temporal enhancement of turbulent magnetic diffusivity does not need to suppress the small-scale dynamo, which in reality is actually rather expected to be vividly operating on much shorter timescales, largely independently of the large-scale process. Furthermore, as stated above, the problem is greatly simplified from the start by the assumption of isotropy and neglection of the strongly anisotropic effect of the background rotation and density stratification. At first, a rough estimate of the anisotropy leads to a separation between horizontal and vertical turbulent $\tilde{\alpha}$ and $\tilde{\eta}$ coefficients. It is also known that in the weak seed field limit (linear stage of growth of the mean magnetic field) the mean induction equation with anisotropic $\tilde{\alpha}_{ij} = \tilde{\alpha}_h \delta_{ij} + (\tilde{\alpha}_v - \tilde{\alpha}_h)\delta_{i3}\delta_{j3}$ and $\tilde{\eta}_{ij} = \tilde{\eta}_h \delta_{ij} + (\tilde{\eta}_v - \tilde{\eta}_h)\delta_{i3}\delta_{j3}$ (say rotation is along the z-axis) separates the evolution of even and odd modes, thus dipolar parity evolves independently with the quadrupolar one. Moreover, one should bear in mind that the reality is even more complex—not only non-stationary but also inhomogeneous, leading to spatially dependent turbulent coefficients, which further differentiates the evolution of different spatial modes. Summarising, the small-scale dynamo is by no means excluded during the periods of enhanced turbulent diffusion, and the simple isotropic, homogeneous model obviously does not grasp the complexity of the strongly anisotropic Earth's core turbulence and the mean-field evolution; and it is the anisotropy and inhomogeneity (rotation, stratification, buoyancy, inhomogeneous wave interactions, etc.) that lead to significant differences in the evolution of the mean dipolar and quadrupolar fields. Although theoretical models, including non-stationarity, inhomogeneity and anisotropy, may be too cumbersome to investigate, it could be of interest to utilise DNS to study the long time evolution of turbulent coefficients in spherical, rapidly rotating shells and its dynamical connection to magnetic excursions.

Furthermore, there is a number of other important models, which identify and study other effects as underlying mechanisms of the geomagnetic reversals/excursions and present a more general approach to the problem. The dynamics of reversals has been modelled with sets of nonlinear ODE's involving supercritical bifurcations and describing the evolution of field modes with different parities, e.g., by [30] or [31]. Later, ref. [32] attempted to grasp the main features of evolution of the planetary field by ODE models with a stochastic noise. A saddle-node-bifurcation model capturing the dynamical features of the geomagnetic field evolution and those of the field obtained in the VKS experiment (cf. [33]) was developed by [34]. A group of nonlinear, one-dimensional evolutional models, which also reported Earth-like features were represented by [35–37]. In short, the described models point to nonlinearities causing chaotic behaviour of the system as the mechanism of generation of reversals/excursions in the system evolution. Of course this does not exclude non-stationarity of the mean-field coefficients as a trigger of the excursion events.

7. Conclusions

Although obviously no decisive conclusion about predictions of geomagnetic excursions can be made from the presented analysis, the paper points to the non-stationarity of the turbulent transport coefficients $\bar{\alpha}$ and $\bar{\eta}$, which are likely to slowly evolve in time due to the non-stationarity of the entire background turbulence in the core. Their evolution was shown to be out of phase, implying the existence of periods of enhanced and suppressed turbulent dynamo process, which may correlate or not with periods of enhanced/suppressed diffusion. In particular, enhancement of the α-effect may coincide with suppression of diffusion (stable field) and vice versa, enhancement of diffusion may coincide with suppression of the dynamo effect (field decay). It would be instructive to study such temporal characteristics of the turbulent diffusivity and the α-effect in numerical simulations, in particular the correlations between the value of the ratio $\bar{\eta}/\bar{\alpha}$ and the occurrences of excursions.

The magnetic diffusion in the Earth's core is thought to be $\eta = 2\,\mathrm{m}^2/\mathrm{s}$ (cf. e.g., [38]), which implies the magnetic diffusion time of about 10^4 years. The non-stationary turbulence can, however, enhance the effective magnetic diffusion of the large-scale dipole to make it a few times larger, which decreases the magnetic diffusion time to a value of a few thousand years, which is in line with the realistic time scales of the geomagnetic excursions/reversals. As shown, in a non-stationary turbulence, such an enhancement of diffusion can be correlated with simultaneous suppression of the α-effect, in which case it is likely to result in a sharp drop of the geomagnetic field intensity, i.e., the excursion. With such a picture, the geomagnetic excursions are manifestations of the chaotic core turbulence and their occurrences are also bound to be chaotic, and no characteristic, significant alterations in the core large-scale flow are expected during excursions, in accordance with results of numerical simulations [16–18]. Perhaps with the development of the available computational abilities and power, in the future, it may become possible to numerically study the Earth-like parameter regime and construct reliable fits to the geomagnetic data in order to provide estimates of the turbulent magnetic diffusivity and the α-effect, which could then be monitored since a drop in the value of the ratio $\bar{\alpha}/\bar{\eta}$ for the outer core could indicate a higher possibility for a geomagnetic excursion.

Funding: The support of the National Science Centre of Poland (grant no. 2017/26/E/ST3/00554) is gratefully acknowledged. This work was partially financed as a part of the statutory activity from the subvention of the Ministry of Science and Higher Education in Poland.

Conflicts of Interest: The author declares no conflict of interest.

Appendix A

The integrals that appear in Equation (22) take the form

$$I_1^{(\alpha)} = \frac{8\pi}{3}\int \mathrm{d}k \int \mathrm{d}\omega \frac{k^6 \mathtt{h}(\omega^2,k)}{|\gamma_\eta|^2}, \quad I_2^{(\alpha)} = \frac{8\pi}{3}\int \mathrm{d}k \int \mathrm{d}\omega \frac{k^6 \mathtt{h}(\omega^2,k)\left(\tilde{\omega}^2+\omega^2+\eta^2k^4\right)}{\left(\omega^2-\tilde{\omega}^2-\eta^2k^4\right)^2+4\omega^2\eta^2k^4}, \tag{A1}$$

$$I_3^{(\alpha)} = \frac{8\pi}{3}\int \mathrm{d}k \int \mathrm{d}\omega \frac{k^4 \mathtt{h}(\omega^2,k)\left(\omega^2-\tilde{\omega}^2-\eta^2k^4\right)}{\left(\omega^2-\tilde{\omega}^2-\eta^2k^4\right)^2+4\omega^2\eta^2k^4}, \tag{A2}$$

$$I_1^{(\eta)} = \frac{8\pi}{3}\int \mathrm{d}k \int \mathrm{d}\omega \frac{k^4 \mathtt{e}(\omega^2,k)}{|\gamma_\eta|^2}, \quad I_2^{(\eta)} = \frac{8\pi}{3}\int \mathrm{d}k \int \mathrm{d}\omega \frac{k^4 \mathtt{e}(\omega^2,k)\left(\tilde{\omega}^2+\omega^2+\eta^2k^4\right)}{\left(\omega^2-\tilde{\omega}^2-\eta^2k^4\right)^2+4\omega^2\eta^2k^4}, \tag{A3}$$

$$I_3^{(\eta)} = \frac{8\pi}{3}\int \mathrm{d}k \int \mathrm{d}\omega \frac{k^2 \mathtt{e}(\omega^2,k)\left(\omega^2-\tilde{\omega}^2-\eta^2k^4\right)}{\left(\omega^2-\tilde{\omega}^2-\eta^2k^4\right)^2+4\omega^2\eta^2k^4}. \tag{A4}$$

Appendix B

Here we present the explicit calculation of the large-scale EMF (30) given in Section 4. By the use of (15) and (28) we write down

$$\begin{aligned}
\epsilon_{ljk}\left\langle \hat{u}_j(\mathbf{q})\hat{b}_k(\mathbf{q}')\right\rangle =& \mathrm{i}\frac{k'_n\langle B\rangle_n}{\gamma'_\eta}\epsilon_{ljk}\frac{1}{\gamma_u\gamma'_u}\left\langle \hat{f}_j\hat{f}'_k\right\rangle \\
&+ \langle B\rangle_n\langle B\rangle_r\epsilon_{ljk}\frac{1}{\gamma_u\gamma'^2_u\gamma'^2_\eta}\frac{k'_nk'_rk'_kk'_s}{k'^2}\frac{\partial\langle B\rangle_s}{\partial x_p}\left\langle \hat{f}_j\hat{f}'_p\right\rangle \\
&+ \langle B\rangle_n\langle B\rangle_r\epsilon_{ljk}\frac{1}{\gamma^2_u\gamma_\eta\gamma'_\eta\gamma'_u}\frac{k'_nk_rk_jk_s}{k^2}\frac{\partial\langle B\rangle_s}{\partial x_p}\left\langle \hat{f}_p\hat{f}'_k\right\rangle \\
&- \epsilon_{ljk}\frac{\partial\langle B\rangle_k}{\partial x_p}\frac{1}{\gamma_u\gamma'_u\gamma'_\eta}\left\langle \hat{f}_j\hat{f}'_p\right\rangle \\
&- \mathrm{i}\frac{k'_n\langle B\rangle_n}{\gamma'_\eta}\epsilon_{ljk}\left[\frac{1}{\gamma_u\gamma'^2_u}P'_{ks}\frac{\partial\langle U\rangle_s}{\partial x_p} + \frac{(\mathbf{k}'\cdot\langle\mathbf{B}\rangle)^2}{\gamma_u\gamma'^2_u\gamma'^2_\eta}\frac{\partial\langle U\rangle_k}{\partial x_p}\right]\left\langle \hat{f}_j\hat{f}'_p\right\rangle \\
&- \mathrm{i}\frac{k'_n\langle B\rangle_n}{\gamma'_\eta}\epsilon_{ljk}\left[\frac{1}{\gamma^2_u\gamma'_u}P_{js}\frac{\partial\langle U\rangle_s}{\partial x_p} + \frac{(\mathbf{k}\cdot\langle\mathbf{B}\rangle)^2}{\gamma^2_u\gamma^2_\eta\gamma'_u}\frac{\partial\langle U\rangle_j}{\partial x_p}\right]\left\langle \hat{f}_p\hat{f}'_k\right\rangle \\
&+ \mathrm{i}\epsilon_{ljk}\langle B\rangle_n\frac{\partial\langle U\rangle_k}{\partial x_p}\frac{k'_n}{\gamma_u\gamma'_u\gamma'^2_\eta}\left\langle \hat{f}_j\hat{f}'_p\right\rangle,
\end{aligned} \tag{A5}$$

where, again, higher-order terms in the gradients $\mathbf{G}$ and $\overline{\mathbf{G}}$ have been neglected. We assume the force correlation tensor in the following isotropic and homogeneous form

$$\left\langle \hat{f}_i(\mathbf{k},\omega)\hat{f}_j(\mathbf{k}',\omega')\right\rangle = \left[F_0(\omega,\omega',k)P_{ij}(\mathbf{k}) + \mathrm{i}F_1(\omega,\omega',k)\epsilon_{ijk}k_k\right]\delta(\mathbf{k}+\mathbf{k}'). \tag{A6}$$

After the change of variables $\omega \to \omega - \mathbf{k}\cdot\langle\mathbf{U}\rangle$, we return to the real space to obtain

$$
\begin{aligned}
&\epsilon_{ljk}\int \mathrm{d}^4q \mathrm{e}^{\mathrm{i}(\mathbf{k}\cdot\mathbf{x}-\omega t)}\int \mathrm{d}^4q'\mathrm{e}^{\mathrm{i}(\mathbf{k}'\cdot\mathbf{x}-\omega' t)}\left\langle \hat{u}_j(\omega,\mathbf{k})\hat{b}_k(\omega',\mathbf{k}')\right\rangle \\
&= 2\langle B\rangle_n \int \mathrm{d}^4q\int \mathrm{d}\omega' \mathrm{e}^{-\mathrm{i}(\omega+\omega')t}\frac{k_n k_l}{\gamma_u\gamma'_u\gamma'_\eta}F_1 \\
&- \epsilon_{ljk}\frac{\partial\langle B\rangle_k}{\partial x_p}\int \mathrm{d}^4q\int \mathrm{d}\omega' \mathrm{e}^{-\mathrm{i}(\omega+\omega')t}\frac{1}{\gamma_u\gamma'_u\gamma'_\eta}F_0 P_{jp}(\mathbf{k}) \\
&+ \langle B\rangle_n\langle B\rangle_r\frac{\partial\langle B\rangle_s}{\partial x_j}\epsilon_{ljk}\int \mathrm{d}^4q\int \mathrm{d}\omega' \mathrm{e}^{-\mathrm{i}(\omega+\omega')t}\frac{1}{\gamma_u\gamma'^2_u\gamma'^2_\eta}\frac{k_n k_r k_k k_s}{k^2}F_0 \\
&+ \langle B\rangle_n\langle B\rangle_r\frac{\partial\langle B\rangle_s}{\partial x_j}\epsilon_{ljk}\int \mathrm{d}^4q\int \mathrm{d}\omega' \mathrm{e}^{-\mathrm{i}(\omega+\omega')t}\frac{1}{\gamma^2_u\gamma'_u\gamma_\eta\gamma'_\eta}\frac{k_n k_r k_k k_s}{k^2}F_0 \\
&+ \langle B\rangle_n\int \mathrm{d}^4q\int \mathrm{d}\omega' \mathrm{e}^{-\mathrm{i}(\omega+\omega')t}\left[\frac{k_n k_l k_k k_s}{k^2\gamma_u\gamma'^2_u\gamma'_\eta}\frac{\partial\langle U\rangle_s}{\partial x_k} + \frac{k_n k_k k_r k_w}{\gamma_u\gamma'^2_u\gamma'^3_\eta}\langle B\rangle_r\langle B\rangle_w\frac{\partial\langle U\rangle_k}{\partial x_l}\right]F_1 \\
&+ \langle B\rangle_n\int \mathrm{d}^4q\int \mathrm{d}\omega' \mathrm{e}^{-\mathrm{i}(\omega+\omega')t}\left[\frac{k_n k_l k_k k_s}{k^2\gamma^2_u\gamma'_u\gamma'_\eta}\frac{\partial\langle U\rangle_s}{\partial x_k} + \frac{k_n k_k k_r k_w}{\gamma^2_u\gamma^2_\eta\gamma'_u\gamma'_\eta}\langle B\rangle_r\langle B\rangle_w\frac{\partial\langle U\rangle_k}{\partial x_l}\right]F_1 \\
&- \langle B\rangle_n\frac{\partial\langle U\rangle_k}{\partial x_l}\int \mathrm{d}^4q\int \mathrm{d}\omega' \mathrm{e}^{-\mathrm{i}(\omega+\omega')t}\frac{k_n k_k}{\gamma_u\gamma'_u\gamma'^2_\eta}F_1.
\end{aligned} \tag{A7}
$$

Next, using the following formulae

$$\int \frac{k_j}{k} f\left(\cos^2\theta\right)\mathrm{d}\mathring{\Omega} = 0, \qquad \int \frac{k_i k_j k_k}{k^3} f(\cos\theta)\mathrm{d}\mathring{\Omega} = 0, \tag{A8a}$$

$$\int \frac{k_j k_n}{k^2} f\left(\cos^2\theta\right)\mathrm{d}\mathring{\Omega} = \pi\int_{-1}^{1} f(X^2)\left\{\delta_{jn}\left(1-X^2\right) + \delta_{j3}\delta_{n3}\left(3X^2-1\right)\right\}\mathrm{d}X, \tag{A8b}$$

$$
\begin{aligned}
\int \frac{k_i k_j k_m k_n}{k^4} f\left(\cos^2\theta\right)\mathrm{d}\mathring{\Omega} &= \frac{\pi}{4}\left(\delta_{ij}\delta_{mn} + \delta_{im}\delta_{jn} + \delta_{in}\delta_{jm}\right)\int_{-1}^{1} f(X^2)\left(1-X^2\right)^2\mathrm{d}X \\
&- \frac{\pi}{4}\left(\delta_{ij}\delta_{m3}\delta_{n3} + \delta_{im}\delta_{j3}\delta_{n3} + \delta_{in}\delta_{j3}\delta_{m3} + \delta_{jm}\delta_{i3}\delta_{n3}\right. \\
&\left. + \delta_{jn}\delta_{i3}\delta_{m3} + \delta_{mn}\delta_{i3}\delta_{j3}\right)\int_{-1}^{1} f(X^2)\left(5X^4 - 6X^2 + 1\right)\mathrm{d}X \\
&+ \frac{\pi}{4}\delta_{i3}\delta_{j3}\delta_{m3}\delta_{n3}\int_{-1}^{1} f(X^2)\left(35X^4 - 30X^2 + 3\right)\mathrm{d}X,
\end{aligned} \tag{A8c}
$$

where $\mathring{\Omega}$ denotes the solid angle and the spherical coordinates (k, θ, ϕ) have been used (with a substitution $X = \cos\theta$) one obtains

$$
\begin{aligned}
&\epsilon_{ljk}\int \mathrm{d}^4q \mathrm{e}^{\mathrm{i}(\mathbf{k}\cdot\mathbf{x}-\omega t)}\int \mathrm{d}^4q'\mathrm{e}^{\mathrm{i}(\mathbf{k}'\cdot\mathbf{x}-\omega' t)}\left\langle \hat{u}_j(\omega,\mathbf{k})\hat{b}_k(\omega',\mathbf{k}')\right\rangle \\
&= \mathcal{I}_1\left(\nu,\eta,\langle B\rangle^2\right)\langle B\rangle_l - \left[\mathcal{I}_2\left(\nu,\eta,\langle B\rangle^2\right) + \langle B\rangle^2\mathcal{I}_3\left(\nu,\eta,\langle B\rangle^2\right)\right][\nabla\times\langle\mathbf{B}\rangle]_l \\
&- \mathcal{I}_4\left(\nu,\eta,\langle B\rangle^2\right)\left[\nabla\langle B\rangle^2\times\langle\mathbf{B}\rangle\right]_l \\
&+ \left\{\mathcal{I}_6\left(\nu,\eta,\langle B\rangle^2\right) + \mathcal{I}_8\left(\nu,\eta,\langle B\rangle^2\right)\langle B\rangle^2 - \mathcal{I}_5\left(\nu,\eta,\langle B\rangle^2\right)\right\}\left[(\nabla\langle\mathbf{U}\rangle)^s\cdot\langle\mathbf{B}\rangle\right]_l \\
&+ \mathcal{I}_7\left(\nu,\eta,\langle B\rangle^2\right)\langle B\rangle_l\frac{\langle\mathbf{B}\rangle\cdot(\nabla\langle\mathbf{U}\rangle)^s\cdot\langle\mathbf{B}\rangle}{\langle B\rangle^2} \\
&- \left\{\mathcal{I}_8\left(\nu,\eta,\langle B\rangle^2\right)\langle B\rangle^2 - \mathcal{I}_5\left(\nu,\eta,\langle B\rangle^2\right)\right\}[(\nabla\times\langle\mathbf{U}\rangle)\times\langle\mathbf{B}\rangle]_l,
\end{aligned} \tag{A9}
$$

with

$$\mathcal{I}_1\left(\nu,\eta,\langle B\rangle^2\right) = 4\pi\int \mathrm{d}k\int_{-1}^{1}\mathrm{d}X\int \mathrm{d}\omega\int \mathrm{d}\omega' \mathrm{e}^{-\mathrm{i}(\omega+\omega')t}\frac{k^4 X^2 F_1(\omega,\omega',k)}{\gamma_u\gamma'_u\gamma'_\eta}, \tag{A10a}$$

$$\mathcal{I}_2\left(\nu,\eta,\langle B\rangle^2\right) = \pi\int dk\int_{-1}^{1} dX\int d\omega\int d\omega' e^{-i(\omega+\omega')t}\frac{k^2(1+X^2)F_0}{\gamma_u\gamma'_u\gamma'_\eta}, \tag{A10b}$$

$$\mathcal{I}_3\left(\nu,\eta,\langle B\rangle^2\right) = \frac{\pi}{2}\int dk\int_{-1}^{1} dX\int d\omega\int d\omega' e^{-i(\omega+\omega')t}\frac{k^4F_0}{\gamma_u\gamma'_u\gamma'_\eta}\left[\frac{1}{\gamma'_u\gamma'_\eta}+\frac{1}{\gamma_u\gamma_\eta}\right]g_3(X), \tag{A10c}$$

$$\mathcal{I}_4\left(\nu,\eta,\langle B\rangle^2\right) = \frac{\pi}{4}\int dk\int_{-1}^{1} dX\int d\omega\int d\omega' e^{-i(\omega+\omega')t}\frac{k^4F_0}{\gamma_u\gamma'_u\gamma'_\eta}\left[\frac{1}{\gamma'_u\gamma'_\eta}+\frac{1}{\gamma_u\gamma_\eta}\right]g_4(X), \tag{A10d}$$

$$\mathcal{I}_5\left(\nu,\eta,\langle B\rangle^2\right) = 2\pi\int dk\int_{-1}^{1} dX\int d\omega\int d\omega' e^{-i(\omega+\omega')t}\frac{k^4X^2F_1}{\gamma_u\gamma'_u\gamma'^2_\eta}, \tag{A10e}$$

$$\mathcal{I}_6\left(\nu,\eta,\langle B\rangle^2\right) = 2\pi\int dk\int_{-1}^{1} dX\int d\omega\int d\omega' e^{-i(\omega+\omega')t}\frac{k^4F_1}{\gamma_u\gamma'_u\gamma'_\eta}\left(\frac{1}{\gamma'_u}+\frac{1}{\gamma_u}\right)g_6(X), \tag{A10f}$$

$$\mathcal{I}_7\left(\nu,\eta,\langle B\rangle^2\right) = \pi\int dk\int_{-1}^{1} dX\int d\omega\int d\omega' e^{-i(\omega+\omega')t}\frac{k^4F_1}{\gamma_u\gamma'_u\gamma'_\eta}\left(\frac{1}{\gamma'_u}+\frac{1}{\gamma_u}\right)g_7(X), \tag{A10g}$$

$$\mathcal{I}_8\left(\nu,\eta,\langle B\rangle^2\right) = 2\pi\int dk\int_{-1}^{1} dX\int d\omega\int d\omega' e^{-i(\omega+\omega')t}\frac{k^6X^4F_1}{\gamma_u\gamma'_u\gamma'_\eta}\left(\frac{1}{\gamma'_u\gamma'^2_\eta}+\frac{1}{\gamma_u\gamma^2_\eta}\right), \tag{A10h}$$

where

$$g_3(x) = -13X^4 + 10X^2 - 1, \tag{A11a}$$

$$g_4(X) = 9X^4 - 10X^2 + 1, \tag{A11b}$$

$$g_6(X) = X^2\left(1 - X^2\right), \tag{A11c}$$

$$g_7(X) = X^2\left(5X^2 - 3\right). \tag{A11d}$$

The mean velocity gradients within the considered model correspond to effects such as the cross-helicity dynamo or the shear-current effect [19–21], but for clarity, we will exclude them here. Hence, we now set $\langle \mathbf{U}\rangle = 0$ and concentrate only on the α-effect and the effects of turbulent diffusion. For simplicity, we will also consider the case when $\nabla\langle B\rangle^2 \times \langle \mathbf{B}\rangle$ is negligible, hence the mean EMF is reduced to

$$\begin{aligned}\epsilon_{ljk}\int d^4q e^{i(\mathbf{k}\cdot\mathbf{x}-\omega t)}\int d^4q' e^{i(\mathbf{k}'\cdot\mathbf{x}-\omega' t)}\left\langle \hat{u}_j(\omega,\mathbf{k})\hat{b}_k(\omega',\mathbf{k}')\right\rangle =& \mathcal{I}_1\left(\nu,\eta,\langle B\rangle^2\right)\langle B\rangle_l \\ &- \left[\mathcal{I}_2\left(\nu,\eta,\langle B\rangle^2\right) + \langle B\rangle^2\mathcal{I}_3\left(\nu,\eta,\langle B\rangle^2\right)\right][\nabla\times\langle \mathbf{B}\rangle]_l.\end{aligned} \tag{A12}$$

Appendix C

The calculation of the explicit relations between the coefficients $\bar{\alpha}_{st}$, $\bar{\eta}_{st}$ and $\alpha_{\Delta\omega}$, $\bar{\eta}_{\Delta\omega}$ is presented here. The integrals (A10a)–(A10c) take the following form

$$\begin{aligned}\mathcal{I}_1\left(\nu,\eta,\langle B\rangle^2\right) =& \mathcal{I}_1^{(st)}\left(\nu,\eta,\langle B\rangle^2\right) \\ &- i2\pi\Gamma\int dk\int_{-1}^{1} dX\int d\omega\frac{k^4X^2\mathcal{F}_1(\omega^2,k)}{\gamma_u}\mathcal{W}(\omega;\tilde{\omega},\nu,\eta),\end{aligned} \tag{A13a}$$

$$\begin{aligned}\mathcal{I}_2\left(\nu,\eta,\langle B\rangle^2\right) =& \mathcal{I}_2^{(st)}\left(\nu,\eta,\langle B\rangle^2\right) \\ &- i\frac{\pi}{2}\Gamma\int dk\int_{-1}^{1} dX\int d\omega\frac{k^2(1+X^2)\mathcal{F}_0(\omega^2,k)}{\gamma_u}\mathcal{W}(\omega;\tilde{\omega},\nu,\eta),\end{aligned} \tag{A13b}$$

$$
\begin{aligned}
\mathcal{I}_3\left(\nu,\eta,\langle B\rangle^2\right) = &\mathcal{I}_3^{(st)}\left(\nu,\eta,\langle B\rangle^2\right)\\
&-\mathrm{i}\frac{\pi}{4}\Gamma\int \mathrm{d}k\int_{-1}^{1}\mathrm{d}X\int \mathrm{d}\omega\frac{k^4\mathcal{F}_0(\omega^2,k)}{\gamma_u^2\gamma_\eta}\left(-13X^4+10X^2-1\right)\mathcal{W}(\omega;\tilde{\omega},\nu,\eta)\\
&-\mathrm{i}\frac{\pi}{4}\Gamma\int \mathrm{d}k\int_{-1}^{1}\mathrm{d}X\int \mathrm{d}\omega\frac{k^4\mathcal{F}_0(\omega^2,k)}{\gamma_u}\left(-13X^4+10X^2-1\right)\overline{\mathcal{W}}(\omega;\tilde{\omega},\nu,\eta),
\end{aligned} \tag{A13c}
$$

where

$$
\mathcal{W}(\omega;\tilde{\omega},\nu,\eta) = \frac{\mathrm{e}^{\mathrm{i}\tilde{\omega}t}}{\gamma_u(-\omega-\tilde{\omega})\gamma_\eta(-\omega-\tilde{\omega})} - \frac{\mathrm{e}^{-\mathrm{i}\tilde{\omega}t}}{\gamma_u(-\omega+\tilde{\omega})\gamma_\eta(-\omega+\tilde{\omega})}, \tag{A14a}
$$

$$
\overline{\mathcal{W}}(\omega;\tilde{\omega},\nu,\eta) = \frac{\mathrm{e}^{\mathrm{i}\tilde{\omega}t}}{\gamma_u^2(-\omega-\tilde{\omega})\gamma_\eta^2(-\omega-\tilde{\omega})} - \frac{\mathrm{e}^{-\mathrm{i}\tilde{\omega}t}}{\gamma_u^2(-\omega+\tilde{\omega})\gamma_\eta^2(-\omega+\tilde{\omega})}, \tag{A14b}
$$

and $\mathcal{I}_3^{(st)}\left(\nu,\eta,\langle B\rangle^2\right)$ is given in (33c). To make analytical progress, we now consider the following asymptotic limit

$$
\forall_k\ \tilde{\omega} \ll \nu k^2,\ \eta k^2, \quad Pm.\mathscr{M}^2(k) \ll 1, \quad Pm = \frac{\nu}{\eta} \ll 1, \tag{A15}
$$

where we have defined

$$
\mathscr{M}^2(k) = \frac{\langle B\rangle^2}{\nu\eta k^2}, \tag{A16}
$$

so that

$$
\begin{aligned}
&\gamma_u(-\omega-\tilde{\omega})\gamma_\eta(-\omega-\tilde{\omega})\\
&= \left\{\nu k^2 + \eta k^2\frac{(\mathbf{k}\cdot\langle\mathbf{B}\rangle)^2}{|\gamma_\eta|^2} + \mathrm{i}(-\omega-\tilde{\omega})\left[\frac{(\mathbf{k}\cdot\langle\mathbf{B}\rangle)^2}{|\gamma_\eta|^2}-1\right]\right\}\left(\eta k^2 - \mathrm{i}(-\omega-\tilde{\omega})\right)\\
&= \nu\eta k^4 - (\omega+\tilde{\omega})^2 + (\mathbf{k}\cdot\langle\mathbf{B}\rangle)^2 + \mathrm{i}(\omega+\tilde{\omega})(\nu+\eta)k^2\\
&\approx \nu\eta k^4 - \omega^2 + (\mathbf{k}\cdot\langle\mathbf{B}\rangle)^2 + \mathrm{i}\omega(\nu+\eta)k^2 + \tilde{\omega}\left[\mathrm{i}(\nu+\eta)k^2 - 2\omega\right],
\end{aligned} \tag{A17}
$$

$$
\frac{1}{\gamma_u(-\omega-\tilde{\omega})\gamma_\eta(-\omega-\tilde{\omega})} \approx \frac{1}{\gamma_u(-\omega)\gamma_\eta(-\omega)} - \tilde{\omega}\frac{\mathrm{i}(\nu+\eta)k^2-2\omega}{\gamma_u^2(-\omega)\gamma_\eta^2(-\omega)}. \tag{A18}
$$

On introducing the following new variable

$$
\omega = \sqrt{\nu\eta}k^2\varpi, \tag{A19}
$$

we obtain

$$
\begin{aligned}
|\gamma_u(\omega)|^2|\gamma_\eta(\omega)|^2 =&\omega^4 + \omega^2\left[k^4\left(\nu^2+\eta^2\right) - 2(\mathbf{k}\cdot\langle\mathbf{B}\rangle)^2\right] + \left(\nu\eta k^4 + (\mathbf{k}\cdot\langle\mathbf{B}\rangle)^2\right)^2\\
=&\nu^2\eta^2k^8\left\{\varpi^4 + \varpi^2\left[\left(Pm+Pm^{-1}\right) - 2\mathscr{M}^2X^2\right] + \left(1+\mathscr{M}^2X^2\right)^2\right\},
\end{aligned} \tag{A20}
$$

and hence with the use of

$$
\left\langle \hat{f}_i(\mathbf{k},\omega)\hat{f}_j(\mathbf{k}',\omega')\right\rangle = \left[\frac{D_0}{k^3}P_{ij}(\mathbf{k}) + \mathrm{i}\frac{D_1}{k^5}\epsilon_{ijk}k_k\right]\delta(\mathbf{k}+\mathbf{k}')\Delta(\omega,\omega';\tilde{\omega},\Gamma), \tag{A21}
$$

where D_0, D_1 and Γ are constants, one obtains

$$
\mathcal{I}_1\left(\nu,\eta,\langle B\rangle^2\right) = \tilde{\alpha}_{st} + \tilde{\alpha}_{st}\Gamma\sin(\tilde{\omega}t) - \tilde{\alpha}_{\Delta\omega}\Gamma\cos(\tilde{\omega}t) \tag{A22a}
$$

$$\mathcal{I}_2\left(\nu, \eta, \langle B\rangle^2\right) = \bar{\eta}_{1\,st} + \bar{\eta}_{1\,st}\Gamma\sin(\tilde{\omega}t) - \bar{\eta}_{1\,\Delta\omega}\Gamma\cos(\tilde{\omega}t), \tag{A22b}$$

$$\langle B\rangle^2\mathcal{I}_3\left(\nu, \eta, \langle B\rangle^2\right) = -\bar{\eta}_{2\,st} - \bar{\eta}_{2\,st}\Gamma\sin(\tilde{\omega}t) + \bar{\eta}_{2\,\Delta\omega}\Gamma\cos(\tilde{\omega}t), \tag{A22c}$$

where

$$\begin{aligned}\bar{\alpha}_{st} &= \frac{4\pi D_1}{\nu\sqrt{\nu\eta}}\int\frac{\mathrm{d}k}{k^5}\int_{-1}^{1}\mathrm{d}X\int\mathrm{d}\varpi\frac{X^2}{\varpi^4+\varpi^2[(Pm+Pm^{-1})-2\mathscr{M}^2X^2]+(1+\mathscr{M}^2X^2)^2}\\ &\approx \frac{4\pi D_1}{\nu\sqrt{\nu\eta}}\int\frac{\mathrm{d}k}{k^5}\int_{-1}^{1}\mathrm{d}X\int\mathrm{d}\varpi\frac{X^2}{\varpi^2(Pm^{-1}-2\mathscr{M}^2X^2)+(1+\mathscr{M}^2X^2)^2},\end{aligned} \tag{A23}$$

$$\begin{aligned}\bar{\alpha}_{\Delta\omega} =& -4\pi\tilde{\omega}\int\mathrm{d}k\int_{-1}^{1}\mathrm{d}X\int\mathrm{d}\omega\frac{k^4X^2\mathcal{F}_1(\omega^2,k)}{|\gamma_u|^4|\gamma_\eta|^4}\Big\{2\omega^4+\omega^2\Big[\nu(\nu-\eta)k^4-2k^2\langle B\rangle^2X^2\Big]\\ &-(\nu+\eta)\eta k^4\Big[\nu\eta k^4+k^2\langle B\rangle^2X^2\Big]\Big\}\\ \approx&\frac{4\pi D_1\tilde{\omega}}{(\nu\eta)^{3/2}}\int\frac{\mathrm{d}k}{k^7}\int_{-1}^{1}\mathrm{d}X\int\mathrm{d}\varpi\frac{X^2\left[Pm^{-1}(1+\mathscr{M}^2X^2)+2\varpi^2\mathscr{M}^2X^2\right]}{\left\{\varpi^2(Pm^{-1}-2\mathscr{M}^2X^2)+(1+\mathscr{M}^2X^2)^2\right\}^2},\end{aligned} \tag{A24}$$

$$\begin{aligned}\bar{\eta}_{st} \approx&\,\pi\eta\langle B\rangle^2\int\mathrm{d}k\int_{-1}^{1}\mathrm{d}X\int\mathrm{d}\omega\frac{k^6g_3(X)\mathcal{F}_0(\omega^2,k)}{|\gamma_u|^4|\gamma_\eta|^4}\left(\omega^2-\nu\eta k^4-k^2\langle B\rangle^2X^2\right)\\ \approx&\frac{\pi D_0\langle B\rangle^2}{\nu(\nu\eta)^{3/2}}\int\frac{\mathrm{d}k}{k^7}\int_{-1}^{1}\mathrm{d}X\int\mathrm{d}\varpi\frac{(13X^4-10X^2+1)(1+\mathscr{M}^2X^2)}{\left\{\varpi^2(Pm^{-1}-2\mathscr{M}^2X^2)+(1+\mathscr{M}^2X^2)^2\right\}^2},\end{aligned} \tag{A25}$$

$$\begin{aligned}\bar{\eta}_{\Delta\omega} \approx& -\frac{3\pi}{2}\tilde{\omega}\langle B\rangle^2\int\mathrm{d}k\int_{-1}^{1}\mathrm{d}X\int\mathrm{d}\omega\frac{k^4\mathcal{F}_0(\omega^2,k)}{|\gamma_u|^4|\gamma_\eta|^4}g_3(X)\left[2\omega^2+(\nu+\eta)\eta k^4\right]\\ &+2\pi(\nu+\eta)(\nu+3\eta)\tilde{\omega}\langle B\rangle^2\int\mathrm{d}k\int_{-1}^{1}\mathrm{d}X\int\mathrm{d}\omega\frac{k^8\omega^2\mathcal{F}_0(\omega^2,k)}{|\gamma_u|^6|\gamma_\eta|^4}g_3(X)\\ &+2\pi\left(\nu^2-\eta^2\right)\tilde{\omega}\langle B\rangle^2\int\mathrm{d}k\int_{-1}^{1}\mathrm{d}X\int\mathrm{d}\omega\frac{k^8\omega^2\mathcal{F}_0(\omega^2,k)}{|\gamma_u|^6|\gamma_\eta|^6}g_3(X)\Big[2(\nu+\eta)\eta k^4\\ &+k^2\langle B\rangle^2X^2\Big]\\ \approx&\frac{3\pi D_0\tilde{\omega}\langle B\rangle^2}{2(\nu\eta)^{5/2}Pm}\int\frac{\mathrm{d}k}{k^9}\int_{-1}^{1}\mathrm{d}X\int\mathrm{d}\varpi\frac{(13X^4-10X^2+1)}{\left\{\varpi^2(Pm^{-1}-2\mathscr{M}^2X^2)+(1+\mathscr{M}^2X^2)^2\right\}^2}\\ &+\frac{2\pi D_0\tilde{\omega}\langle B\rangle^2}{(\nu\eta)^{5/2}Pm}\int\frac{\mathrm{d}k}{k^9}\int_{-1}^{1}\mathrm{d}X\int\mathrm{d}\varpi\frac{g_3(X)\varpi^2(Pm^{-1}-\mathscr{M}^2X^2)}{\left\{\varpi^2(Pm^{-1}-2\mathscr{M}^2X^2)+(1+\mathscr{M}^2X^2)^2\right\}^3}.\end{aligned} \tag{A26}$$

The above integrals can be evaluated in the asymptotic limit (A15) in a straightforward manner, leading to

$$\frac{\bar{\alpha}_{st}}{\bar{\alpha}_{\Delta\omega}} \approx \frac{8\langle B\rangle\sqrt{Pm}\int_{k_{\min}}^{K}\frac{\mathrm{d}k}{k^3}}{\pi\tilde{\omega}\int_{k_{\min}}^{K}\frac{\mathrm{d}k}{k^4}} \approx \frac{12\langle B\rangle\sqrt{Pm}k_{\min}}{\pi\tilde{\omega}}, \tag{A27a}$$

$$\frac{\bar{\eta}_{st}}{\bar{\eta}_{\Delta\omega}} \approx \frac{4\nu\int_{k_{\min}}^{K}\frac{\mathrm{d}k}{k^6}}{3\tilde{\omega}\int_{k_{\min}}^{K}\frac{\mathrm{d}k}{k^8}} \approx \frac{28\nu k_{\min}^2}{15\tilde{\omega}}, \tag{A27b}$$

where K denotes the short-wavelength dissipative cut-off of the Fourier spectrum and we have assumed $k_\ell \ll K$. The induction equation can be cast in the following form

$$\frac{\partial\langle\mathbf{B}\rangle}{\partial t} = \nabla\times(\bar{\alpha}_{ns}\langle\mathbf{B}\rangle) - \nabla\times(\bar{\eta}_{ns}\nabla\times\langle\mathbf{B}\rangle), \tag{A28}$$

with

$$\bar{\alpha}_{ns} = \bar{\alpha}_{st} + \bar{\alpha}_{st}\Gamma\sin(\tilde{\omega}t) - \bar{\alpha}_{\Delta\omega}\Gamma\cos(\tilde{\omega}t), \tag{A29a}$$

$$\bar{\eta}_{ns} = \eta + \bar{\eta}_{st} + \bar{\eta}_{st}\Gamma\sin(\tilde{\omega}t) - \bar{\eta}_{\Delta\omega}\Gamma\cos(\tilde{\omega}t). \tag{A29b}$$

The Equations (A29a) and (A29b) can be easily rearranged into

$$\bar{\alpha}_{ns} = \bar{\alpha}_{st} + \frac{\bar{\alpha}_{st}\Gamma}{\cos\phi_\alpha}\sin(\tilde{\omega}t - \phi_\alpha), \tag{A30a}$$

$$\bar{\eta}_{ns} = \eta + \bar{\eta}_{st} + \frac{\bar{\eta}_{st}\Gamma}{\cos\phi_\eta}\sin(\tilde{\omega}t - \phi_\eta), \tag{A30b}$$

where

$$\tan\phi_\alpha = \frac{\bar{\alpha}_{\Delta\omega}}{\bar{\alpha}_{st}}, \qquad \tan\phi_\eta = \frac{\bar{\eta}_{\Delta\omega}}{\bar{\eta}_{st}}. \tag{A31}$$

References

1. Roberts, P.H.; Soward, A.M. Magnetohydrodynamics of the Earth's core. *Ann. Rev. Fluid Mech.* **1972**, *4*, 117–154. [CrossRef]
2. Dormy, E.; Soward, A.M. *Mathematical Aspects of Natural Dynamos*; Chapman & Hall/CRC Taylor & Francis Group: Boca Raton, FL, USA, 2007.
3. Roberts, P.H.; King, E.M. On the genesis of the Earth's magnetism. *Rep. Prog. Phys.* **2013**, *76*, 096801. [CrossRef] [PubMed]
4. Steenbeck, M.; Krause, F.; Radler, K.-H. A calculation of the mean electromotive force in an electrically conducting fluid in turbulent motion, under the influence of Coriolis forces. *Z. Naturforsch.* **1966**, *21*, 369–376. [CrossRef]
5. Roberts, P.H. Fundamentals of dynamo theory. In *Lectures on Solar and Planetary Dynamos*; Proctor, M., Gilbert, A., Eds.; Publications of the Newton Institute; Cambridge University Press: Cambridge, UK, 1994; pp. 1–58.
6. Moffatt, H.K.; Dormy, E. *Self-Exciting Fluid Dynamos*; Cambridge University Press: Cambridge, UK, 2019.
7. Aldridge, K.D.; Lumb, L.I. Inertial waves identified in the Earth's fluid outer core. *Nature* **1987**, *325*, 421–423. [CrossRef]
8. Braginsky, S.I. Magnetic waves in the core of the Earth. *Geomagn. Aeron.* **1967**, *7*, 851–859. [CrossRef]
9. Braginsky, S.I. Magnetic waves in the core of the Earth. II. *Geophys. Astrophys. Fluid Dyn.* **1980**, *14*, 189–208. [CrossRef]
10. Hori, K.; Jones, C.A.; Teed, R.J. Slow magnetic Rossby waves in the Earth's core. *Geophys. Res. Lett.* **2015**, *42*, 6622–6629. [CrossRef]
11. Schubert, G. (Ed.) Core dynamics. In *Treatise on Geophysics*; Elsevier: Amsterdam, The Netherlands, 2015; Volume 8.
12. Mizerski, K.A. Large-scale hydromagnetic dynamo by Lehnert waves in nonresistive plasma. *SIAM J. Appl. Math.* **2018**, *78*, 1402–1421. [CrossRef]
13. Mizerski, K.A. Large-scale dynamo action driven by forced beating waves in a highly conducting plasma. *J. Plasma Phys.* **2018**, *84*, 735840405. [CrossRef]
14. Mizerski, K.A. Renormalization group analysis of the turbulent hydromagnetic dynamo: The effect of nonstationarity. *Astroph. J. Supp. Ser.* **2020**, *251*, 21. [CrossRef]
15. Mizerski, K.A.; Bajer, K.; Moffatt, H.K. The mean electromotive force generated by elliptic instability. *J. Fluid Mech.* **2012**, *707*, 111–128. [CrossRef]
16. Driscoll, P.; Olson, P. Polarity reversals in geodynamo models with core evolution. *Earth Planet. Sci. Lett.* **2009**, *282*, 24–33. [CrossRef]
17. Driscoll, P.; Olson, P. Effects of buoyancy and rotation on the polarity reversal frequency of gravitationally driven numerical dynamos. *Geophys. J. Int.* **2009**, *178*, 1337–1350. [CrossRef]
18. Olson, P. The geodynamo's unique longevity. *Phys. Today* **2013**, *66*, 30–35. [CrossRef]
19. Yokoi, N. Cross helicity and related dynamo. *Geophys. Astrophys. Fluid Dyn.* **2013**, *107*, 114–184. [CrossRef]
20. Krause, F.; Rädler, K.-H. *Mean-Field Magnetohydrodynamics and Dynamo Theory*; Pergamon Press: Oxford, UK, 1980.
21. Rogachevskii, I.; Kleeorin, N. Nonlinear theory of a "shear-current" effect and mean-field magnetic dynamos. *Phys. Rev. E* **2004**, *70*, 046310. [CrossRef]
22. Yakhot, V.; Orszag, S.A. Renormalization group analysis of turbulence. I. Basic theory. *J. Sci. Comput.* **1986**, *1*, 3–51. [CrossRef]
23. Mizerski, K.A. Renormalization group analysis of the magnetohydrodynamic turbulence and dynamo. *J. Fluid Mech.* **2021**, *926*, A13. [CrossRef]
24. Chandrasekhar, S. On force-free magnetic fields. *Proc. Nat. Acad. Sci. USA* **1956**, *42*, 1–5. [CrossRef] [PubMed]
25. Chandrasekhar, S.; Kendall, P.C. On force-free magnetic fields. *Astroph. J.* **1957**, *126*, 457–460. [CrossRef]

26. Rädler, K.-H.; Brandenburg, A. Contributions to the theory of a two-scale homogeneous dynamo experiment. *Phys. Rev. E* **2003**, *67*, 026401. [CrossRef]
27. Tassi, E.; Pegoraro, F.; Cicogna, G. Solutions and symmetries of force-free magnetic fields. *Phys. Plasmas* **2008**, *15*, 092113. [CrossRef]
28. Brandenburg, A.; Subramanian, K. Astrophysical magnetic fields and nonlinear dynamo theory. *Phys. Rep.* **2005**, *417*, 1–209. [CrossRef]
29. Tobias, S.M. The turbulent dynamo. *J. Fluid Mech.* **2021**, *912*, P1. [CrossRef] [PubMed]
30. Melbourne, I.; Proctor, M.R.E.; Rucklidge, A.M. A heteroclinic model of geodynamo reversals and excursions. In *Dynamo and Dynamics, a Mathematical Challenge*; Chossat, P., Armbruster, D., Oprea, I., Eds.; Kluwer: Dordrecht, The Netherlands, 2001; pp. 363–370.
31. Hoyng, P.; Ossendrijver, M.A.J.H.; Schmidt, D. The geodynamo as a bistable oscillator. *Geophys. Astrophys. Fluid Dyn.* **2001**, *94*, 263–314. [CrossRef]
32. Hoyng, P.; Duistermaat, J.J. Geomagnetic reversals and the stochastic exit problem. *Europhys. Lett.* **2004**, *68*, 177–183. [CrossRef]
33. Berhanu, M.; Monchaux, R.; Fauve, S.; Mordant, N.; Petrelis, F.; Chiffaudel, A.; Daviaud, F.; Dubrulle, B.; Marie, L.; Ravelet, F.; et al. Magnetic field reversals in an experimental turbulent dynamo. *Europhys. Lett.* **2007**, *77*, 59001. [CrossRef]
34. Petrelis, F.; Fauve, S.; Dormy, E.; Valet, J.-P. Simple Mechanism for Reversals of Earth's Magnetic Field. *Phys. Rev. Lett.* **2009**, *102*, 144503. [CrossRef] [PubMed]
35. Meinel, R.; Brandenburg, A. Behaviour of highly supercritical alpha-effect dynamos. *Astron Astrophys.* **1990**, *238*, 369–376.
36. Stefani, F.; Gerbeth, G.; Günther, U.; Xu, M. Why dynamos are prone to reversals. *Earth Planet. Sci. Lett.* **2006**, *143*, 828–840. [CrossRef]
37. Stefani, F.; Xu, M.; Sorriso-Valvo, L.; Gerbeth, G.; Günther, U. Oscillation or rotation: A comparison of two simple reversal models. *Geophys. Astrophys. Fluid Dyn.* **2007**, *101*, 227–248. [CrossRef]
38. Braginsky, S.I.; Roberts, P.H. Equations governing convection in Earth's core and the geodynamo. *Geophys. Astrophys. Fluid Dyn.* **1995**, *79*, 1–97. [CrossRef]

symmetry

Article

Anchored and Lifted Diffusion Flames Supported by Symmetric and Asymmetric Edge Flames

Zhanbin Lu [1,2,3] **and Moshe Matalon** [4,*]

1 Shanghai Institute of Applied Mathematics and Mechanics, School of Mechanics and Engineering Science, Shanghai University, Shanghai 200072, China; zblu@i.shu.edu.cn
2 Shanghai Key Laboratory of Mechanics in Energy Engineering, Shanghai 200072, China
3 Shanghai Institute of Aircraft Mechanics and Control, Zhangwu Road, Shanghai 200092, China
4 Department of Mechanical Science and Engineering, University of Illinois at Urbana-Champaign, Urbana, IL 61801, USA
* Correspondence: matalon@illinois.edu

Abstract: Numerous combustion applications are concerned with the stabilization of diffusion flames formed by injecting gaseous fuels into a co-flowing stream containing an oxidizer. The smooth operation of these devices depends on the attachment and lift-off characteristics of the edge flame at the base of the diffusion flame. In this paper, we address fundamental issues pertinent to the structure and dynamics of edge flames, which have attributes of both premixed and diffusion flames. The adopted configuration is the mixing layer established in the wake of a splitter plate where two streams, one containing fuel and the other oxidizer, merge. The analysis employs a diffusive-thermal model which, although it excludes effects of gas expansion, systematically includes the influences of the overall flow rate, unequal strain rates in the incoming streams, stoichiometry, differential and preferential diffusion, heat loss and gas–solid thermal interaction, and their effect on the edge structure, speed, and temperature. Conditions when the edge flame is anchored to the plate, lifted-off and stabilized in the flow, or blown-off, are identified. Two stable modes of stabilization are observed for lifted flames; the edge flame either remains stationary at a specified location or undergoes spontaneous oscillations along a direction that coincides with the trailing diffusion flame.

Keywords: edge flames; lifted diffusion flames; tribrachial flames; triple flames; edge speed; oscillating flames; gas–solid thermal interactions; diffusive-thermal model; diffusive-thermal instabilities

Citation: Lu, Z.; Matalon, M. Anchored and Lifted Diffusion Flames Supported by Symmetric and Asymmetric Edge Flames. *Symmetry* **2023**, *15*, 1547. https://doi.org/10.3390/sym15081547

Academic Editors: Andrzej Herczyński and Roberto Zenit

Received: 6 July 2023
Revised: 27 July 2023
Accepted: 4 August 2023
Published: 6 August 2023

1. Introduction

Combustion phenomena are commonly classified as premixed or non-premixed, depending on whether the fuel and oxidizer are already mixed at the molecular level when introduced into the combustion chamber or are individually supplied from different origins. When ignition occurs in a premixed system, a *premixed flame* propagates throughout the mixture, consuming the reactants and generating heat. In a non-premixed system, the fuel and oxidizer are introduced from separate streams to a common region where, upon ignition, mixing and reaction take place simultaneously. Since diffusion plays an essential role in mixing the reactants, the flame in non-premixed systems is referred to as a *diffusion flame*. The distinction between premixed and diffusion flames is a useful way to globally characterize combustion systems. There are circumstances, however, in which burning occurs in a hybrid mode; fuel and oxidizer enter separately but partially mix at the outset and combustion takes place in a stratified ambience once the mixture is ignited. Consider, for example, a jet of fuel burning into an environment containing an oxidizer, reproduced in Figure 1a from the experiments of Chung and Lee [1]. At low speeds, chemical reaction occurs in the immediate vicinity of the injection port and extends downstream along a diffusion flame that separates a region where there is primarily fuel from a region containing mainly oxidizer. At higher speeds, the flame lifts off and stabilizes within the jet away from the injector. The base of the lifted flame, referred to as an *edge flame*, separates a burning from a non-burning state; it has a distinct tribrachial structure that combines characteristics of both premixed and diffusion flames as illustrated in Figure 1b. The highly curved lower part is a premixed flame, which is sustained by the stratified mixture of fuel and oxidizer that has been generated near the injection port. It has two arms: a fuel-rich branch extending towards the fuel side and a fuel-lean branch extending towards the oxidizer side. The diffusion flame that stretches out downstream emanates from the point where the mixture is locally in stoichiometric proportion. This structure has been often referred to in the literature as a *tribrachial flame* or a *triple flame*.

Edge flames play a crucial role in the stabilization of lifted diffusion flames. In large industrial boilers that typically run at high flow rates, the diffusion flame is lifted off the injector, which is favorable because it prevents thermal contact that could lead to erosion of the burner material. The disadvantage is that the freely standing edge flame may be subject to instabilities and possible blow-off. Another example is the diffusion flame separating streams of gaseous hydrogen and liquid oxygen in a liquid rocket engine. The diffusion flame seen in the vicinity of the high-speed gaseous stream further downstream is actually stabilized in the neighborhood of the oxygen injector lip. Once established, the diffusion flame cannot be easily extinguished because the strain rate encountered in practical operation conditions is typically too small and cannot cause its extinction [2]. The main problem, therefore, reduces to anchoring the edge flame near the injector lip. Anchoring and stabilizing lifted diffusion flames, conditioned on the properties of the edge flames at their base, are essential for the safe, efficient, and smooth operation of combustion devices. Poor stabilization may lead to disastrous consequences.

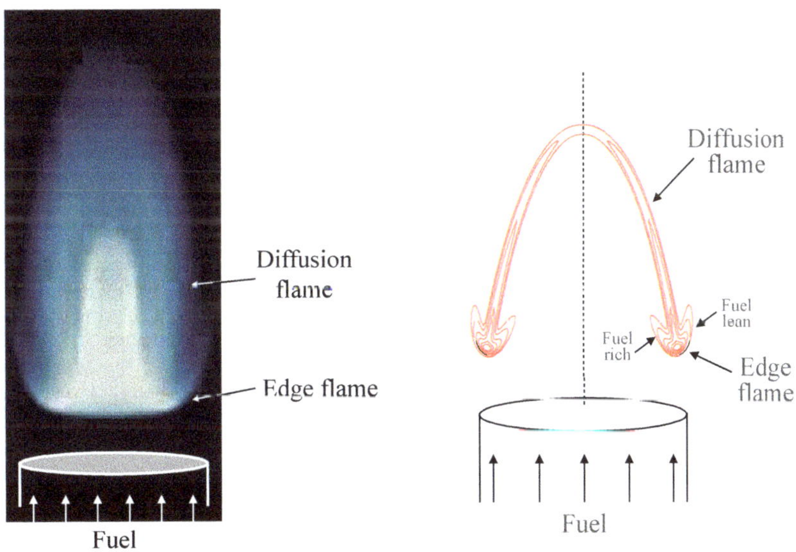

Figure 1. Lifted jet diffusion flame; the fuel issued from a narrow tube is burning in an environment containing an oxidizer. The photograph on the left is taken from [1]; the sketch on the right illustrates the structure of the partially premixed edge flames formed at the base of the diffusion flame.

The symmetric flame structure illustrated in Figure 1 would result under idealized conditions. Various factors, such as streams of unequal strain rates, fuel and oxidizer of unequal molecular diffusivity or supplied in off-stoichiometric proportions, heat losses, and unequal reactant consumption rates would all lead to asymmetric flames. In this study, we focus on the various factors that affect the structure and dynamics of the edge flames supporting the lifted diffusion flames. To this end, we examine the combustion field in the wake of a thin splitter plate where two initially separated streams, one containing fuel and the other an oxidizer, merge. Despite its apparent simplicity, the mathematical problem is rather complex; the flow and combustion fields are strongly coupled due to the heat released by the chemical reactions and the equations governing this intricate process must be solved in an infinite domain that mimics the region ahead and behind the plate trailing edge in order to properly allow for upstream diffusion and accurately capture the far-field asymptotic behavior in the wake region. To simplify the description, we adopt a diffusive-thermal model that filters out the hydrodynamic disturbances induced by gas expansion, practically adopting a constant-density model. Accordingly, the flow field is determined a priori by solving the Navier–Stokes equations with a constant (average) density and used thereafter in the heat and mass transport equations to determine the combustion field. The merging shear flow model considered here, which as formulated by Liñán [3] represents the near-wake flow structure for large Reynolds numbers, has successfully captured the intricate edge flame structure sustained near the plate, the attachment and lift-off behavior of the diffusion flame, and a revelation of the flame stability properties leading to oscillations, blow-off, and/or extinction. In addition, we have provided a comprehensive parametric study describing the influence of a wide range of practical parameters, including mass-flow rate, diffusion properties of the fuel and oxidizer, stoichiometry, radiative heat loss, and thermal characteristics of the splitter plate. Comparison with experimental observations confirms the relevance of our results to the anchoring and stabilization of lifted diffusion flames in various practical settings.

Edge flames have also been studied in other contexts, such as weakly curved premixed flames in stratified mixtures [4–7] and in strained mixing layers in premixed [8–11] and non-premixed [12,13] systems, and have been addressed in a number of model problems [14–16]. They are also relevant to studies of turbulent diffusion flames acting as an agent for re-establishing combustion in a hole formed on the flame surface [17]. Although they share some common features with edge flames examined in the present article, the mathematical problems are fundamentally different. Further details can be found in the review articles of Buckmaster [18], Chung [19], Lyons [20] and Matalon [21].

The mathematical formulation is presented next. In the subsequent sections, we provide a review and extension of studies concerned with edge flames in mixing layers.

2. Model and Formulation

The oxidation of practical fuels involves a complex network of chemical reactions with a multitude of parameters that are not all well known. Although reduced mechanisms have been suggested for common fuels, they typically involve many elementary reactions and a large number of intermediate species that need to be tracked over a wide range of time scales. This complicates analysis of the governing equations appreciably and could only be used in numerical simulations intended to address the burning of a particular fuel under specified conditions. For fundamental understanding, it is preferable to model the chemical activity by an overall step of the form

$$\text{Fuel} + s\ \text{Oxidizer} \rightarrow (1+s)\,\text{Products} + \{Q\},$$

implying that a mass s of an oxidizer is consumed for every unit mass of fuel, producing a mass $1+s$ of products and liberating an amount Q of heat. The fuel consumption, or reaction rate, is then given by

$$\omega = B(\rho Y_{\mathrm{F}})^{n_{\mathrm{F}}}(\rho Y_{\mathrm{O}})^{n_{\mathrm{O}}}\,\mathrm{e}^{-E/RT},$$

where ρ and T are the density and temperature of the mixture, Y_{F} and Y_{O} are the mass fractions of the fuel and oxidizer, E is the activation energy, R the universal gas constant, B is a pre-exponential factor, and $n_{\mathrm{F}}, n_{\mathrm{O}}$ are reaction orders that are empirically determined to accommodate different mixture combinations; see, for example, [22]. To minimize the number of parameters and refrain from dealing with a particular fuel under specified conditions, we assume below that $n_F = n_O = 1$. The implication of this assumption will be outlined below.

We consider below the two-dimensional merging-flow configuration, which is shown in Figure 2 for a symmetric setup. Two streams are separated far upstream by an infinitesimally thin semi-infinite splitter plate; one contains a fuel, with initial mass fraction Y_{F_0}, and the other an oxidizer, with initial mass fraction Y_{O_0}. The ratio between the initial mass fraction of fuel and that of oxidizer, normalized by their stoichiometric proportion, represents the *initial mixture strength* of the system, $\phi = sY_{F_0}/Y_{O_0}$ (similar to the equivalence ratio in a fuel and oxidizer premixture). A lean system corresponds to $\phi < 1$, and a rich system to $\phi > 1$. The ambient temperature and the temperature in both streams are constant and equal to T_0. The incoming streams begin to merge beyond the tip of the plate, bringing the fuel and oxidizer together and forming in the wake of the plate a velocity shear layer and a much wider mixing layer that extends upstream. A Cartesian coordinate system (Oxy) is established with the origin coinciding with the tip of the plate, the x-axis parallel to the plate, and the y-axis perpendicular to the plate pointing toward the fuel stream. For large Reynolds number, the incoming flow along the plate is asymptotically equivalent to two infinite streams of uniform but not necessarily equal strain rates, corresponding to the local velocity gradients of the Blasius boundary layers generated by the plate. The flow extends laterally towards the lower deck of the triple-deck structure that characterizes the flow in the near wake region; for more details, see the description in [3,23]. Hence, the velocity field $\mathbf{v} = (u, v)$ as $x \to -\infty$ is given by

$$v = 0, \qquad u = \begin{cases} Ay, & \text{for } y > 0\,, \\ -\alpha Ay, & \text{for } y < 0\,, \end{cases}$$

where A and αA are the (constant) shear strain rates in the incoming fuel and oxidizer streams, respectively. The parameter α characterizes the skewness of the merging flow field; the case $\alpha \leq 1$ corresponds to a co-flowing setup, which is often used in experiments to stabilize the flame [24], while $\alpha \to 0$ mimics a fuel jet injecting into a quiescent gas [1], where the oxidizer is totally entrained into the mixing layer from the surroundings. Upon successful ignition, a diffusion flame stabilized by an edge flame at its base is established in the flow field.

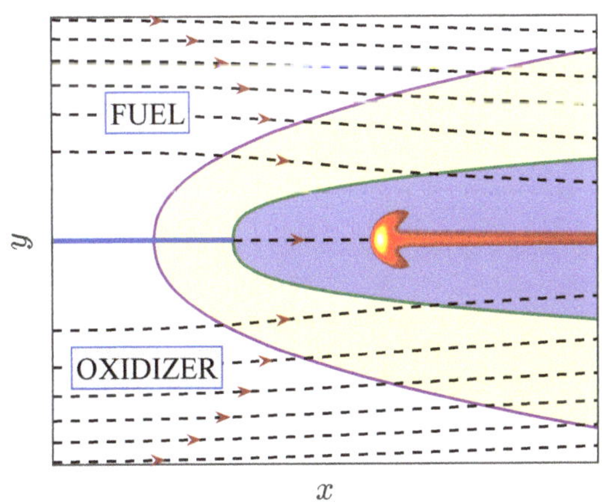

Figure 2. A symmetric edge flame stabilized in the wake of two merging streams of initially uniform and constant strain rates. The edge flame is represented by reaction-rate contours and the flow field is illustrated by equally spaced streamlines. The shaded areas correspond to the velocity shear layer (inner region) and the mixing layer (outer region), respectively.

The combustion process is described by conservation statements of mass, momentum, and energy of the entire mixture and mass balance equations for the fuel and oxidizer. For constant properties, these equations take the form

$$\begin{aligned}
\frac{\partial \rho}{\partial t} + \nabla\cdot(\rho\mathbf{v}) &= 0,\\
\rho(\frac{\partial \mathbf{v}}{\partial t} + \mathbf{v}\cdot\nabla\mathbf{v}) &= -\nabla p + \mu\nabla^2\mathbf{v},\\
\rho c_p(\frac{\partial T}{\partial t} + \mathbf{v}\cdot\nabla T) &= \lambda\nabla^2 T + Q\omega - Q_\mathrm{R},\\
\rho(\frac{\partial Y_\mathrm{F}}{\partial t} + \mathbf{v}\cdot\nabla Y_\mathrm{F}) &= \rho D_\mathrm{F}\nabla^2 Y_\mathrm{F} - \omega,\\
\rho(\frac{\partial Y_\mathrm{O}}{\partial t} + \mathbf{v}\cdot\nabla Y_\mathrm{O}) &= \rho D_\mathrm{O}\nabla^2 Y_\mathrm{O} - s\omega,
\end{aligned}$$

where t is time and p is the dynamic pressure. The coefficients μ, c_p, and λ are the viscosity, specific heat (at constant pressure) and conductivity of the mixture, and $D_\mathrm{F}, D_\mathrm{O}$ are the mass diffusivities of the fuel and oxidizer. The volumetric heat loss rate by gas radiation, for an optically thin gas, takes the form

$$Q_\mathrm{R} = 4\sigma l_p^{-1}(T^4 - T_0^4)$$

where σ is the Stefan–Boltzmann constant and l_p is the Planck mean-absorption length. These equations must be supplemented by an appropriate equation of state, relating the density of the mixture to the temperature. Deflagration waves are low-Mach-number processes and nearly isobaric, such that the density is inversely proportional to the temperature. The small pressure variations from the constant ambient pressure are only necessary to balance the equally small momentum changes.

The aforementioned equations display a strong coupling between the flow and combustion fields because of the density variations associated with the increase in temperature that results from the heat released by the chemical reactions. The gas expansion modifies the velocity field which, in turn, affects the transport of energy and chemical species. A notable simplification results if the density is assumed constant by adopting, for example, an average value. The *constant-density*, or *diffusive-thermal*, model can be obtained formally by assuming that the heat release is small compared to the initial enthalpy of the mixture. As a result, the temperature rise is small and the density in the fluid-mechanic equations is, with respect to leading order, the constant ambient value. A small heat release, however, is not a characteristic of combustion systems. Using instead the constant-density assumption with an average value for the density is more appealing. Using this approximation, the flow field is determined a priori and then used in the energy- and mass-balance equations to determine the combustion field. The results presented below will be examined within this framework.

To express the equations in dimensionless form, we scale velocities with respect to a characteristic speed of an edge flame chosen as the laminar flame speed corresponding to a stoichiometric premixture, namely a mixture consisting of $Y_{\mathrm{F}_0}/(1+\phi)$ fuel and $\phi Y_{\mathrm{O}_0}/(1+\phi)$ oxidizer, of unity fuel and oxidizer Lewis numbers. An expression for laminar flame speed S_L of a stoichiometric mixture (of arbitrary Lewis numbers) derived in the limit of large activation energies [25] is given by

$$S_\mathrm{L} = \sqrt{\frac{4B\rho D_\mathrm{T}\,\phi Y_{\mathrm{O}_0}}{\beta^3 \mathrm{Le}_\mathrm{F}^{-1}\mathrm{Le}_\mathrm{O}^{-1}(1+\phi)}}\ \mathrm{e}^{-E/2RT_\mathrm{a}}, \tag{1}$$

where $D_\mathrm{T} = \lambda/\rho c_p$ is the thermal diffusivity of the mixture, $T_\mathrm{a} = T_0 + QY_{\mathrm{F}_0}/c_p(1+\phi)$ is the adiabatic flame temperature, $\mathrm{Le}_\mathrm{F} = D_\mathrm{T}/D_\mathrm{F}$ and $\mathrm{Le}_\mathrm{O} = D_\mathrm{T}/D_\mathrm{O}$ are the Lewis numbers, representing the ratios of the thermal diffusivity of the mixture to the molecular diffusivities of the fuel and oxidizer, and $\beta = E(T_a - T_0)/RT_a^2$ is the activation-energy parameter, or Zeldovich number. The laminar flame speed corresponding to unity fuel and oxidizer Lewis numbers is identified with superscript 0, i.e., denoted S_L^0. Using S_L^0 as a unit speed, length and time are scaled by the diffusion length D_T/S_L^0 and diffusion time $D_T/S_L^{0\,2}$, and pressure by $\rho S_L^{0\,2}$. The steady flow resulting from the merging streams illustrated in Figure 2 is described by

$$\begin{aligned} \nabla \cdot \mathbf{v} &= 0, \\ \mathbf{v} \cdot \nabla \mathbf{v} &= -\nabla p + \mathrm{Pr}\, \nabla^2 \mathbf{v}, \end{aligned} \tag{2}$$

using the same symbols for the dimensionless velocity and pressure, where $\mathrm{Pr} = \mu c_p/\lambda$ is the Prandtl number, or the reciprocal of the corresponding Reynolds number. For the remaining equations, it is convenient to normalize the fuel and oxidizer mass fractions by their initial supply values, Y_{F_0} and Y_{O_0}, and define a scaled temperature $\theta = (T - T_0)/(T_a - T_0)$. The combustion field is then described by

$$\begin{aligned} \frac{\partial \theta}{\partial t} + \mathbf{v} \cdot \nabla \theta &= \nabla^2 \theta + (1+\phi)\,\Omega - H\left[(1+q\theta)^4 - 1\right], \\ \frac{\partial Y_F}{\partial t} + \mathbf{v} \cdot \nabla Y_F &= \mathrm{Le}_F^{-1} \nabla^2 Y_F - \Omega, \\ \frac{\partial Y_O}{\partial t} + \mathbf{v} \cdot \nabla Y_O &= \mathrm{Le}_O^{-1} \nabla^2 Y_O - \phi\,\Omega, \end{aligned} \tag{3}$$

using the same symbols for the normalized mass fractions. The parameters in these equations, in addition to the initial mixture strength ϕ and the Lewis numbers Le_F and Le_O, are the heat release or thermal expansion parameter $q = (T_a - T_0)/T_0$, and the volumetric heat-loss coefficient $H = 4\sigma T_0^4/\rho c_p l_p (T_a - T_0) S_L^{0\,2}$. The dimensionless reaction rate is given by

$$\Omega = \mathbb{D}\beta^3 Y_F Y_O \exp\left\{\frac{\beta(1+q)(\theta-1)}{(1+q\theta)}\right\}, \tag{4}$$

where the factor β^3 was introduced solely for convenience, being the proper scaling in the asymptotic treatment of both premixed [25] and diffusion flames [26]. The coefficient

$$\mathbb{D} = \frac{1}{\beta^3} \frac{D_T/S_L^{0\,2}}{\left(\rho Y_{O_0} B\, \mathrm{e}^{-E/RT_a}\right)^{-1}} = \frac{1+\phi}{4\phi}, \tag{5}$$

which represents the ratio of the diffusion to chemical reaction times, depends only on the initial mixture strength and may be considered as the local Damköhler number that characterizes the conditions near the edge flame. The ratio of the flow to the chemical reaction times,

$$D = \frac{\mathrm{Pr}}{\beta^3} \frac{A^{-1}}{\left(\rho Y_{O_0} B\, \mathrm{e}^{-E/RT_a}\right)^{-1}} \tag{6}$$

is the global Damköhler number, which serves in this study as a primary parameter that controls the flow conditions; decreasing D corresponds to increasing the strain rate A or, equivalently, increasing the overall mass-flow rate. Below, we refer to D as the Damköhler number. The Prandtl number was included in (6) for consistency with the definition used in previous studies.

The boundary conditions associated with these equations, which confirm the description shown in Figure 2 are:

$$\begin{aligned} \text{as } x \to -\infty: \quad & u = \tilde{A}y, \quad v = 0, \quad Y_F = 1, \quad Y_O = 0, \quad \theta = 0, \quad \text{for } y > 0, \\ & u = -\alpha\tilde{A}y, \quad v = 0, \quad Y_F = 0, \quad Y_O = 1, \quad \theta = 0, \quad \text{for } y < 0, \end{aligned}$$

$$\text{as } y \to +\infty : \qquad u = \widetilde{A}y, \quad Y_F = 1, \quad Y_O = 0, \quad \theta = 0,$$
$$\text{as } y \to -\infty : \qquad u = -\alpha\widetilde{A}y, \quad Y_F = 0, \quad Y_O = 1, \quad \theta = 0,$$

where
$$\widetilde{A} = \frac{AD_T}{S_L^{0\,2}} = \frac{1+\phi}{4\phi}\frac{\mathrm{Pr}}{D}$$

is the dimensionless strain rate, inversely proportional to the Damköhler number D. We note that although the shear flow extends to infinity in the lateral directions, the vertical velocity component v can be made to vanish only far upstream. Elsewhere, we require

$$\text{as } y \to \pm\infty : \qquad \partial v/\partial y = \partial p/\partial y = 0,$$

a constraint that results from the pressure gradient induced by the displacement effect of the boundary layer along the plate, as discussed in [27–29]. The conditions along the splitter plate are:

$$\text{along } y = 0 \text{ for } x < 0 : \qquad u = v = 0, \quad \frac{\partial Y_F}{\partial y} = \frac{\partial Y_O}{\partial y} = 0, \quad \theta = \theta_s,$$

where θ_s is the temperature of the solid plate; they correspond to no-slip conditions, the impermeability of the solid plate to fuel and oxidizer, and local thermal equilibrium between the solid and gas phases. The temperature θ_s depends on the nature of the plate and the thermal interaction between the solid and gaseous phases. Energy conservation, for a thin plate of thickness h, yields the following heat-conduction equation

$$\frac{1}{r_D}\frac{\partial\theta_s}{\partial t} = \frac{\partial^2\theta_s}{\partial x^2} + \frac{1}{r_\lambda h}\left[\!\left[\frac{\partial\theta}{\partial y}\right]\!\right],$$

where r_λ and r_D are, respectively, the ratios of thermal conductivity and thermal diffusivity between the solid and the gas, and $[\![\partial\theta/\partial y]\!] = \partial\theta/\partial y|_{y=0^+} - \partial\theta/\partial y|_{y=0^-}$ is the heat conducted laterally from the plate to the gaseous mixture. The limit $r_\lambda \to \infty$, which corresponds to infinite conductivity of the plate's material, or vanishing thermal resistance, implies that $\theta_s = 0$ and represents a (cold) isothermal plate. The other extreme, $r_\lambda \to 0$, which corresponds to the vanishing conductivity of the plate's material, or infinite thermal resistance, implies that $[\![\partial\theta/\partial y]\!] = 0$ and represents an adiabatic plate. At the tip of the plate, $x = y = 0$, we impose

$$u = v = 0, \quad \frac{\partial Y_F}{\partial x} = \frac{\partial Y_O}{\partial x} = 0, \quad \theta = \theta_s, \quad \frac{\partial\theta}{\partial x} = r_\lambda\frac{\partial\theta_s}{\partial x}.$$

Finally, far downstream the temperature and mass fractions are expected to evolve into a uniform state such that

$$\text{as } x \to \infty : \qquad \frac{\partial\theta}{\partial x} = \frac{\partial Y_F}{\partial x} = \frac{\partial Y_O}{\partial x} = 0.$$

The conditions on the velocity field will be discussed below.

3. The Fast Chemistry Limit

We begin with an examination of the diffusion flame in the fast chemistry limit, first envisaged by Burke and Schumann [30]. We also assume that the combustion system is adiabatic, by neglecting heat losses, i.e., $H = 0$, and considering an adiabatic plate, i.e., $r_\lambda = 0$. For the sake of analytical treatment, we momentarily put aside the merging shear flow and replace it with a simple uniform flow along the x-direction. If the uniform axial velocity U is used as a reference speed, the temperature and mass fraction equations for steady conditions simplify to

$$\begin{aligned}\frac{\partial \theta}{\partial x} &= \nabla^2\theta + (1+\phi)\,\Omega\,,\\ \frac{\partial Y_F}{\partial x} &= \mathrm{Le}_F^{-1}\,\nabla^2 Y_F - \Omega\,,\\ \frac{\partial Y_O}{\partial x} &= \mathrm{Le}_O^{-1}\,\nabla^2 Y_O - \phi\,\Omega\,,\end{aligned} \tag{7}$$

where the numerator in the definition of $\mathbb{D}$ appears in the reaction rate term Ω, and similarly in the definition of D, it is replaced by the characteristic flow time D_T/U^2. The fast chemistry limit corresponds to an infinitely large Damköhler number, namely when a chemical reaction proceeds much faster than the flow. Formally, $D \to \infty$ implies that $Y_F Y_O = 0$ throughout the combustion field, which can only be achieved if both reactants are consumed instantaneously along a sheet that separates a region where there is fuel but no oxidizer, from a region where there is only oxidizer. Then $\Omega = 0$ on either side of the reaction sheet (also known as the Burke–Schumann flame sheet), and the resulting reaction-free equations must be solved subject to the following jump conditions across the sheet:

$$\begin{gathered}[\![\theta]\!] = [\![Y_F]\!] = [\![Y_O]\!] = 0\,,\\ \left[\!\!\left[\frac{\partial\theta}{\partial n}\right]\!\!\right] = -\frac{1+\phi}{\mathrm{Le}_F}\left[\!\!\left[\frac{\partial Y_F}{\partial n}\right]\!\!\right] = -\frac{1+\phi}{\phi \mathrm{Le}_O}\left[\!\!\left[\frac{\partial Y_O}{\partial n}\right]\!\!\right],\end{gathered} \tag{8}$$

where $\partial/\partial n$ is the derivative along the normal to the sheet, and $[\![\bullet]\!] = \bullet|_{n=0^+} - \bullet|_{n=0^-}$ is the jump operator. These conditions state that all variables are continuous across the sheet, that fuel and oxidizer flow into the sheet from opposite sides in stoichiometric proportions, and that heat is conducted proportionately to the fuel and oxidizer regions. They are sufficient for the mathematical description of the combustion field, including the shape and location of the reaction sheet, as discussed by Cheatham and Matalon [26].

The distribution of the reactant mass fractions is often determined in terms of the mixture fraction, defined as

$$Z = \frac{1+\phi Y_F - Y_O}{1+\phi}\,, \tag{9}$$

which varies from $Z = 1$ in the fuel region (where there is no oxidizer) to $Z = 0$ in the oxidizer region (where there is no fuel). In the present case, Z varies smoothly from $Z = 0$ to $Z = 1$ when y extends from $-\infty$ to $+\infty$. The mixture fraction is a useful property when the fuel and oxidizer have equal Lewis numbers, i.e., $\mathrm{Le}_F = \mathrm{Le}_O = \mathrm{Le}$ because, according to Equation (7), it satisfies

$$\frac{\partial Z}{\partial x} = \mathrm{Le}^{-1}\nabla^2 Z\,, \tag{10}$$

which contains no source terms, so that Z is a conserved scalar; namely, a quantity that is neither created nor destroyed by chemical reactions. The mixture fraction can then be determined a priori, with the mass fractions expressed in terms of Z from the definition (9). The reaction sheet, along which $Y_F = Y_O = 0$, corresponds to $Z = Z_{st}$, where $Z_{st} = 1/(1+\phi)$ is the stoichiometric value of the mixture fraction. The distributions of the fuel and oxidizer mass fractions are given by

$$Y_F = \begin{cases} \phi^{-1}[(1+\phi)Z - 1]\,, & Z_{st} < Z < 1\\ 0\,, & 0 < Z \le Z_{st}\end{cases} \tag{11}$$

$$Y_O = \begin{cases} 0\,, & Z_{st} < Z < 1\\ 1-(1+\phi)Z\,, & 0 < Z \le Z_{st}\,,\end{cases} \tag{12}$$

where (here and below) the upper/lower expressions correspond to the fuel/oxidizer regions. Note that the stoichiometric surface $Z(x,y) = Z_{st}$ is also the surface prescribed by the stoichiometric condition $\phi Y_F = Y_O$ in a non-reacting environment.

The temperature field cannot be simply expressed in terms of the mixture fraction, unless $\mathrm{Le} = 1$. For unity Lewis numbers, the combinations $\theta + (1+\phi)Y_{\mathrm{F}}$ and $\phi\theta + (1+\phi)Y_{\mathrm{O}}$ are also conserved scalars that can be expressed in terms of the mixture fraction as follows:

$$\begin{aligned} \theta + (1+\phi)Y_{\mathrm{F}} &= (1+\phi)Z, \\ \phi\theta + (1+\phi)Y_{\mathrm{O}} &= (1+\phi)(1-Z). \end{aligned}$$

Consequently, the temperature is given by

$$\text{for Le} = 1: \qquad \theta = \begin{cases} \phi^{-1}(1+\phi)(1-Z)], & Z_{\mathrm{st}} < Z < 1 \\ (1+\phi)Z, & 0 < Z \leq Z_{\mathrm{st}}, \end{cases} \tag{13}$$

with $\theta = 1$ along the stoichiometric surface. Indeed, for unity Lewis numbers, the temperature along the Burke–Schumann flame sheet is the adiabatic temperature $T = T_a$.

It is important to note that the mass fractions admit the simple form (11) and (12) only for equal Lewis numbers, and the temperature can be simply expressed in terms of the mixture fraction Z if, in addition, $\mathrm{Le} = 1$. The usefulness of the mixture fraction formulation, however, is limited to situations where the combustion field depends on one coordinate which can be easily transformed to Z. Otherwise, one needs to solve an appropriate equation, such as (10) in the present case, to determine Z in terms of the physical coordinates.

We proceed with the case of equal but non-unity Lewis numbers and introduce the following parabolic coordinate transformation:

$$\xi = \left\{\frac{1}{2}\left(\sqrt{x^2+y^2}+x\right)\right\}^{\frac{1}{2}}, \qquad \eta = \pm\left\{\frac{1}{2}\left(\sqrt{x^2+y^2}-x\right)\right\}^{\frac{1}{2}},$$

where $\eta \lesseqgtr 0$ corresponds to $y \lesseqgtr 0$, respectively. Conversely, the Cartesian coordinates are related to the parabolic coordinates through the relations

$$x = \xi^2 - \eta^2, \qquad y = 2\xi\eta,$$

for $0 \leq \xi < \infty$, and $-\infty < \eta < \infty$. In terms of the parabolic coordinates (ξ, η), the splitter plate lies along $\xi = 0$, and Equation (10) transforms into

$$2\xi\frac{\partial Z}{\partial \xi} - 2\eta\frac{\partial Z}{\partial \eta} = \mathrm{Le}^{-1}\left(\frac{\partial^2 Z}{\partial \xi^2} + \frac{\partial^2 Z}{\partial \eta^2}\right),$$

with the boundary conditions:

$$\begin{aligned} &\frac{\partial Z}{\partial \xi} = 0, \quad \text{for } \xi = 0, && \left|\frac{\partial Z}{\partial \xi}\right| < \infty, \quad \text{as } \xi \to \infty, \\ &Z = 0, \quad \text{as } \eta \to -\infty, && Z = 1, \quad \text{as } \eta \to \infty. \end{aligned}$$

We note that the boundary conditions are identically satisfied when $\partial Z/\partial \xi = 0$ and, consequently, the solution that depends on η only, is

$$Z = \frac{1}{2}\left[1 + \mathrm{erf}\left(\sqrt{\mathrm{Le}}\,\eta\right)\right]. \tag{14}$$

The mass fractions (11) and (12) can now be easily expressed in terms of (x, y). Solution (14) implies that the contours of $Z(x, y)$ coincide with the coordinate lines η, and for a constant η are parabolas of the form $x = y^2/4\eta^2 - \eta^2$. The stoichiometric surface, or Burke–Schumann flame-sheet location, is obtained from $Z = Z_{st}$, and given by

$$\eta_{st} = \frac{1}{\sqrt{\text{Le}}}\,\text{erf}^{-1}\left(\frac{1-\phi}{1+\phi}\right).$$

When the fuel and oxidizer are supplied in stoichiometric proportions, i.e., $\phi = 1$, the flame sheet coincides with the axis $y = 0$ and the combustion field is symmetric with respect to the axis. An asymmetric flame results otherwise; for a lean system ($\phi < 1$) it lies in the fuel region and for a rich system ($\phi > 1$) in the oxidizer region.

Although the temperature cannot be simply expressed in terms of the mixture fraction, a solution can be obtained by recognizing that θ on either side of the reaction sheet satisfies an equation similar to Z, with Le^{-1} in front of the Laplacian replaced by 1. The solution that vanishes as $\eta \pm \infty$ is

$$\theta = \begin{cases} \dfrac{\theta_{st}}{1-\text{erf}\,\eta_{st}}(1-\text{erf}\,\eta), & \text{for } \eta > \eta_{st} \\[2ex] \dfrac{\theta_{st}}{1+\text{erf}\,\eta_{st}}(1+\text{erf}\,\eta), & \text{for } \eta < \eta_{st}\,, \end{cases} \quad (15)$$

where θ_{st} is the temperature along the diffusion flame sheet $\eta = \eta_{st}$. The jump condition (8) relating the gradients of temperature and mass fractions yields

$$\theta_{st} = \frac{(1+\phi)^2}{4\phi\sqrt{\text{Le}}}(1 - \text{erf}^2\eta_{st})\,\text{e}^{(1-\text{Le})\eta_{st}^2}\,. \quad (16)$$

For unity Lewis numbers, $\theta_{st} = 1$ as noted earlier, and the temperature along the Burke–Schumann flame sheet is the adiabatic flame temperature T_a. For non-unity Lewis numbers, the temperature along the Burke–Schumann flame sheet, or the stoichiometric temperature, differs from the adiabatic flame temperature and depends on the Lewis number, as first recognized by Cheatham and Matalon [26] and discussed in [31] for a counterflow diffusion flame. This remains true even for $\phi = 1$, namely, when the flame sheet lies along the y-axis and the combustion field is symmetric with respect to the centerline. The result $\theta_{st} = \text{Le}^{-1/2}$ in this case implies that for the same stoichiometric mixture, the flame temperature of a diffusion flame consuming all the fuel and oxidizer is larger than the flame temperature of a premixed flame by a factor $\text{Le}^{-1/2}$. It will be demonstrated in later sections that this distinction plays a significant role in determining the temperature and propagation speed of edge flames. The insight gained from the explicit Formula (16) helps in understanding the more complex effects of Lewis number and initial mixture strength on edge flames in merging shear flows.

The dependence of the stoichiometric temperature on the Lewis number and initial mixture strength is illustrated in Figure 3 by contours of θ_{st} with respect to Le and $\phi/(1+\phi)$. The initial mixture strength has been normalized as suggested by Law [32] for the equivalence ratio in premixed combustion. In terms of $\phi/(1+\phi)$, lean/rich conditions are distributed in a symmetric way below/above the stoichiometric value $\phi/(1+\phi) = 0.5$ (dashed horizontal line). In general, the stoichiometric temperature increases on decreasing the Lewis number, namely when the mass diffusivities of the fuel and oxidizer are significantly larger than the thermal diffusivity of the mixture. The symmetric distribution of θ_{st} relative to stoichiometry is expected, because the flow is uniform and for equal Lewis numbers the roles of fuel and oxidizer are interchangeable.

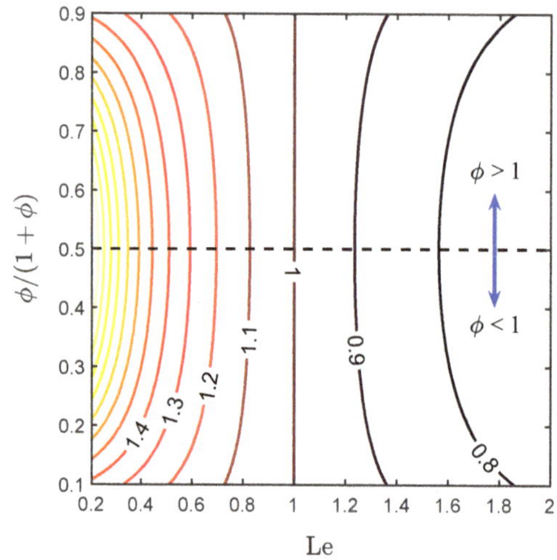

Figure 3. Contours of the Burke–Schumann diffusion-flame temperature, or stoichiometric temperature θ_{st}, in the parameter plane characterized by the Lewis number Le of the mixture and the normalized initial mixture strength, $\phi/(1+\phi)$, for a constant uniform flow.

4. The Merging Shear Flow

The merging of two uniform shear flows downstream of the trailing edge of a semi-infinite flat plate has been the subject of past investigations [33–35], primarily in the aerodynamic field. For large Reynolds numbers, the flow is described by the boundary layer equations and admits a similarity solution which, when expressed in terms of the streamfunction, takes the form

$$\psi(x,y) = x^{2/3}F(\eta),$$

where $\eta = y/x^{1/3}$ is the similarity variable (not to be confused with the notation in the previous section). For unequal initial strain rates, $F(\eta)$ is the solution of the boundary value problem

$$3F''' + 2FF'' - F'^2 = 2c,$$

$$F(\eta) \sim \frac{1}{2}\eta^2 + c, \qquad \text{as } \eta \to +\infty,$$

$$F(\eta) \sim -\frac{1}{2}\alpha\eta^2 - \frac{c}{\alpha}, \qquad \text{as } \eta \to -\infty,$$

where primes denote differentiation with respect to η and c is an eigenvalue. The corresponding velocity components are given by

$$u = x^{1/3}F', \qquad v = \frac{1}{3x^{1/3}}(\eta F' - 2F). \tag{17}$$

The solution of this problem was reexamined recently by Lu and Matalon [23] evaluating the constant c and displaying profiles of the $u(\eta)$ and $v(\eta)$ for various values of the strain rate ratio α. The similarity solution exhibits a singularity at the plate's trailing edge, as attested by the behavior of the vertical velocity v as $x \to 0$, which invalidates the boundary layer approximation exactly in the region where fuel and oxidizer begin mixing and the base of the diffusion flame is expected to be stabilized. The underlying flow that supports the edge flame must therefore be based on the Navier–Stokes Equation (2), and the solution must asymptotically match the profiles (17) far downstream, namely with $u \sim x^{1/3}$ and $v \sim x^{-1/3}$ as $x \to \infty$. The flow field that adheres to this condition was determined numerically for various values of α using a vorticity-streamfunction formulation. The equations were discretized in a rectangular domain on a uniform mesh with $\Delta x = \Delta y = 0.05$, which were found sufficient to meet the requirements of grid independence. Derivatives were approximated via a second-order central difference scheme, and the resulting difference equations were solved using a Gauss–Seidel iteration with successive over-relaxation.

Figure 4 shows the flow field for two representative values of α, for $\mathrm{Pr} = 0.72$. When $\alpha = 1$, the flow field is symmetric with respect to the horizontal centerline, which coincides with the dividing streamline that emanates from the trailing edge of the plate. When $\alpha = 0.1$, the imbalance in vorticity in the incoming streams causes the dividing streamline to deviate from the centerline toward the fuel side, indicative of fluid entrainment from the oxidizer region. Results for different α's indicate that the deviation of the dividing streamline increases as α decreases, implying an enhancement in the oxidizer entrainment. Evidently, in both cases shown in the figure, the flow past the trailing edge of the plate undergoes acceleration in the near-wake region. For a symmetric structure, the stabilization of the edge flame in the mixing layer is significantly affected by the streamwise-velocity gradient but when α decreases, the role of diffusion becomes more pronounced, as discussed below.

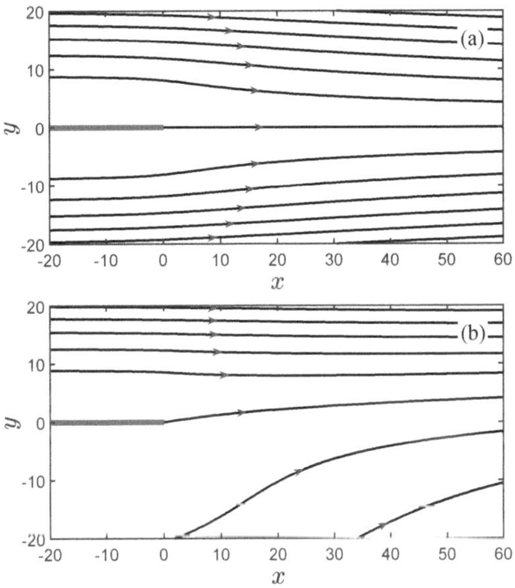

Figure 4. The flow field of two merging shear flows for two representative strain rate ratios: (**a**) $\alpha = 1$, and (**b**) $\alpha = 0.1$, both at Damköhler number $D = 10$, represented by streamlines that correspond to equally spaced streamfunction values.

5. The Edge-Flame Structure

Given the velocity field $\mathbf{v}(x, y)$, as determined in the preceding section, we now address Equation (3) for the description of the combustion field. Among all the parameters involved, we have fixed $q = 5$ and $\beta = 10$, which correspond to the combustion reaction of typical hydrocarbon fuels under standard conditions, and examined the influence of the remaining parameters. These include the Damköhler number D, the fuel and oxygen Lewis numbers $\mathrm{Le_F}$ and $\mathrm{Le_O}$, the initial strain rate ratio α, the initial mixture strength ϕ, the thermo-physical properties of the splitter plate, r_λ and r_D, and the volumetric heat-loss coefficient H, which will be examined separately. For consistency, the combustion field was determined numerically in the same rectangular domain and on the same mesh as the flow field. The governing equations and boundary conditions were discretized via a second-order finite-difference scheme and solved using the Gauss–Seidel iteration method with successive over-relaxation. For steady solutions, an intermediate temperature value was chosen and fixed at a specified grid point to capture all possible solutions, stable and unstable. The stability of the computed states was assessed by examining the time evolution of small perturbations superimposed to a given steady state. The time-dependent calculations were carried out with appropriate time steps Δt, ensuring that the required temporal precision was satisfied. Further details can be found in [23] and the references cited therein.

5.1. Symmetric Edge Flames

Figure 5 illustrates the structure of a symmetric flame, stabilized in the near wake of two merging streams of equal strain rates, $\alpha = 1$, and a flow rate corresponding to $D = 10$, with the parameter values: $\mathrm{Le_F} = \mathrm{Le_O} = 1$, $\phi = 1$, $H = 0$, $r_\lambda = \infty$. We refer to this case as the *baseline case* and, unless otherwise stated, hereinafter when varying one of the parameters it is implicitly assumed that all the other parameters remain unchanged. The flame structure in the figure is represented by contours of reaction rate Ω, temperature θ, and fuel and oxidizer mass fractions, Y_F and Y_O, respectively. Because the flow field is symmetric with respect to the x-axis, and the fuel and oxidizer of equal diffusivities are supplied in the incoming streams in stoichiometric proportion, the resulting edge flame formed in the wake of the splitter plate is also symmetric. The flame has an apparent tribrachial structure; it consists of a curved premixed flame at the front with fuel-rich and fuel-lean branches extending above and below, which is referred to as the *edge flame*, and a straight diffusion flame trailing downstream along the centerline. An enlarged picture of an edge flame can be seen in one of the figures shown below. The temperature and mass fraction distributions shown in the figure illustrate the extent and width of the thermal and mixing layers. For the isothermal plate considered here, the thermal layer develops beyond the trailing edge of the plate, while the mixing layer extends behind the plate as a result of upstream diffusion.

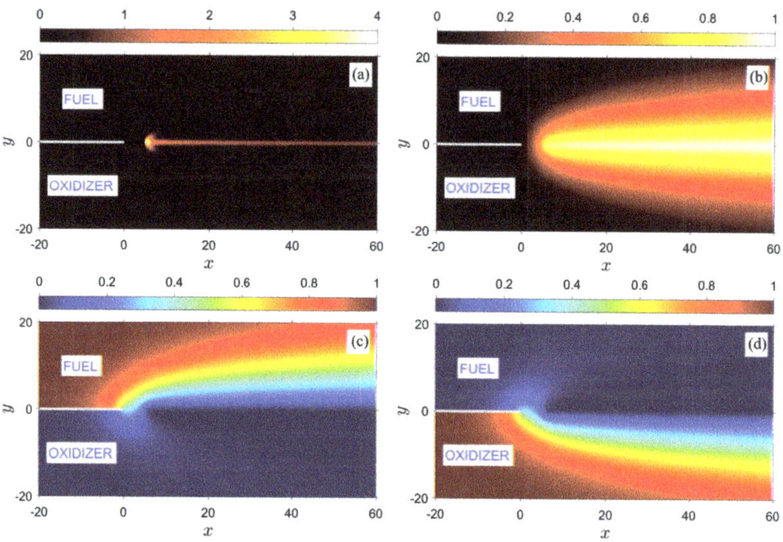

Figure 5. An edge flame stabilized in the near wake of two merging streams of equal initial strain rates, for unity Lewis numbers, stoichiometry, and adiabatic conditions—the baseline case. The structure of the edge flame is illustrated by contours of (**a**) reaction rate Ω, (**b**) temperature θ, (**c**) fuel mass fraction Y_F, and (**d**) oxidizer mass fraction Y_O.

The discussion below necessitates a precise and unambiguous definition of the edge-flame position. To this end, we define its coordinates, x_e and y_e, as the location where the reaction rate Ω attains its maximum value. The axial distance between the edge flame and the tip of the plate, i.e., x_e, will be referred to as the *flame-standoff distance*.

Figure 6 shows that the flame-standoff distance is strongly affected by the overall flow rate, i.e., variations in the Damköhler number D, and by the disparity between heat and mass diffusivities, or variations in the Lewis number Le. In general, when decreasing D from relatively large values, the edge flame moves towards the plate, i.e., down the velocity gradient, in order to reach a new balance with the local flow velocity for its stabilization. The standoff distance, however, reaches a minimum value because heat loss to the cold plate prevents the edge flame from getting closer. The response of the edge flame when approaching the plate and the minimum standoff distance depend on the Lewis number. To retain the symmetry, the fuel and oxidizer Lewis numbers were assumed to have a common value, i.e., $\mathrm{Le_F} = \mathrm{Le_O} = \mathrm{Le}$. The figure shows the dependence of the standoff distance x_e of steady solutions on the Damköhler number, for $0.8 \leq \mathrm{Le} \leq 1.6$. Depending on the Lewis number, the response curves exhibit one of two behaviors: a C-shaped response for relatively low Lewis numbers, as exemplified by Le = 0.8, 1.0, and 1.2, and a U-shaped response for higher values, as exemplified by Le = 1.4 and 1.6. The C-shaped curve consists of two branches with a turning point corresponding to a minimum Damköhler number, $D = D_{\mathrm{ext}}$, below which no solution exists. For $D > D_{\mathrm{ext}}$, the solution is multivalued. Steady solutions along the lower branch (solid curve) are stable with respect to small perturbations; when perturbed, the flame recedes back in time to its initial state. Solutions along the upper branch (dotted curve) are unstable; when perturbed, the flame advances or retreats from its initial equilibrium position indefinitely. Since only stable states are realized physically, the turning point D_{ext} defines an *adiabatic extinction limit*. In practical situations, when increasing the mass-flow rate by decreasing the Damköhler number starting with large values, the edge flame will first approach the plate along the stable lower branch, reach a minimum standoff distance, and then lift slightly up to D_{ext}. When further increasing the mass-flow rate, the flame gets blown off. Both the Damköhler number and the flame-standoff distance at extinction show a decreasing trend when Le decreases, suggesting that under otherwise similar conditions flames with lower Lewis numbers (e.g., light fuels) can survive higher flow rates.

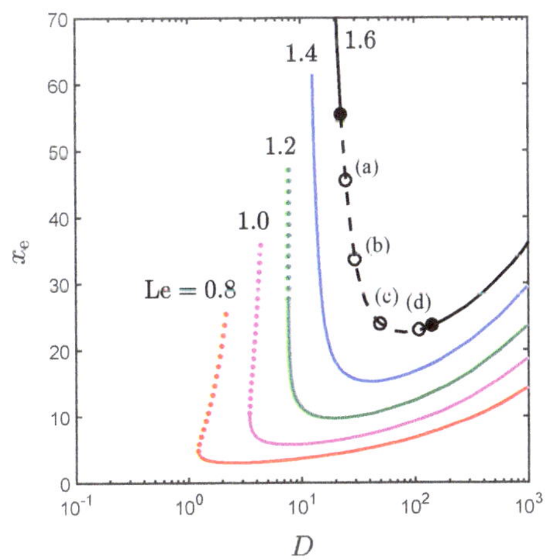

Figure 6. Response curves showing the dependence of the edge-flame-standoff distance x_e on the Damköhler number D for five representative values of Le. The solid segments correspond to stable steady states, the dotted segments to unstable steady states, and the dashed segment to oscillatory states. The filled circles '•' mark the marginal stability boundaries between stable steady states and stable oscillatory states, and the open circles '∘' marked by (**a–d**), which correspond to Damköhler numbers $D = 25, 30, 50$, and 110, respectively, are oscillatory states described in Figure 7.

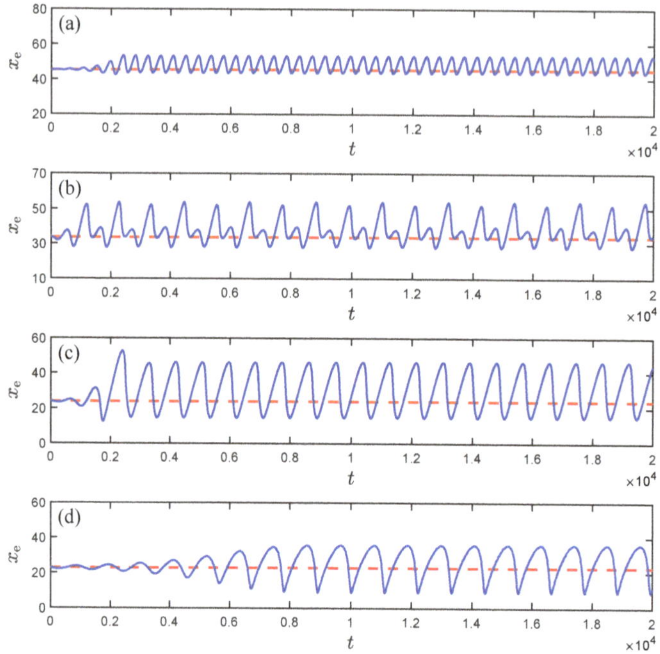

Figure 7. Edge-flame oscillation represented by time evolution of the flame-standoff distance x_e, for four representative Damköhler numbers marked by (**a–d**) on the response curve corresponding to Le = 1.6 in Figure 6. The dashed line marks the initial equilibrium standoff distance of the edge flame.

Different from the C-shaped response, the U-shaped curve associated with high-Lewis-number flames shows that a single steady solution exists for all D; the response curve does not have a turning point or an explicit extinction limit. In practical situations, when increasing the mass-flow rate by decreasing the Damköhler number starting with a large value, the edge flame will first approach the plate, reach a minimum standoff distance, and then lift off and stabilize further downstream. The flame lifts off to significantly large distances by increasing the mass-flow rate only slightly. Both the minimum standoff distance and the lift-off distance show an increasing trend when Le increases, suggesting that, under otherwise similar conditions, flames with higher Lewis numbers (e.g., heavy fuels) can stabilize sufficiently far from the plate. The stabilization mode, however, depends on the Lewis number. For Le = 1.4, the entire response curve corresponds to stable steady states. For Le = 1.6, an intermediate range of Damköhler numbers associated with unstable states exist (dashed curve) between the stable states corresponding to smaller and larger values of D. When a steady state within this range is slightly perturbed, the edge flame is observed to move back and forth along the centerline relative to its initial equilibrium position. Because these oscillations persist in time, the flame remains practically stabilized near the plate but not in a steady fashion. The marginal states separating stable steady states from stable oscillatory states are marked along the response curve by a '•' symbol. Figure 7 shows typical oscillations for four representative values of the Damköhler number marked in Figure 6 by (a)–(d). The amplitude and frequency, as well as the nature of the oscillations, vary with D. For example, oscillations of smaller amplitude are seen in (a), while a complex oscillation pattern of multiple frequencies is seen in (b). A closer examination shows that during its motion, the edge flame drags along the trailing diffusion flame, but oscillations along the diffusion flame decay quickly when moving further downstream [23,36]. The oscillations, which begin in the premixed segment of the edge flame, are presumed to share a similar underlying mechanism with the well-known large-

Lewis-number pulsation of premixed flames attributed to a diffusive-thermal instability. Further results show that the ranges of both the Damköhler number D and the flame-standoff distance x_e corresponding to oscillatory states expand on increasing the Lewis number [23].

The influence of the Lewis number on the structural characteristics of edge flames is further illustrated in Figure 8, where temperature profiles along the centerline are plotted for the same five representative Lewis numbers used earlier in Figure 6. For the sake of comparison, the Damköhler number corresponding to each of these five cases was selected such that the temperature decays to the ambient value ($\theta = 0$) at approximately the same position ($x \approx 11$). We observe that when increasing Le the temperature on the trailing diffusion flame falls consistently while the thickness of the preheat zone preceding the edge flame increases. The reason lies in the different effects that the Lewis number has on the temperature of diffusion and premixed flames, as discussed in Section 3. Unlike the temperature of a premixed flame which remains the adiabatic flame temperature for all Lewis numbers, the flame temperature of a diffusion flame, or the stoichiometric temperature, decreases when increasing Le. The structure of an edge flame, which combines characteristics of both premixed and diffusion flames, is influenced by the Lewis number through the thermal interaction between the leading premixed segment and the trailing diffusion flame. When Le $= 1$, the temperature of both the edge flame and the diffusion flame, is the adiabatic flame temperature. When Le < 1, the temperature of the diffusion flame exceeds the adiabatic flame temperature; the inherent heat transfer from the diffusion flame toward the premixed segment yields a super-adiabatic edge temperature. The opposite is true when Le > 1; heat is directed from the premixed segment towards the diffusion flame resulting in a sub-adiabatic edge temperature.

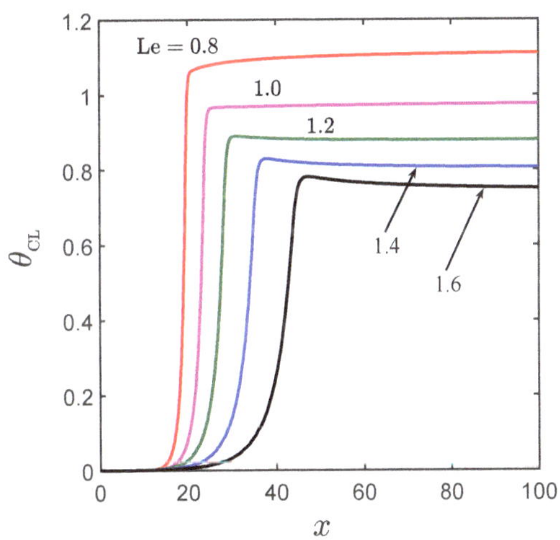

Figure 8. Temperature profiles for Le $= 0.8, 1.0, 1.2, 1.4, 1.6$, at $D = 2.004, 4.073, 7.853, 14.495, 25.360$, respectively. The different values of D were judiciously selected to enable comparison of the preheat zone and trailing diffusion-flame temperature between the various profiles.

The significant variations in edge temperature associated with the Lewis number have implications on both the edge propagation speed S_e and the thickness of the preheat zone preceding the edge flame. The edge speed, similar to the edge temperature, is expected to increase with decreasing Le. Because a stable stationary edge flame assumes the location where the edge speed S_e balances the incoming flow, a larger flow rate (smaller D) would be required to stabilize the low-Lewis-number flames corresponding to a smaller standoff distance. Indeed, the flow rate required to stabilize the edge flame with Le $= 0.8$ is significantly higher than that required to stabilize an edge flame with Le $= 1.6$, as indicated in the caption of Figure 8. The same reasoning applies to the thickness of the preheat zone preceding the edge flame. Assuming an exponential decay $\theta \sim \exp\{S_e(x - x_e)\}$, which is the typical profile of a premixed flame, the characteristic thickness of the preheat zone proportional to S_e^{-1} increases on increasing the Lewis number, as observed in Figure 8.

The insights gained from the two distinct types of responses identified for edge flames and the role played by the Lewis number may be exploited to understand the stabilization of laminar jet diffusion flames relative to the nozzle rim. As evident from Figure 6, the flame-standoff distance of low-Lewis-number edge flames characterized by C-shaped response curves is generally small in magnitude, on the order of the flame thickness. From a practical perspective, the edge flame is effectively attached to the tip of the plate. This attachment behavior persists when continuously increasing the flow rate until blow-off occurs. This implies that substantial lift-off is virtually impossible for low-Lewis-number edge flames. By contrast, for high-Lewis-number flames characterized by U-shaped response curves, the edge flame may be lifted and stabilized at substantially large distances from the plate by gradually increasing the flow rate. Note that both stabilization modes, namely the stationary mode which corresponds to an edge flame held stationary at a well-defined distance, and the oscillatory mode which corresponds to an edge flame moving back and forth relative to a mean position, can be realized in practice and thus are considered *globally* stable. Such distinct attachment and lift-off behaviors of edge flames have been experimentally observed by Chung and Lee [1], who proposed a Schmidt-number-based criterion to predict the transition between the attached and lifted regimes of jet diffusion flames. Their prediction based on flow and concentration profiles of cold jets, naturally involves a Schmidt number. When converted to a Lewis number, using the relation $\mathrm{Sc} = \mathrm{Pr} \cdot \mathrm{Le}$, their condition is in good agreement with the transition between attached and lifted edge flames predicted in our mixing layer model, as discussed in [23]. Choosing the Lewis number as the primary control parameter describing the stabilization of jet diffusion flames seems more appropriate, considering the fundamental and direct role it plays in determining the characteristics of edge flames, as discussed in the preceding paragraph. It is also worth mentioning that experiments on jet diffusion flames have also observed U-shaped responses of flame-standoff distance with respect to the flow rate [37] and sustained flame oscillations [38,39]. These experimental findings have been attributed to buoyancy effects, which are fundamentally different from the present numerical predictions exclusively associated with diffusive-thermal influences. Systematic microgravity experiments in conjunction with numerical simulations accounting for buoyancy convection are therefore desired to elucidate the unique influence of these different mechanisms.

5.2. Asymmetric Edge Flames

The symmetric structure of the edge and trailing diffusion flames described above was obtained for selected conditions that constitute the baseline case and its symmetric variants. Streams of unequal initial strain rates ($\alpha \neq 1$), fuel and oxidizer supplied in off-stoichiometric proportions ($\phi \neq 1$), preferential diffusion of fuel and oxidizer ($\mathrm{Le_F} \neq \mathrm{Le_O}$) would all lead to asymmetric flames. In the following, we examine some of these cases.

Figure 9 illustrates an asymmetric edge flame stabilized within the flow and resulting purely from preferential diffusion, i.e., unequal diffusivities of fuel and oxidizer, with $\mathrm{Le_F} = 1.8$ and $\mathrm{Le_O} = 1.0$. The significantly larger diffusivity of the oxidizer offsets the balanced transport of the two reactants such that the trailing diffusion flame as a whole deviates from the centerline toward the fuel side. For the same reason, the premixed flame branches of the tribrachial edge structure are no longer symmetrically distributed on the two sides of the diffusion flame; the fuel-rich branch appears longer and the entire edge, measured relative to the normal direction, is deflected toward the fuel side. Meanwhile, in line with the trends identified in Figures 6 and 8, for a given overall mass-flow rate (same Damköhler number), an increase in the fuel Lewis number leads to an appreciable increase in the flame-standoff distance and decrease in flame temperature.

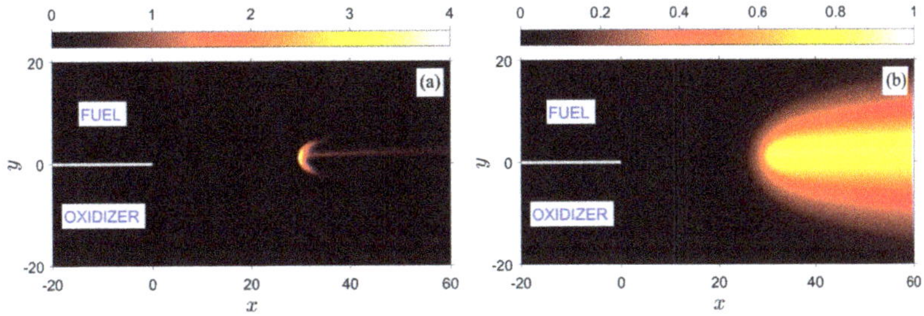

Figure 9. Asymmetric edge and trailing diffusion flames resulting from preferential diffusion, with $Le_F = 1.8$ and $Le_O = 1.0$. The edge flame is illustrated in (**a**) by contours of the reaction rate Ω, while (**b**) shows the temperature distribution θ.

Figure 10 shows an asymmetric edge flame stabilized within the flow and resulting purely from unequal strain rates in the incoming streams, i.e., with $\alpha = 0.1$. As the characteristic flow velocity originating from the oxidizer stream is considerably lower than that originating from the fuel stream, the diffusion flame deviates from the centerline toward the oxidizer stream to facilitate its intake of oxygen and the two premixed flame branches of the tribrachial structure are, for the same reason, no longer symmetric. The flame-standoff distance and temperature distribution along the diffusion flame appear, for this equal-Lewis-number case, to be scarcely influenced by a change in the flow field at the given Damköhler number. However, from an overall perspective, diffusion will play a more significant role when decreasing the strain rate ratio α, leading to a leftward shift of the entire edge flame response curve, and implying a delay in the blow-off conditions for low-Lewis-number flames and in the lift-off conditions of high-Lewis-number flames [23].

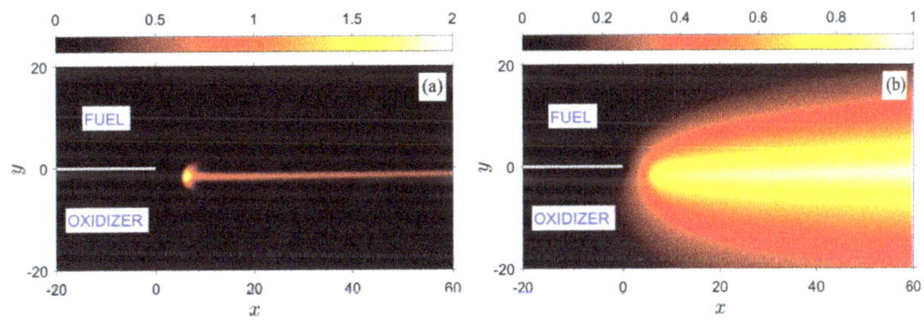

Figure 10. Asymmetric edge and trailing diffusion flames resulting from incoming streams of unequal strain rates, with $\alpha = 0.1$. The edge flame is illustrated in (**a**) by contours of the reaction rate Ω, while (**b**) shows the temperature distribution θ.

The extent to which unequal strain rates combined with inter-diffusion of reactants influence the edge-flame position is shown in Figure 11 for $\alpha = 0.1$ and unequal Lewis numbers. The oxidizer Lewis number is assumed equal to one, as appropriate for air, and two values of the fuel Lewis number are considered; a low value $Le_F = 0.4$ corresponding to a relatively light fuel, and a high value $Le_F = 2.0$ corresponding to a relatively heavy fuel. It should be noted that diluting one or both streams with an inert one can significantly affect the corresponding Lewis number due to changes in the mixture thermal diffusivity; see, for example, [24]. In addition to the edge flame, illustrated in the figure by reaction-rate contours, the stoichiometric surface $Z = Z_{st}$ is shown as a dashed line. Although for $\alpha = 0.1$ considered here the oxidizer is entrained into the mixing layer and the dividing streamline is deflected towards the fuel region, the edge and diffusion flames deviate towards the oxidizer region when $Le_F = 0.4$ due to the high mobility of the fuel. When $Le_F = 2.0$, the inclination is towards the fuel region. The diffusion flame trailing behind the edge flame always aligns itself with the stoichiometric surface. We note parenthetically that due to preferential diffusion ($Le_F \neq Le_O$) the mixture fraction Z is not a conserved scalar and, as a result, $Z = Z_{st}$ is not a smooth curve; its determination that accounts for the sharp deflection observed across the curved part of the edge flame will be discussed below in Section 6.

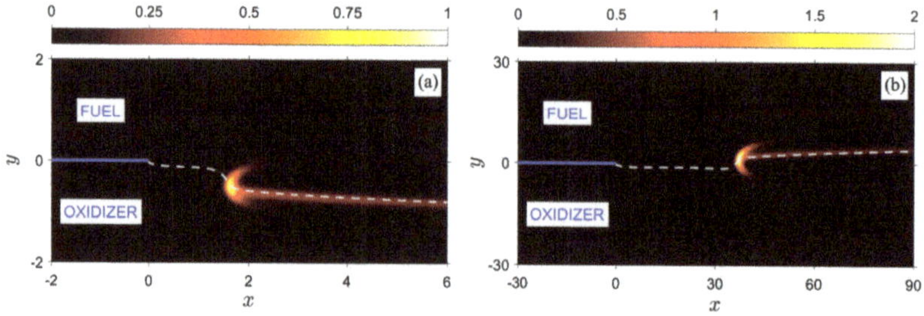

Figure 11. Asymmetric edge and trailing diffusion flames resulting from incoming streams of unequal strain rates, with $\alpha = 0.1$, and preferential diffusion. In both cases, $Le_O = 1.0$, with (**a**) corresponding to $Le_F = 0.4$ and $D = 0.0332$, and (**b**) to $Le_F = 2.0$ and $D = 13.782$. The edge flame is illustrated by contours of the reaction rate Ω, and the stoichiometric surface $Z = Z_{st} = 0.5$ is delineated by the dashed line.

Figure 12 illustrates a situation of asymmetry resulting from fuel and oxidizer supplied in the incoming streams in off-stoichiometric proportions, with an initial mixture strength $\phi = 5$. Because the fuel is in excess, the reaction is dominated by the supply of oxygen to the reaction zone. As a response, the diffusion flame adjusts its shape and position and finally stabilizes itself in the oxidizer stream, in an effort to balance the transport of the two reactants. A notable difference compared to the previous two non-symmetric cases is that now the edge flame loses its tribrachial structure; instead, it takes on a hook-like shape with only the fuel-rich branch retained, while the fuel-lean branch appears to completely degenerate into the diffusion flame.

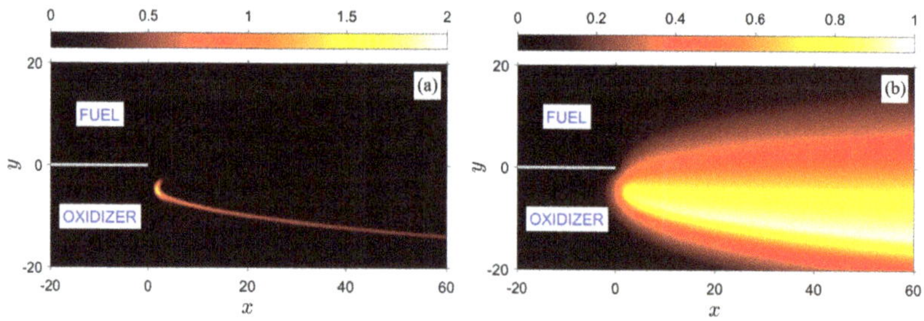

Figure 12. Asymmetric edge and trailing diffusion flames in a rich system corresponding to an initial mixture strength $\phi = 5$. The edge flame is illustrated in (**a**) by contours of the reaction rate Ω, while (**b**) shows the temperature distribution θ.

Asymmetry in the combustion field can also result from chemistry imbalance, for example, unequal reaction orders as demonstrated by Juanos and Sirignano [7] for a propane–air mixture with $n_F = 0.1$ and $n_O = 1.65$, or from the interaction of two adjacent flames as illustrated by Kurdyumov and Jimenez [40].

6. The Edge-Flame Speed

Similar to the definition of flame speed, the *edge speed* S_e is defined as the propagation of the edge flame along its normal, relative to the flow. When held stationary, S_e is equal to the velocity component of the incoming gas normal to the edge flame, namely normal to the curved premixed flame front formed at the base of the diffusion flame. The edge speed is expected to depend on the reactant concentrations immediately ahead of the flame front, due to their direct influence on the edge-flame temperature, and on their distributions in the transverse direction which determine the local curvature of the premixed flame surface. The Lewis number is expected to play a significant role because, as seen in Section 5, the edge temperature is influenced by the heat transfer to/from the trailing diffusion flame wherein the Lewis number serves as a key parameter. Before attempting to provide an accurate definition of the edge speed, we first examine means by which the local distribution of the reactants near a general edge flame (not necessarily symmetric) can be quantified.

The distribution of the fuel and oxidizer mass fractions ahead of the edge flame can be conveniently described in terms of the mixture fraction Z defined in Equation (9) which, using the mass fraction Equation (3), must satisfy

$$\frac{\partial Z}{\partial t} + \mathbf{v} \cdot \nabla Z - \mathrm{Le}_{\mathrm{F}}^{-1} \nabla^2 Z = (\mathrm{Le}_{\mathrm{F}}^{-1} - \mathrm{Le}_{\mathrm{O}}^{-1}) \nabla^2 Y_{\mathrm{O}} . \tag{18}$$

For equal Lewis numbers $\mathrm{Le}_{\mathrm{F}} = \mathrm{Le}_{\mathrm{O}}$, the source term on the right-hand side vanishes and the mixture fraction Z is a conserved scalar that satisfies a reaction-free equation; namely, it is neither created nor destroyed by chemical reactions. The stoichiometric surface $Z(x,y) = Z_{\mathrm{st}}$ can therefore be determined a priori throughout the combustion field. The trailing diffusion flame, located where the fuel and oxidizer fluxes are in stoichiometric proportions, lies approximately along this surface. Exact overlapping occurs in the asymptotic limit $D \to \infty$. For unequal Lewis numbers, the mixture fraction Z is no longer a conserved scalar; it may no longer be determined a priori and must be solved for as part of the overall solution. The stoichiometric surface $Z(x,y) = Z_{\mathrm{st}}$, which still approximates the location of the trailing diffusion flame, is no longer coincident with the equivalent surface in the reaction-free mixing region ahead of the edge flame. As a consequence, the stoichiometric surface originating near the tip of the plate will be deflected when passing through the edge flame in order to make a smooth connection with the trailing diffusion flame. This situation is exemplified in Figure 13 by the case corresponding to $\mathrm{Le}_{\mathrm{F}} = 0.4$ and $\mathrm{Le}_{\mathrm{O}} = 1.0$, which shows contours of the mixture fraction Z computed a posteriori from the definition (9) after solving for the mass fractions Y_{F} and Y_{O}. Because of the larger diffusivity of fuel, the diffusion flame as a whole deviates from the centerline toward the oxidizer stream, and the stoichiometric surface shown by a dashed curve bends down from its initial location ahead of the edge flame to align with the diffusion flame further downstream. One also notes the asymmetry in the distribution of the mixture fraction Z on both sides of the diffusion flame; the contours are sparser on the fuel side and denser on the oxidizer side, indicating a sharper gradient of oxidizer towards the flame which is necessary to achieve the proper mass fluxes of fuel and oxidizer into the diffusion flame.

The accurate evaluation of the edge speed necessitates unambiguously identifying the direction normal to the curved premixed flame front at the base of the diffusion flame. This may be done by determining the plane tangential to the premixed flame front, but this choice is ambivalent because one or both wings of the edge flame often degenerate into the diffusion flame, making the tangential plane hard to identify. A more satisfactory method, which is applicable to edge flames of various structures, with two or one arms or even a single nucleus, is to identify the normal to the edge flame with the direction of steepest descent of the reaction rate contours [23,41]. Consequently, the unit normal $\mathbf{n}$ is defined as

$$\mathbf{n} = -\frac{(\nabla \Omega)_{\max}}{|(\nabla \Omega)_{\max}|} \tag{19}$$

at the edge position $(x_{\mathrm{e}}, y_{\mathrm{e}})$, and the edge speed is given by $S_{\mathrm{e}} = -\mathbf{v}_{\mathrm{e}} \cdot \mathbf{n}$, where $\mathbf{v}_{\mathrm{e}}$ is the local gas velocity. Likewise, the tangential component of the local mixture fraction gradient is given by

$$(\nabla Z)_{\mathrm{e}_\perp} = (\nabla Z)_{\mathrm{e}} - \left[(\nabla Z)_{\mathrm{e}} \cdot \mathbf{n}\right] \mathbf{n} , \tag{20}$$

where the subscript '⊥' represents the direction tangential to the flame front. Using these prescriptions, we show in Figure 14 an enlargement of the edge flame corresponding to Figure 13 along with the unit normal **n** and the associated edge speed S_e, as well as the local mixture fraction gradient $(\nabla Z)_e$ and its tangential component $(\nabla Z)_{e\perp}$.

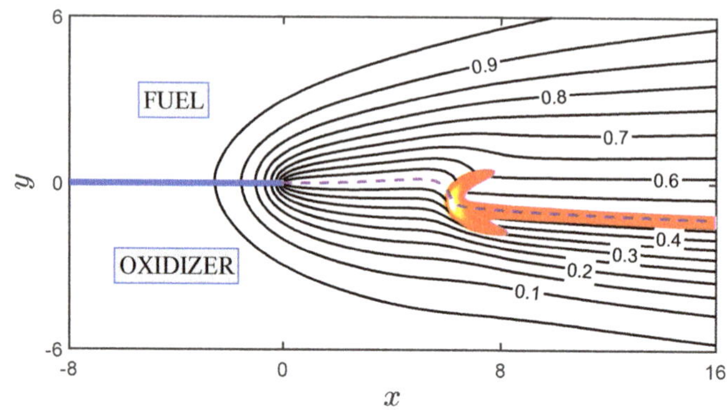

Figure 13. Contours of the mixture fraction Z, for an asymmetric edge flame resulting from unequal fuel and oxidizer Lewis numbers, $Le_F = 0.4$ and $Le_O = 1.0$, at Damköhler number $D = 0.49$. Numbers labelled on the contour lines indicate the values of Z. The dashed line corresponds to the value $Z = Z_{st} = 0.5$. The flame is represented by reaction rate contours.

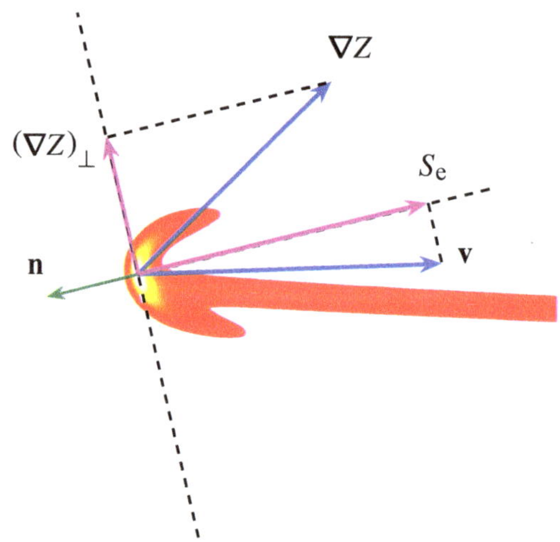

Figure 14. A schematic illustration of the local properties of the edge flame corresponding to Figure 13, showing the unit normal **n**, the local mixture fraction gradient ∇Z and its tangential component $(\nabla Z)_\perp$, and the edge speed S_e, all evaluated at the edge location (x_e, y_e).

To quantitatively delineate the dynamical properties of edge flames, we seek a relationship that exhibits the dependence of the edge speed S_e on $(\nabla Z)_{e\perp}$, which characterizes the transverse distribution of the reactants ahead of the edge flame, and on the Lewis number, which influences the edge speed through the edge temperature. To this end, we first examine the distribution of the edge location (x_e, y_e) when varying the Damköhler number D, as shown in Figure 15a for representative values of the fuel Lewis number Le_F and $Le_O = 1$. The curves in this figure correspond to the phase portraits of the edge location when the Damköhler number is varied along the response curve that traces the dependence of the edge location on D. The overall trend when D is decreased from large values is for the edge to converge toward the tip of the plate, a change manifested by a decrease in y_e from positive values when $Le_F > 1$ and an increase in y_e from negative values when $Le_F < 1$, and in both cases a decrease in x_e (except for very small Le_F). Once a minimum standoff distance is reached and the flame lifts off, the trajectory exhibits large variations in x_e with relatively small changes in y_e. The phase portrait data of Figure 15a is used to extract the local edge speed and its dependence on the tangential component of the local mixture fraction gradient.

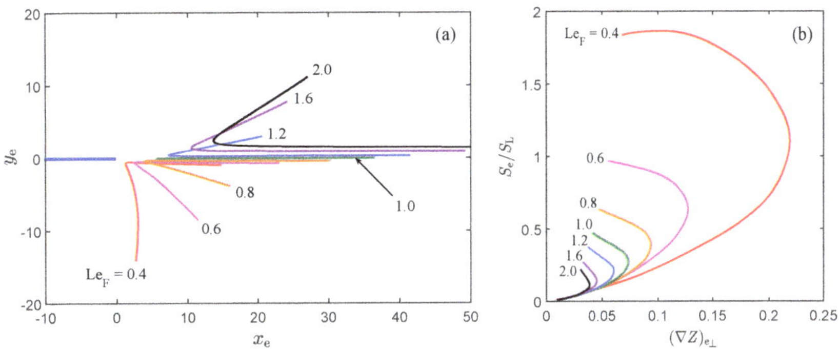

Figure 15. (**a**) The distributions of the edge location (x_e, y_e), and (**b**) the dependence of the normalized edge-flame speed S_e/S_L on the tangential component of the local mixture fraction gradient $(\nabla Z)_{e\perp}$, for several representative values of the fuel Lewis number Le_F and $Le_O = 1$.

Figure 15b shows the dependence of the edge speed on the tangential component of the local mixture fraction gradient, after normalizing the edge speed S_e by the laminar flame speed corresponding to a stoichiometric mixture equivalent to the one in the mixing zone preceding the edge flame of the same Lewis number, namely $S_L = \sqrt{Le_F}\, S_L^0$ as per Equation (1). For each Le_F, the dependence of S_e on $(\nabla Z)_{e\perp}$ consists of two branches. The lower branch corresponds to relatively small standoff distances, where the edge flames are subjected to substantial conductive heat loss to the plate. Note that the curves corresponding to different fuel Lewis numbers collapse onto a common envelope when $(\nabla Z)_{e\perp} \to 0$, implying that within that range conductive heat losses to the plate dominate over Lewis-number effects. This collapse is rather remarkable considering that the edge locations corresponding to different Lewis numbers are rather scattered in space, as seen in Figure 15a. The common limit corresponds to the fast chemistry limit, $D \to \infty$, when the flame becomes attached to the plate and the edge-flame speed approaches zero. The upper branch tracing the dependence of S_e on $(\nabla Z)_{e\perp}$ corresponds to edge flames at comparatively large standoff distances, which are essentially free of thermal interaction with the plate. Evidently, Lewis number effects here are dominant. As can be seen, the normalized edge speed corresponding to such freely standing flames displays an overall increasing trend with decreasing Le_F. This is expected, because a decrease in the Lewis number leads to an enhancement in the edge temperature and concurrently in the edge speed S_e, along with a diminution in the corresponding laminar flame speed S_L. The approach $(\nabla Z)_{e\perp} \to 0$ along the upper branch is realized when the edge-flame-standoff distance is infinitely large, a limit not amenable in calculations performed thus far. Nevertheless, it can be presumed that in this limit the edge flame degenerates into a planar premixed flame, such that for all Lewis numbers $S_e/S_L \to 1$.

7. Volumetric Heat Loss

In this section, we examine the effect of heat loss on the structure of the edge flame and its stabilization. Figure 16 depicts an edge flame with parameters corresponding to the baseline case, but with $H = 2.71 \times 10^{-4}$. The figure shows the edge flame depicted by reaction contours, and the distributions of temperature and mass fractions. When compared to the adiabatic case shown in Figure 5, we observe that the reaction rate and temperature decay rapidly in the downstream direction, the edge flame is stabilized at a distance further away from the tip of the plate, and the trailing diffusion flame has a finite length. The asymptotic behavior for large x, examined by Kurdyumov and Matalon [42] in a similar configuration, shows that far downstream the solution is necessarily the frozen state ($\theta = 0$), which confirms that the diffusion flame cannot extend to infinity.

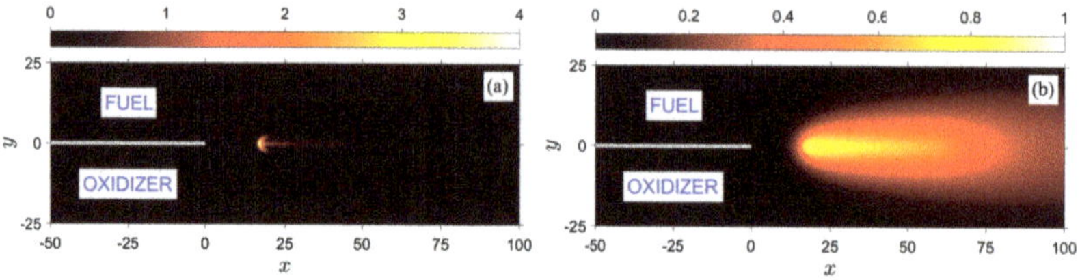

Figure 16. An edge flame corresponding to the baseline case and subject to volumetric heat loss with $H = 2.71 \times 10^{-4}$. The structure of the edge flame is illustrated by contours of (**a**) reaction rate Ω, and (**b**) temperature θ.

Figure 17 illustrates the response of steady edge flames to heat loss by displaying the dependence of the flame-standoff distance x_e on the heat-loss coefficient H, for three representative values of the Lewis number Le at a fixed Damköhler number $D = 10$. The response curve consists of two distinct branches with a turning point at $H = H_{ext}$. Two steady states exist for $H < H_{ext}$ and none for $H > H_{ext}$, so that the maximum value of the heat-loss coefficient defines a *non-adiabatic extinction limit*. We observe that the extinction limit increases when decreasing the Lewis number, suggesting that, under otherwise similar conditions, low-Lewis-number edge flames are generally less vulnerable to external heat loss. This tendency is in line with the trend exhibited by the adiabatic extinction limit shown in Figure 6, both limits being attributed to the influence of differential diffusion manifested by the Lewis number.

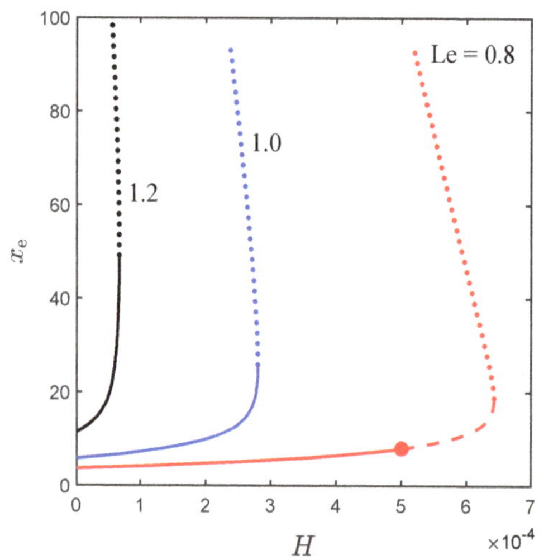

Figure 17. The dependence of the flame-standoff distance x_e on the heat-loss coefficient H for three representative values of the Lewis number, at a fixed Damköhler number $D = 10$. Steady states on the lower branch of the response curves (solid curves) are stable and those on the upper branch (dotted curves) are unstable; solutions on the lower branch beyond the '•' symbol (dashed segment) correspond to stable oscillatory states.

Of greater importance is the stability of the steady states, which shows dissimilar inferences for the high and low Lewis numbers considered. For $\mathrm{Le} = 1.0$ and 1.2, steady states on the lower branch of the response curve are stable with respect to small perturbations. When solutions on the upper branch are slightly perturbed, the flame-standoff distance increases or decreases indefinitely; the corresponding states are unstable and therefore physically unrealistic. The marginal stability state corresponds to the turning point $H = H_{\mathrm{ext}}$. Hence, in practice, when the heat-loss intensity is gradually increased for a given flow rate (constant Damköhler number), the edge flame will continuously retreat from the plate and extinguish when H exceeds H_{ext}.

Steady states on the upper branch of the response curve remain unstable and physically unrealistic, for low-Lewis-number flames as well. But unlike the high-Lewis-number flames, steady states on the lower branch of the response curve are stable only for H smaller than a critical value $H_c < H_{\mathrm{ext}}$; for $\mathrm{Le} = 0.8$, the critical value $H_c \approx 5.006 \times 10^{-4}$ is marked in the figure by a '•' symbol. For $H > H_c$, when a steady solution is slightly perturbed, the edge flame evolves into a stable state exhibiting sustained oscillations of constant amplitude and frequency, relative to the initial equilibrium position, as exemplified in Figure 18. The amplitude of the oscillations is found to grow rapidly on increasing H, presumably becoming infinitely large when $H \to H_{\mathrm{ext}}$. The nature of the oscillations observed here is different than those predicted in Section 5; due to substantial heat loss, the edge flame strives to survive, moving from a burning state to one which is nearly extinguished, while at high Lewis number (no heat loss) by moving back and forth the edge flame simply attempts to find an equilibrium position.

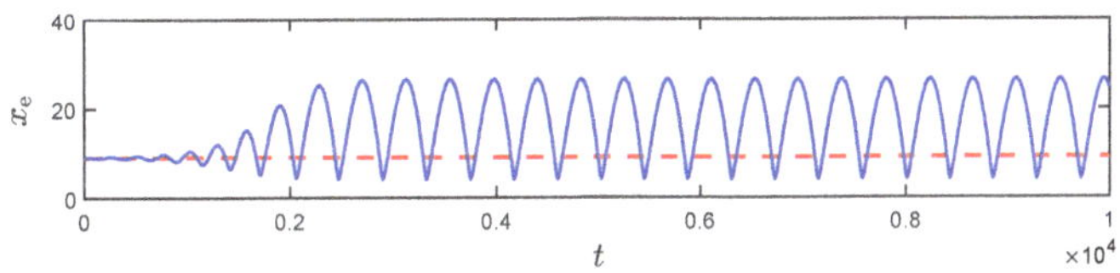

Figure 18. Time evolution of a slightly perturbed steady state (dashed line) corresponding to $\mathrm{Le} = 0.8$ and $H = 5.487 \times 10^{-4}$. The curve shows the flame-standoff distance x_e as a function of time.

8. Thermally Active Splitter Plate

The discussion so far has assumed that the splitter plate separating the two streams is retained at a constant temperature equal to the temperature of the two streams. This ideal situation corresponds to a plate with infinitely large thermal conductivity. In practical circumstances, however, the thermal conductivity of the plate is finite and there is a thermal interaction between the plate and the gaseous mixture which may significantly influence the dynamical properties of the edge flame stabilized in its wake. Figure 19 illustrates the temperature distributions of edge flames stabilized in the vicinity of four plates of various thermal conductivity, under otherwise identical conditions. In addition to the "cold isothermal" plate corresponding to $r_\lambda \to \infty$, and the "adiabatic" plate which constitutes the other extreme with $r_\lambda = 0$, two cases with finite values of r_λ are displayed. The case corresponding to $r_\lambda = 35$ mimics a plate made of glass, while the case corresponding to $r_\lambda = 1$, i.e., the "virtual-gas" plate, is a hypothetical case of a plate with the same thermo-physical properties as the gas phase. Commonly used materials, such as aluminum and stainless steel, have considerable large values of r_λ and have not been analyzed due to the high computational cost, as discussed in [36]. The edge flames in the wake of these four plates are stabilized at different distances, resulting from the extent of thermal interactions with the plates. Figure 20 shows the temperature profiles along $y = 0$ for the four plates considered in Figure 19, with the temperature for $x \leq 0$ corresponding to θ_s. The cold isothermal plate, which retains a constant temperature $\theta_s = 0$, acts as a heat sink extracting heat from the combustion field. The adiabatic plate precludes any thermal interaction with the flame but holds the heat conducted to it from the edge flame so as to raise its temperature. In the other two cases, the heat conducted from the edge flame to the plate elevates the temperature θ_s to a different extent and a different distance upstream. The heated plate then relays the heat it receives to the fuel and oxidizer streams through its lateral surfaces. The preheated reactants raise the temperature in the mixing region ahead of the edge flame, resulting in an edge temperature that exceeds its stoichiometric value.

The heat-recirculation cycle, absent in the cold and adiabatic plates, would support stabilization and delay possible flame blow-off. Such a mechanism may explain the favorable effect of certain materials on the stabilization of jet diffusion flames [43,44]. In [36], we provided a quantitative measure that characterizes the heat recirculation and its efficiency, relying not only on the total recirculated heat but also on its distribution along the plate. As an experimentally measurable quantity, it can be exploited in practice for the selection of appropriate nozzle materials that optimize the stabilization of jet diffusion flames.

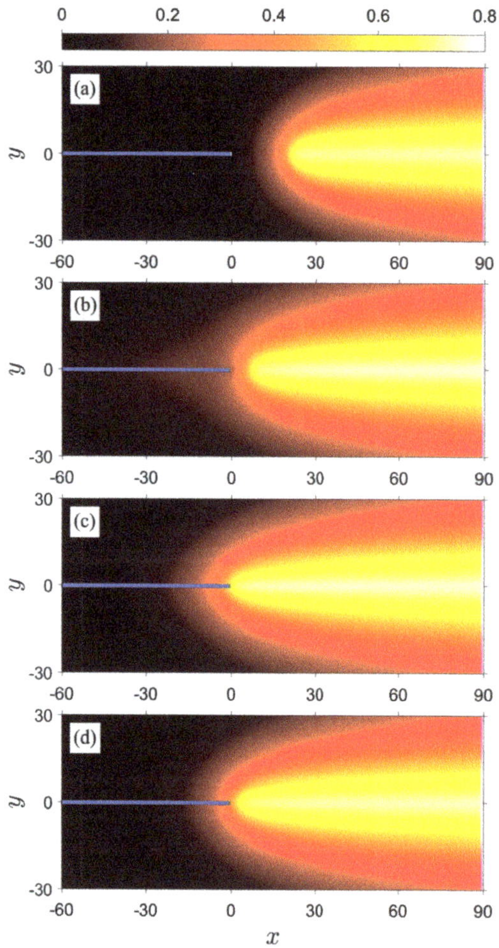

Figure 19. Temperature distribution corresponding to edge flames stabilized in the vicinity of four distinct plates: (**a**) isothermal cold plate ($r_\lambda = \infty$), (**b**) glass plate ($r_\lambda = 35$), (**c**) virtual-gas plate ($r_\lambda = 1$), (**d**) adiabatic plate ($r_\lambda = 0$), under otherwise identical conditions, i.e., $D = 81.615$ and $\mathrm{Le} = 1.6$. The flame-standoff distances corresponding to these plates are $x_e = 22.64$, 8.94, 2.30, and 4.84, respectively.

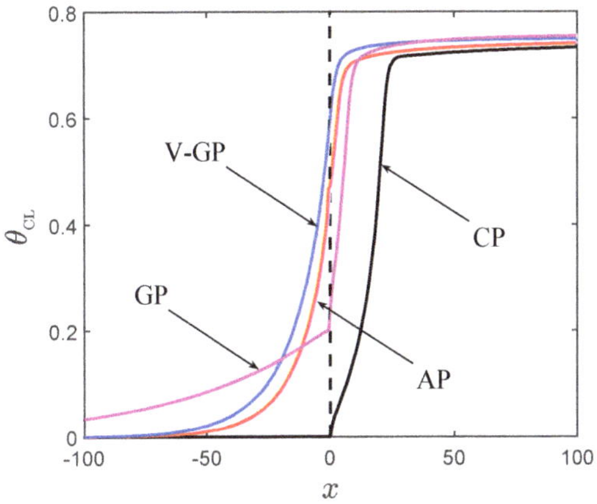

Figure 20. Temperature profiles along the centerline ($y = 0$) corresponding to the four plates considered in Figure 19: isothermal cold plate (CP), glass plate (GP), virtual-gas plate (V-GP), and adiabatic plate (AP), at $D = 81.615$ and $\text{Le} = 1.6$.

9. Concluding Remarks

The combustion behavior in a mixing layer that develops when two separate streams of reactants merge is a canonical problem that provides fundamental insight into practical problems such as jet diffusion flames, or diffusion flames in co-flowing systems. Although the diffusion flame itself in such applications is typically robust and may not be easily extinguished, its endurance depends crucially on the stabilization of the edge flame at its base and the interaction of the edge flame with the fuel injector or nozzle rim. Poor stabilization will have consequences on the entire flame. Using a diffusive-thermal framework that decouples the flow and combustion fields for mathematical simplicity, we have systematically investigated in this paper the structural characteristics and dynamical properties of an edge flame sustained in the wake of a plate separating two merging shear flows, one with fuel and the other with an oxidizer, and supporting the diffusion flame that extends downstream. We provide a comprehensive discussion of the relevant physical parameters affecting the sustenance and stability of the edge flame, including the overall mass-flow rate and the strain rates in the separate streams, the initial fuel and oxidizer stoichiometric proportions, the distinct diffusion properties of the fuel and oxidizer compared to the thermal diffusivity of the mixture, influences of volumetric heat loss, and thermal interaction between the gaseous mixture and the splitter plate. Conditions leading to idealized symmetric flames and those resulting in asymmetric structures were delineated.

In addition to a substantial flow, which is necessary to bring the fuel and oxidizer together, the fuel Lewis number representing the ratio of the thermal diffusivity of the mixture to the molecular diffusivity of the fuel, is perhaps the most important parameter that controls the stabilization of the diffusion flame. Taking combustion in air as an example, the Lewis number of oxygen is near one and has a limited influence. Due to their high molecular mobility, light fuels (low Lewis numbers) combine and react with the oxidizer in the immediate vicinity of the splitter plate. For moderate flow rates, the resulting edge flame remains practically attached to the plate, supporting the diffusion flame in its wake. At high flow rates, the edge and diffusion flames get blown off by the flow, which precludes flame liftoff. By contrast, heavy fuels (high Lewis numbers) which diffuse slowly travel further into the mixing region to meet and react with the oxidizer, and the resulting edge flame is established at a distance further away from the plate. When increasing the flow rate, the edge and diffusion flames gradually lift from the plate and stabilize in the flow at larger and larger distances. Two modes of stabilization have been identified for sufficiently high Lewis numbers: a stationary mode, where the edge flame is held at a well-defined distance, and an oscillatory mode, where the edge flame undergoes sustained oscillations relative to a mean location. The dependence of flame attachment and liftoff on the Lewis number may also be attributed to its effect on the edge temperature and consequently on the edge speed that must balance the velocity of the incoming flow. The edge temperature is dictated by the stoichiometric temperature of the diffusion flame, which increases on decreasing the Lewis number, resulting in super-adiabatic temperatures for low-Lewis-number flames and sub-adiabatic temperatures for high-Lewis-number flames. Faster propagating edge flames can therefore balance a higher velocity in the accelerating flow that emerges immediately beyond the splitter plate.

Evidently, the most significant simplification adopted in this study is the constant-density assumption. Gas expansion resulting from the large change in density that results from the heat released by the chemical reaction has two main effects on the edge flame. It causes a deflection of the streamlines crossing the curved frontal segment of the edge flame, leading to an increase in the edge speed [45,46], and it induces a lateral growth of the mixing layer, which affects the local mixture fraction gradient near the edge flame. These effects will have a quantitative effect on the edge position and edge speed, but will not undo the fundamental physics of diffusive-thermal nature discovered in this study.

Author Contributions: This research is a result of an ongoing collaboration between Z.L. and M.M. Both authors have contributed into the conceptualization, methodology, data analysis, and writing of the manuscript. All authors have read and agreed to the published version of the manuscript.

Funding: This research was funded partially by the National Natural Science Foundation of China (NSFC) under grant number 52176119 to Zhanbin Lu, and by the US National Science Foundation under grant number CBET 19-11530 to Moshe Matalon.

Data Availability Statement: The data presented in this study are available on request from both authors.

Conflicts of Interest: The authors declare no conflict of interest. The funders had no role in the design of the study; in the collection, analyses, or interpretation of data; in the writing of the manuscript; or in the decision to publish the results.

References

1. Chung, S.H.; Lee, B.J. On the characteristics of laminar lifted flames in a nonpremixed jet. *Combust. Flame* **1991**, *86*, 62–72. [CrossRef]
2. Juniper, M.; Candel, S. Edge diffusion flame stabilization behind a step over a liquid reactant. *J. Propuls. Power* **2003**, *19*, 332–341. [CrossRef]
3. Liñán, A. Ignition and flame spread in laminar mixing layers. In *Combustion in High-Speed Flows;* Buckmaster, J., Jackson, T.L., Kumar, A., Eds.; Kluwer Academic Publishers: Alphen aan den Rijn, The Netherlands, 1994; pp. 461–476.
4. Buckmaster, J.; Matalon, M. Anomalous Lewis number effects in tribrachial flames. *Proc. Combust. Inst.* **1988**, *22*, 1527–1535. [CrossRef]
5. Dold, J.W. Flame propagation in a nonuniform Mixture: Analysis of a slowly varying triple flame. *Combust. Flame* **1989**, *76*, 71–88. [CrossRef]
6. Hartley, L.J.; Dold, J.W. Flame propagation in a nonuniform mixture: Analysis of a propagating triple-flame. *Combust. Sci. Technol.* **1991**, *80*, 23–46. [CrossRef]
7. Juanós, A.J.; Sirignano, W.A. Triple flame: Inherent asymmetries and pentasectional character. *Combust. Theory Model.* **2014**, *18*, 454–473. [CrossRef]
8. Daou, J.; Linan, A. Ignition and extinction fronts in counterflowing premixed reactive gases. *Combust. Flame* **1999**, *118*, 479–488. [CrossRef]
9. Vedarajan, T.; Buckmaster, J.; Ronney, P. Two-dimensional failure waves and ignition fronts in premixed combustion. *Proc. Combust. Inst.* **1998**, *27*, 537–544. [CrossRef]
10. Liao, K.P.; Matalon, M.; Pantano, C. A flow pattern that sustains an edge flame in a straining mixing layer with finite thermal expansion. *Proc. Combust. Inst.* **2015**, *35*, 1015–1021. [CrossRef]
11. Shields, B.; Freund, J.B.; Pantano, C. Stationary ethylene-air edge flamesin a wedge-shaped region at low and high strain rates. *Combust. Theory Model.* **2021**, *25*, 1039–1063. [CrossRef]
12. Daou, J.; Liñán, A. Triple flames in mixing layers with nonunity Lewis numbers. *Proc. Combust. Inst.* **1998**, *27*, 667–674. [CrossRef]
13. Daou, J.; Matalon, M.; Liñán, A. Premixed Edge-Flames under Transverse Enthalpy Gradients. *Combust. Flame* **2000**, *121*, 107–121. [CrossRef]
14. Buckmaster, J. Edge-flames. *J. Eng. Math.* **1997**, *31*, 269–284. [CrossRef]

15. Nayagam, V.; Williams, F.A. Lewis-number effects on edge-flame propagation. *J. Fluid Mech.* **2002**, *458*, 219–228. [CrossRef]
16. Nayagam, V.; Williams, F. Curvature effects on edge flame propagation Iin the premixed-flame regime. *Combust. Sci. Technol.* **2004**, *176*, 2125–2142. [CrossRef]
17. Knaus, R.; Pantano, C. A computational approach to flame hole dynamics using an embedded manifold approach. *J. Comput. Phys.* **2015**, *296*, 209–240. [CrossRef]
18. Buckmaster, J. Edge-flames. *Prog. Energy Combust. Sci.* **2002**, *28*, 435–475. [CrossRef]
19. Chung, S.H. Stabilization, propagation and instability of tribrachial triple flames. *Proc. Combust. Inst.* **2007**, *31*, 877–892. [CrossRef]
20. Lyons, K.M. Toward an understanding of the stabilization mechanisms of lifted turbulent jet flames: Experiments. *Prog. Energy Combust. Sci.* **2007**, *33*, 211–231. [CrossRef]
21. Matalon, M. Flame dynamics. *Proc. Combust. Inst.* **2009**, *32*, 57–82. [CrossRef]
22. Westbrook, C.E.; Dryer, F. Simplified reaction mechanisms for the oxidation of hydrocarbon fuels in flames. *Combust. Sci. Technol.* **1981**, *27*, 31–43. [CrossRef]
23. Lu, Z.; Matalon, M. Structure and dynamics of edge flames in the near wake of unequal merging shear flows. *Combust. Theory Model.* **2016**, *20*, 258–295. [CrossRef]
24. Füri, M.; Papas, P.; Monkewitz, P.A. Non-premixed jet flame pulsations near extinction. *Proc. Combust. Inst.* **2000**, *28*, 831–838. [CrossRef]
25. Bechtold, J.; Matalon, M. Effects of stoichiometry on stretched premixed flames. *Combust. Flame* **1999**, *119*, 217. [CrossRef]
26. Cheatham, S.; Matalon, M. A general asymptotic theory of diffusion flames with application to cellular instability. *J. Fluid Mech.* **2000**, *414*, 105–144. [CrossRef]
27. Li, T.Y. Effects of free-stream vorticity on the bahavior of a viscous boundary layer. *J. Aeronaut. Sci.* **1956**, *23*, 1128–1129.
28. Murray, J.D. The boundary layer on a flat plate in a stream with uniform shear. *J. Fluid Mech.* **1961**, *11*, 309–316. [CrossRef]
29. Toomre, A.; Rott, N. On the pressure induced by the boundary layer on a flat plate in shear flow. *J. Fluid Mech.* **1964**, *19*, 1–10. [CrossRef]
30. Burke, S.; Schumann, T. Diffusion flames. *Ind. Eng. Chem.* **1928**, *20*, 998. [CrossRef]
31. Koundinyan, S.; Matalon, M.; Stewart, D. Counterflow diffusion flames: Effects of thermal expansion and non-unity Lewis numbers. *Combust. Theory Model.* **2018**, *22*, 585–612. [CrossRef]
32. Law, C.K. *Combustion Physics*; Cambridge University Press: New York, NY, USA, 2006.
33. Goldstein, S. Concerning some solutions of the boundary layer equations in hydrodynamics. *Proc. Camb. Philos. Soc.* **1930**, *26*, 1–30. [CrossRef]
34. Rott, N.; Hakkinen, R.J. Similar solutions for merging shear flows. *J. Aerosp. Sci.* **1962**, *29*, 1134–1135. [CrossRef]
35. Hakkinen, R.J.; Rott, N. Similar solutions for merging shear flows II. *AIAA J.* **1965**, *3*, 1553–1554. [CrossRef]
36. Lu, Z.; Matalon, M. Edge flames in mixing layers: Effects of heat recirculation through thermally active splitter plates. *Combust. Flame* **2020**, *217*, 262–273. [CrossRef]
37. Van, K.; Jung, K.S.; Yoo, C.S.; Oh, S.; Lee, B.J.; Cha, M.S.; Park, J.; Chung, S.H. Decreasing liftoff height behavior in diluted laminar lifted methane jet flames. *Proc. Combust. Inst.* **2019**, *37*, 2005–2012. [CrossRef]
38. Van, K.H.; Park, J.; Yoon, S.H.; Chung, S.H.; Cha, M.S. Mechanism on oscillating lifted flames in nonpremixed laminar coflow jets. *Proc. Combust. Inst.* **2019**, *37*, 1997–2004. [CrossRef]
39. Oh, S.; Van, K.H.; Jung, K.S.; Yoo, C.S.; Cha, M.S.; Chung, S.H.; Park, J. On the oscillating flame characteristics in nonpremixed laminar coflow-jets: An experimental and numerical study. *Proc. Combust. Inst.* **2021**, *38*, 2049–2056. [CrossRef]
40. Kurdyumov, V.N.; Jiménez, C. Lifted jet edge flames: Symmetric and non-symmetric configurations. *Combust. Theory Model.* **2022**, *26*, 1114–1129. [CrossRef]
41. Lu, Z.; Matalon, M. The speed and temperature of an edge flame stabilized in a mixing layer: Dependence on fuel properties and local mixture fraction gradient. *Combust. Sci. Technol.* **2020**, *192*, 1274–1291. [CrossRef]
42. Kurdyumov, V.N.; Matalon, M. Radiation losses as a driving mechanism for flame oscillations. *Proc. Combust. Inst.* **2002**, *29*, 45–52. [CrossRef]
43. Fujiwara, K.; Nakamura, Y. Experimental study on the unique stability mechanism via miniaturization of jet diffusion flames (microflame) by utilizing preheated air system. *Combust. Flame* **2013**, *160*, 1373–1380. [CrossRef]
44. Gao, J.; Hossain, A.; Matsuoka, T.; Nakamura, Y. A numerical study on heat-recirculation assisted combustion for small scale jet diffusion flames at near-extinction condition. *Combust. Flame* **2017**, *178*, 182–194. [CrossRef]
45. Ruetsch, G.R.; Vervisch, L.; Liñán, A. Effects of heat release on triple flames. *Phys. Fluids* **1995**, *7*, 1447–1454. [CrossRef]
46. Kurdyumov, V.N.; Matalon, M. Effects of thermal expansion on the stabilization of an edge-flame in a mixing-layer model. *Proc. Combust. Inst.* **2009**, *32*, 1107–1115. [CrossRef]

symmetry

MDPI

Review

Discrete and Continuous Symmetries of Stratified Flows Past a Sphere

Yuli D. Chashechkin

Laboratory of Fluid Mechanics, Ishlinsky Institute for Problems in Mechanics RAS, 119526 Moscow, Russia; yulidch@gmail.com

Abstract: This study presents the detailed experimental results of fine structures and dynamics in a stratified flow past a sphere, which is towed with constant velocity in a transparent basin. We developed experimental procedures based on the complete solutions of the truncated fundamental fluid equations. These complete solutions describe the waves and fine accompanying ligaments, as well as the vortices and other flow structures. To visualize the flow, a variety of classical schlieren and electrolytic precipitation procedures were used. Ligaments appear in the schlieren images of the flow as fine interfaces and fibers. They strengthen the influence of the relatively weak density gradient in a continuously stratified fluid (CSF). The symmetry in the wake is discrete at small Froude numbers with the domination of buoyancy effects. At increased velocity and high Froude numbers, when the inertial and non-linear effects turn out to be significant, an axial symmetry becomes continuous.

Keywords: stratified fluid; towing sphere; schlieren instrument; electrolytic precipitation; internal waves; prismatic wake; vortex column; rings

Citation: Chashechkin, Y.D. Discrete and Continuous Symmetries of Stratified Flows Past a Sphere. *Symmetry* **2022**, *14*, 1278. https://doi.org/10.3390/sym14061278

Academic Editors: Andrzej Herczyński and Roberto Zenit

Received: 24 May 2022
Accepted: 13 June 2022
Published: 20 June 2022

1. Introduction

Denser particles sink, while lighter particles float under the impact of buoyancy forces. Naturally, a stratification in a heterogeneous liquid or gas occurs (a stratification is a continuous variation of density $\rho(z)$ over depth z).

The stratification is discrete when the density changes abruptly. It happens at the boundaries of immiscible media; for example, between the atmosphere and the hydrosphere.

With density changes, we observe the variations of such physical properties as the velocity of sound propagation, the refractive index of light [1]. They allow observing the patterns of flows in the bulk or on the surface of a liquid either with the naked eye [2] or using optical [3,4] and acoustic instruments [5].

The development of space technologies for remote sensing has opened up new ways for tracking the temporal variability of naturally occurring structures in the atmosphere and the oceans [6]. The received images of natural phenomena are useful for further development of the theory of fluid flows, and for the improvement of their numerical simulation and laboratory modeling.

The natural length scale $\Lambda = |d \ln \rho(z)/dz|^{-1}$, the frequency $N = \sqrt{g/\Lambda}$, or the buoyancy period $T_b = 2\pi/N$ are applied to describe the continuous stratification of incompressible liquids in environmental, industrial, or laboratory settings.

Natural oscillations of stratified media were noticed as early as the 18th century [2] and their frequency was calculated a century later [7]. There are two types of continuous stratification. Strong stratification, with $N \sim 1\ \mathrm{s}^{-1}$, is typical for laboratory conditions, and the weak one with $N \sim 0.01\ \mathrm{s}^{-1}$ characterizes the ocean and planetary atmospheres, as well as two types of uniform density fluids that are nearly homogeneous ($N \to 0 \sim 10^{-5} \neq 0$) or actually homogeneous, with $N \equiv 0$. The condition of an actually uniform density $\rho \equiv 0$ leads to the degeneration of the fundamental equations system and to a confusing of "motion" and "flow" concepts. The motion, which is considered to be a transformation of a metric space into itself, and the flow, which is a physical process of matter, momentum, and energy transfer, become mathematically undistinguishable in approximation of an actually homogeneous liquid [8].

A standard classification of flows, including laminar (layered) and irregular (turbulent) flows [9,10], is based on the characteristic structural features of flow patterns. Specific forms of phenomena at a smaller scale allow distinguishing waves and vortices [9,10]. They were visualized with a pen and brush beginning with the expressive drawings of Leonardo da Vinci and other Renaissance artists [11], and then by optical instruments [12].

Numerous experiments [12] showed that at low velocities the symmetries of homogeneous fluid flows around obstacles depend on the shape of the body even in the case of a spatially uniform flow.

The aesthetic aspect of the symmetric flow patterns influences positively the mathematical description with the reduced dimension of the problem.

Correspondingly, the solutions of classical flow problems are significantly simplified (in particular, for the flows around a sphere or a disk [9,10]). The natural flow pattern in the bulk of a homogeneous liquid does not depend on the directions of the velocity vector with respect to the gravity acceleration vector.

Surprisingly, even if the stratification is weak (i.e., the density variations on the body size can hardly be registered by modern instruments), it is noticed both in the dynamics and in the structure of the flow.

In a heterogeneous medium there are directions, with the gradients of a gravitational force potential, $\mathbf{g} = -\nabla\Phi$, determining the free-fall acceleration $\mathbf{g}$ and the density gradient $\nabla\rho$. The action of noncollinear gradients directly produces vorticity in the liquid flow [13].

The greatest gradients are placed near the surface of the submerged body, where the action of various anisotropic mechanisms of energy and matter transfer is the most pronounced. In a moving fluid the energy is transferred by macroscopic processes (by the flows with the local velocity $\mathbf{v}$ and different waves with the group velocity c_g), as well as by microscopic processes (the slow atomic–molecular diffusion as well as fast processes of direct transfer of internal energy into thin flow components [8]).

The calculations show that the stratification removes the degeneracy in the singular components of the complete solution of the linearized set of fundamental equations (i.e., the multiplicity of the dispersion equation roots for homogeneous liquids [8]).

Simultaneously, it elucidates a number of hydrodynamic paradoxes [14].

The anisotropy of the stratified media affects the redistribution of energy between the structural components in a thin layer near the surface of the submerged body. The purpose of our study is to reveal the influence of a continuous stratification on the symmetry breaking in the flow past a sphere—a perfectly symmetric body.

2. A Brief History of Flow around a Sphere Studies

There were a number of theoretical and practical reasons why scientists took interest in studying the flow around a uniformly moving sphere.

For many centuries, balls made of various materials (from natural stones to metal alloys) served as artillery shells and moveable parts in many transport mechanisms. In the middle of the 18th century, with the development of universities, the problem of flow around a sphere became the object of intensive theoretical and experimental research. The studies were carried out inquisitively, as a consequence of scientific development logic, and under government contracts in Russia [15], UK [16], Germany [17], and France [18]. The result of this research work was the creation of the continuity equation for an incompressible fluid and a compressible gas [19] and the formulation of the first closed system of equations for ideal fluid flows [20]. Paradoxical results of their application to the drag on a sphere problem were established very soon.

New energy transfer equations (for heat transfer [21]) contributed to the development of Navier's theory of fluid motion, incorporating shear stresses caused by viscous friction [22]. G.G. Stokes re-derived Navier's equations of a viscous fluid motion in terms of the continuous medium conception [23].

Analyzing the damping of oscillations of a pendulum gravimeter, he calculated the drag on a sphere in axial symmetric flow [24] in a linear approximation and created the basis of subsequent experimental and theoretical studies.

In experiments, marking impurities in the form of small immiscible liquid drops, solid particles or gas bubbles were introduced to discrete visualization of the flow pattern in a homogeneous liquid [12]. Schlieren instruments, sensitive to variations in the density gradient, were used for continuous visualization of the flow patterns in a stratified medium in addition to separated markers [4,25]. The method was chosen based on the analysis of the system of governing equations and in agreement with traditional approaches.

3. System of Stratified Fluid Mechanics Equations

The effects of compressibility and heat transfer are usually neglected in the case of slow, compared to the velocity of sound, flows of low-viscosity liquids with high heat capacity [10,26].

The equations of state for the Gibbs potential and the thermal diffusivity equation are omitted from the general system of equations as well [8]. The unperturbed (initial) density profile $\rho_0(z)$ replaces the equation of state (in a coordinate frame used further axis z is vertical and opposed to the gravity acceleration $\mathbf{g}$; the body moves along axis x).

The main physical quantities characterizing the state and flow of stratified media are density ($\rho(\mathbf{x},\ t)$), momentum ($\mathbf{p} = \rho\mathbf{v}$), or velocity ($\mathbf{v} = \mathbf{p}/\rho$), defined as the ratio of invariant parameters, concentrations of the stratifying $S = S_0(z) + s$, and visualizing c components. The truncated system of equations for an incompressible stratified fluid with the linearized equation of state includes the equations of continuity, momentum transfer, diffusion for the stratifying component, and visualizing impurity, in the Boussinesq approximation, and takes the following form [8,10,26]

$$\begin{gathered}
\operatorname{div}\mathbf{v} = 0; \quad \rho = \rho_{00}\left(1 - \frac{z}{\Lambda} + s + c\right); \\
\frac{\partial \mathbf{v}}{\partial t} + \mathbf{v}(\nabla\cdot\mathbf{v}) = -\frac{1}{\rho_{00}}\nabla P + \nu\Delta\mathbf{v} - (s + c)\mathbf{g}; \\
\frac{\partial s}{\partial t} + \mathbf{v}\cdot\nabla s = \kappa_S \Delta s + \frac{v_z}{\Lambda}; \quad \frac{\partial c}{\partial t} + \mathbf{v}\cdot\nabla c = \kappa_C \Delta c.
\end{gathered} \tag{1}$$

Here, ν is the kinematic viscosity, κ_S and κ_c are stratifying agent and the impurity diffusion coefficients, ρ_{00} is the density on a reference level, and s and c are the density and impurity perturbations (contraction coefficients are included in the definition). The boundary conditions on the surface of a fixed or a uniformly moving sphere with velocity U are traditional. They are no-slip for velocity and impermeability for substances on the surface of the sphere or the surface flux for a visualizing component $j_n = j_c$ [8,27].

System (1) is characterized by a set of physical parameters, transforming into a number of temporal and spatial scales. The density profile defines the scale of stratification Λ. The ratio of a sphere diameter D to velocity U forms the kinematic time scale $\tau_D = D/U$. The attached internal wavelength is $\lambda\!\!\!^{-}= UT_b$.

Kinetic coefficients determine a relatively large viscous wave scale $L_N^\nu = \sqrt[3]{g\nu}/N$ as well as a fine scale $L_g^\nu = \sqrt[3]{\nu^2/g}$ and transverse scales of the ligaments due to viscosity $\delta_N^\nu = \sqrt{\nu/N}$ or diffusivity of the stratifying $\delta_N^{\kappa_s} = \sqrt{\kappa_s/N}$ and of the visualizing impurity $\delta_N^{\kappa_c} = \sqrt{\kappa_c/N}$.

The thicknesses of an additional group of ligaments depend directly on the body velocity $\delta_U^\nu = \nu/U$, $\delta_U^{\kappa_s} = \kappa_s/U$, $\delta_U^{\kappa_c} = \kappa_c/U$ [8,27]. The dynamics of the fine components is characterized by a small intrinsic time scale, $\tau_U^\nu = \delta_U^\nu/U$, reflecting a possible ability of fast rearrangement of the flow pattern. A large number of small scales (for laboratory conditions $\delta_N^\nu \sim \delta_\omega^\nu \sim 1$ mm, $\delta_N^\kappa \sim 0.05$ mm, $\delta_U^\nu \sim 0.1$ mm for $N \sim 1\ \text{s}^{-1}$, kinematic viscosity $\nu = 0.01$ cm^2/s, and a table salt diffusion coefficient $\kappa \sim 1.4 \cdot 10^{-5}$ cm^2/s) show the complexity of the fine structures of stratified flows. It requires their investigation methods and instruments with a high temporal and spatial resolution.

Ratios of scales define the traditional dimensionless combinations, which are the Reynolds number, $\text{Re} = D/\delta_U^\nu = UD/\nu$ (and its analogues—Peclet number with diffusion coefficients $\text{Pe} = UD/\kappa$); Froude number, $\text{Fr} = (\lambda/2\pi D)^2 = (U/ND)^2$, equaling the reciprocal value of the energetic criterion; Richardson number, $\text{Ri} = \omega_U^2/N^2$, $\omega_U = (\partial U/\partial z)$, which is widely used in the theory of stratified turbulent flows [28]; a large length scale ratio, $\text{C} = \Lambda/D$; Schmidt number, $\text{Sc} = \nu/\kappa_S$; and others not so frequently used.

The classification of flow components is based on a complete solution of the linearized system (1), followed from implementations of the compatibility condition [29]. It includes waves and supplementary ligaments.

Internal waves in the bulk of a CSF are visualized by schlieren instruments [4,25]. Registered variations in the magnitude and gradient of the optical refractive index illustrate the displacements of fluid particles from equilibrium positions [30].

The schlieren images of the internal waves coincide with the calculations based on the linear theory of the amplitude-phase properties of the attached waves. These waves are created by a horizontally moving sphere and along an inclined trajectory [31–33].

Ligaments are described by the singular components of the complete solutions of fundamental equations system [8,10,26] and its reduced version (1). They correspond to thin high-gradient interfaces and fibers in the wake past an obstacle [34]. A special class of fine flows (diffusion-induced flows (DiF)) is formed due to the inhomogeneity of the molecular transfer of the stratifying component and interruption of the transfer on the impermeable surface of a rigid body. We can observe them near a motionless body in a fluid at rest [9,35], but when the motion starts, DiF transform into fine disturbances. These fine disturbances in the vicinity of a moving body are drawn out by the shear flow as interfaces and filaments in the wake. They are extended by the mean flow and slowly widened under the action of molecular processes [36].

The formation of each structural component, which is the wake as a whole, vortices, and internal waves, as well as the general and fine deformation of the density profile, contributes to the drag force. A relatively small contribution of wave effects to a total drag coefficient at large values of the Froude number increases with a decreasing sphere velocity, reaching the maximum in the interval $0.7 < \text{Fr} < 1$ or $4.4 < (\lambda/D) < 6.3$ [31,37,38] and decreases further when the body velocity grows. It follows from the theory of linear internal waves. Moreover, the results agree with measurements [37–40]. Additional simplifications, including the assumption of the absence of diffusion $\kappa_S = 0$ [41] or the equality of dissipative coefficients of kinematic viscosity and diffusivity $\text{Sc} = \nu/\kappa_S = 1$ [42], are made when constructing the numerical solutions of a complex multiscale system (1).

The system (1) turns into the classical Navier–Stokes equations with the approximation of an actually constant density, in which the diffusion equations and the term with the force of gravity in the momentum transfer equation are excluded [10]. The patterns of pressure fields, velocity, and vorticity components determine the laminar or turbulent nature of the flow. The detailed calculations of the streamlines, pressure, and the vorticity fields in the wake past the sphere in the mode of a toroidal rear vortex formation, as well as the shape and position of the separation line, were carried out [43].The determination of the position of singular points on the flow separation line and the classification of the shapes of vortex elements in the wake supplement the calculations of forces acting on the sphere [44]. The calculated flow pattern past the sphere in the range $20 < \text{Re} < 400$ gave the angular position of the circular separation line of the wake with the vortex structure [45].

Fine visualization of the vortex structure of the axisymmetric dyed wake past the sphere at low Reynolds numbers (the soluble dye was washed off the surface of a freely falling body) helps register the attached rear toroidal vortex outer shape. Its axisymmetric envelope gradually changes from a concave form into a conical and convex forms with an increasing velocity [46]. Comparison with the previous experiments shows that the rear vortex exists at Re ~ 10 and even lower values of velocity Re ≈ 3.5.

Difficulties in studying flows with low body velocities are associated with uncontrolled background fluid motions. They occur because of the temperature or atmospheric pressure variations and mechanical movements of the basin excited by street traffic. For the visualization of the liquid flow pattern around a sphere, a soluble dye as well as luminescent compositions have been applied, which glow brightly in ultraviolet light, highlighting hydrogen bubbles and small particles. These small particles allow the recording of the flow velocity and flow patterns with 2D and 3D PIV tools.

Oil slicks, smoke, small particles, drops, and silk threads are used in wind tunnels in approximation that the impurity is "passive"; i.e., it is completely carried by the flow. However, this assumption should be confirmed in each specific case, taking into account the difference in the molecular properties of the studied and visualizing media, the Brownian motion of small particles, as well as the transfer and rotation of free solids in the shear flow.

At high velocities, when compressibility effects are evident, the schlieren instrument [3,4] becomes the most effective tool for the visualization of a general flow pattern and singular fine components such as shock waves, wake envelopes, sharp interfaces, and thin fibers (ligaments), which transfer energy, momentum, and density perturbations from the body boundary into shock waves and the wake [47,48].

4. Visualization of a Stratified Flow around a Sphere

The study of the stratification effects is usually carried out in a transparent basin filled with a sodium chloride solution of variable concentration using the "continuous displacement" method [49,50]. The linear relationship of background density and the optical refractive index [51] allow the application of various optical imaging methods, which accompany the techniques for homogeneous media [3,4,12]. The direct shadow and various versions of the schlieren methods, both the classical ones with a slit diaphragm and the Foucault knife [3,4,51,52] and technically more simple "synthetic" or moiré schlieren methods, have been applied to the study of stratified flows [53,54]. Although classical schlieren instruments have a higher sensitivity and spatial resolution than moiré tools, in practice the latter have become more widespread due to the consistency with modern PC image processing.

The high chemical activity of the NaCl solution was used for visualization in an "electrolytic precipitation" method; i.e., electrochemical anodic oxidation of lead, tin, or their alloys under the action of direct electric current [55,56]. Small particles of metal oxides and chlorates with a size of about one micrometer, formed as a result of a complex of chemical reactions, produce a white translucent suspension. The shape and location of the electrodes determine the position of the source and the density of the suspension The flow pattern resembles the tinting of flows with smoke or microdrops in wind tunnels and soluble dyes in the aquatic environment [12].

In the process of studying the effect of stratification on the flow pattern around a sphere, we mainly paid attention to the description of internal waves, which caused the displacements of fluid particles from equilibrium positions [30,31] as well as high gradient interfaces and fibrous. Fine flow components characterizing ligaments were not identified by most of the traditional imaging methods [12], including moiré or "synthetic schlieren" methods [53,54], nor suspended particles, because of the limited spatial and temporal resolution (the impact of temporal resolution importance was illustrated in [12] as well).

5. Laboratory Experiment Technique

The studies of the flow patterns around a sphere towed in a CSF or homogeneous fluid were carried out on the setups "Wave Fields Fine Structure Modeling (WFF) and "Laboratory Mobile Tank (LMT)", as a part of the Unique Science Facility "Hydrophysical Complex of the IPMech RAS". This complex was used for the modeling of hydrodynamic processes in the environment, their impact on underwater technical objects, as well as the distribution of impurities in the ocean and atmosphere (USF HPC IPMech) [57]. The setups differ in the size of the working basins ($0.7 \times 0.25 \times 0.7$ m^3 for WFF and $2.20 \times 0.40 \times 0.60$ m^3 for LMT) and a number of auxiliary mechanisms. A photo of the WFF setup is shown in Figure 1; a detailed description of the LMT setup is given in [28]. The models are towed by one movable or two thin threads (tight, stabilizing the trajectory, and a towing wire) with constant and variable velocities. Schlieren images of the flows were produced by the IAB-458 instrument (a field view is 23 cm).

Figure 1. Experimental setup WFF USF of the "HPC IPMech RAS".

The basin was filled with an aqueous solution of common salt of variable concentration with a selected density profile by the continuous displacement method [49,50]. Flow patterns were observed around a sphere made of Plexiglas with a diameter of D =1, 2, 3, 4, 5 cm, which was placed ($U = 0$) in the tank or towed at a constant velocity U. A tin belt or segment was placed on some sphere surfaces (Figure 2). Formed particles from monoxide and oxide of tin, tin dichloride, and tin acids with a size of about 1 µm under the action of a direct electric current were transported by the flow. When the electric current increased, the larger particles were formed and the wake containing them began to sink as a whole, which was especially noticeable in a homogeneous liquid. The surface of the sphere was carefully polished before each experiment and the traces of the metal oxidation were removed.

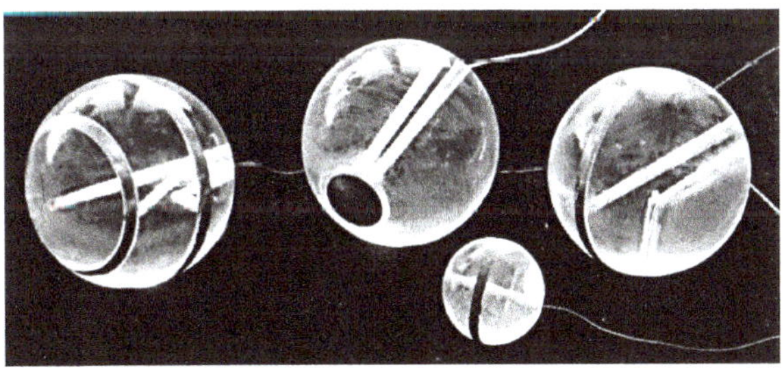

Figure 2. Spheres with anodes of various shapes.

The main stratification parameter, the buoyancy period T_b, was measured with an optical instrument [58] or an electrical conductivity sensor [59] in a short internal wave field around a density marker—a vertical wake past a rising gas bubble or a sinking crystal (salt or sugar). The bubble floated (the crystal sank) vertically if the Reynolds number, determined by its diameter D and velocity U, satisfied the condition $\mathrm{Re} = UD/\nu < 201$. In these experiments, the buoyancy period was in the range $3.5 < T_b < 14$ s.

A vertical slit was installed in the illumination part of the IAB-458, and a Foucault knife or Maksutov's thread was placed in the receiving part of the instrument to visualize the fields of the horizontal component of the refractive index and density gradient, respectively. A colored schlieren method ("natural rainbow" schlieren [52]) with a horizontal slit and a linear grating in the receiving part visualized the vertical component of the refractive index gradient. With auxiliary optics, the spatial resolution of the instrument was better than 0.01 cm.

6. Laboratory Studies of Stratified Flows

The center of mass of the stratified fluid layer is located below the geometric center in the gravity field. It means that a non-equilibrium medium with the density profile has supplies of available potential energy. The latent energy is converted into the kinetic energy of the liquid flow in a thin layer of DiF near an inclined impermeable surface, on which the molecular transfer of the stratifying component is interrupted. There is flow if the gravity field is formed, even in the absence of additional external forcing. The nature of such flows is the difference between atomic–molecular interactions in the bulk of a liquid and near a solid body boundary. The mechanism of fine flow structure formation by interruption of the diffusion flux is universal and exists in all types of non-uniform density liquid and gas flows. Firstly, the theory of stationary DiF on an infinite inclined plane was developed [9,35]. The profiles of salinity and velocity of this stationary flow are similar and characterized by a common combination scale, $\delta = \sqrt[4]{\nu\kappa_S / N^2 \sin^2\alpha}$, where α is the inclination angle of the plane to the horizon. The solution diverges at small values of the angle α of plane inclination.

The salinity and velocity perturbation profiles are characterized by different transverse length scales for salinity and velocity perturbations in asymptotic solutions for small-time approximation, presented firstly in [60] and in first terms of the exact solution expansions constructed in [61].

In the local coordinate frame (ξ, ζ), axis ζ is normal to the plane; the asymptotic solution in the small-time approximation for salinity perturbations is characterized by the length scale $\delta_N^{\kappa_S} = \sqrt{\kappa_S/N}$, where time t is normalized by the buoyancy period $\tau = t/T_b$ [60],

$$s' = -2\frac{\delta_N^{\kappa_S}\sqrt{\tau}}{\Lambda}\mathrm{ierfc}\left(\frac{\zeta}{2\delta_N^{\kappa_S}\sqrt{\tau}}\right), \tag{2}$$

The asymptotic expression for the velocity of the induced velocity is described by both scales $\delta_N^{\kappa_S}$ $\delta_N^{\nu} = \sqrt{\nu/N}$ [60],

$$\begin{gathered} u(\zeta) = \frac{N^2\,\delta_s\,\tau^{3/2}}{\nu-\kappa_s}\left[\mathrm{i}^3\mathrm{erfc}\left(\frac{\zeta}{2\delta_\nu\sqrt{\tau}}\right) - \mathrm{i}^3\mathrm{erfc}\left(\frac{\zeta}{2\delta_s\sqrt{\tau}}\right)\right]\sin 2\alpha, \\ \mathrm{i}^n\mathrm{erfc}(z) = \int_z^\infty \mathrm{i}^{n-1}\mathrm{erfc}(x)dx,\ \mathrm{i}^0\mathrm{erfc}(z) = \frac{2}{\sqrt{\pi}}\int_z^\infty e^{-x^2}dx,\ \mathrm{i}^{-1}\mathrm{erfc}(z) = \frac{2}{\sqrt{\pi}}e^{-z^2}. \end{gathered} \tag{3}$$

The flow components with incommensurable length scale values show the total unsteadiness of the phenomena.

The flow pattern consists of a sequence of counter flows with rapidly decaying amplitude. Due to the smallness of the transverse scales, it does not practically depend on the shape of the body everywhere, except for its poles, which are the extreme points of the obstacle in the vertical plane [62]. Thin DiF are not resolved by the mostly used methods of a flow visualization.

The geometry of the problem and the consistency of the diffusion flux on the overhanging surface (an excess of salt accumulates on this surface) form a directed downward flow.

The flow is directed upward on an open inclined surface where a salt deficit is observed. Accordingly, the flow around the sphere is symmetric with respect to the vertical axis and is antisymmetric with respect to the plane of the horizontal equator. Asymptotic and numerical solutions of the system (1) for DiF on a sphere are presented in [63].

Here, the sections of the sequence of emerging toroidal vortices, which form convergent and divergent jets near the upper pole of the sphere, are shown in Figure 3. Amplitudes of perturbations rapidly decay with distance from the body surface.

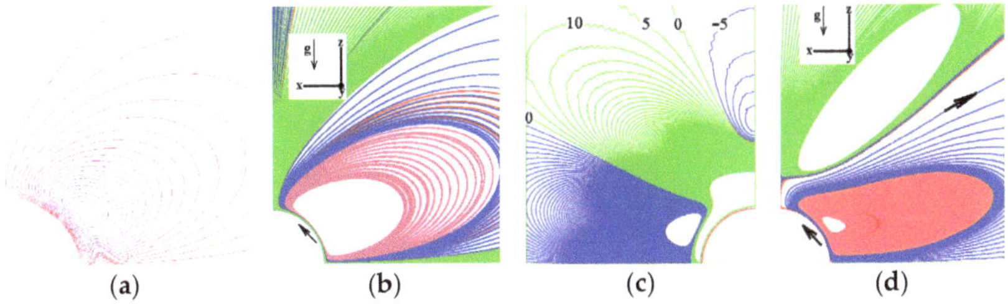

Figure 3. Pattern of the DiF flow on a fixed sphere of a small diameter (d = 2 cm, T_b = 6.34 s): (**a**,**b**)—streamlines from the analytical solution and numerical simulation t = 0.5 T_b; (**c**,**d**)—salinity disturbance isolines and streamlines, $t = T_b$.

The results of the numerical and analytical solution of the system (1) agree quite well. In the non-stationary pose, new circular cells appear subsequently near the pole of the sphere with a buoyancy period. Each new cell shifts the entire system of perturbations that have arisen earlier onto the sphere surface. The general shape of the flow structure does not depend on the size of the sphere; however, the thickness of the cells and the flow velocity increase with the growth of the body diameter.

The comparison of calculations of the flow pattern at short and long times (Figures 3 and 4) shows that the number of cells, as well as the value of the maximum velocity, steadily increases, and the rate of change gradually fades. However, the flow does not have a stationary limit. The free symmetric body remains motionless on the horizon of neutral buoyancy due to the geometry of the flow.

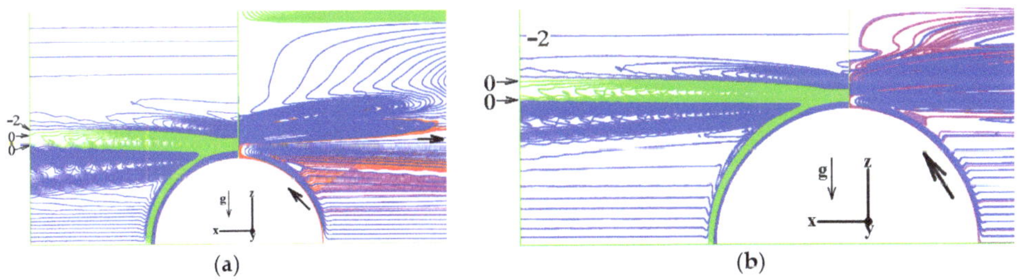

Figure 4. Central section of a DiF pattern on the upper half of the sphere (T_b = 6.34 s): (**a**,**b**)—D = 2, 4 cm, $\tau = t/T_b$ = 1662, 1075; there are salinity disturbances on the left side of the figures, and streamlines on the right side (image scales are different).

In the schlieren photographs of the flow around a motionless sphere, the symmetrical pattern contains a sequence of widening dark and light bands near the poles. The angle of inclination of the bands to the horizon monotonically decreases with distance from the source (Figure 5a,b). The shape of the flow near the poles of a sphere, which indicates a noticeable change in the density gradient in a thin layer, is in good agreement with the calculation [63]. A geometrically similar flow is located at the edge of a cell of thermoconcentration convection over a localized ("point") heat source in a CSF. It was interpreted as "dissipative-gravitational waves" in [64].

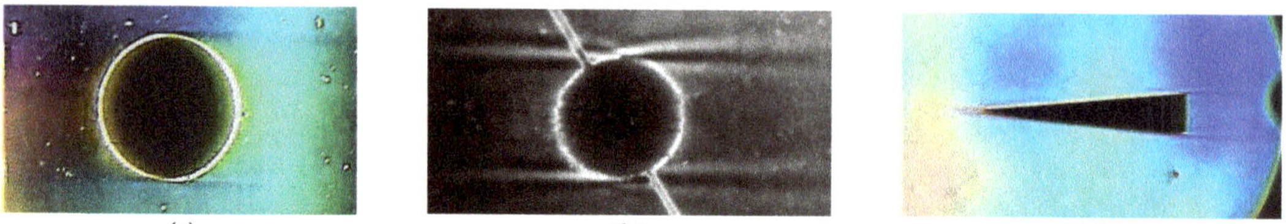

(**a**) (**b**) (**c**)

Figure 5. Schlieren pattern of DiF, $T_b = 6$ s: (**a**,**b**)—on a sphere $D = 5$ cm (Maksutov's thread and Foucault knife), and (**c**) on a wedge of length $L = 10$ cm and the base height $h = 2$ cm suspended on the horizon of neutral buoyancy.

A general change in the density distribution influences the molecular diffusion in the flow fluid near the inclined surface of an impermeable body. The intensification of the density gradient in an almost resting liquid confirms the non-stationarity and spatial inhomogeneity of the process of a flow formation. It transforms a weak mean gradient into a stronger one in the thin layer. Numerical calculations of the DiF were carried out on the basis of the system of Equation (1) near various bodies that are an inclined plate [62], a sphere [63], or a horizontal wedge [65–67].

DiF transport the substance along the side faces of the wedge and creates a deficit of mass ahead of the apex. The asymmetric distribution of pressure in the flow pattern induced by diffusion on a motionless wedge includes the area of pressure deficit in front of its apex (Figure 6a). It causes self-motion of the asymmetric body [65,66].

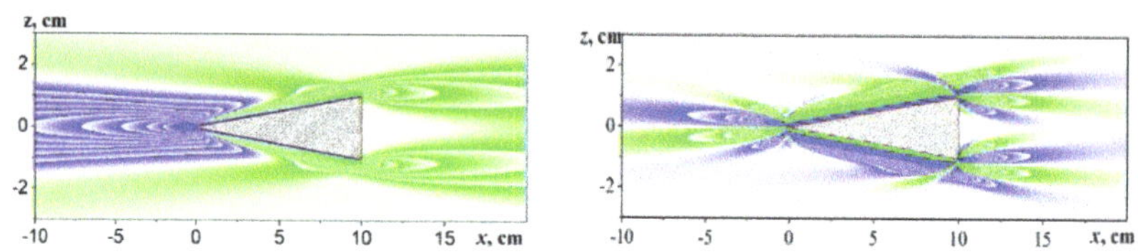

(**a**) (**b**)

Figure 6. Calculated DiF pattern near a fixed wedge (length $L = 10$ cm, shelf height $h = 2$ cm) in CSF at rest, $T_b = 6.3$ s: (**a**,**b**)—disturbances of the pressure and the horizontal component of the density gradient.

The fact that the height of the pressure deficit area even somewhat exceeds the height of the body base, ensures the efficiency of its action. In this case, the variations in velocity, salinity, and the horizontal component of its gradient are concentrated in thin layers adjacent to the lateral surface of the wedge (Figure 6b).

In the experiment, the interfaces are weakly expressed ahead of the body apex and are distinctly expressed at the edges of the base, as is shown in the schlieren image of the flow in Figure 5c. The self-motion of a wedge is experimentally demonstrated in [65,67]; a detailed calculation and schlieren visualization of the flow pattern is given in [66].

7. Rectangular Cross-Section of the Density Wake Past a Sphere at Low Froude Number

As the sphere starts to move in the horizontal plane, the pattern of DiF as a whole and its symmetry changes radically. To visualize the fine structure of the wake past the sphere and its spatial shape we used sensitive schlieren methods and electrolytic precipitation.

Careful examination of the images in Figure 7 shows that the axisymmetric DiF near the poles of the sphere is transformed into unsteady upstream internal waves that are continuations of the attached waves past and above the sphere. Ligaments correspond to extended double light/dark stripes adjoining the poles past the sphere in Figure 7a,c [8].

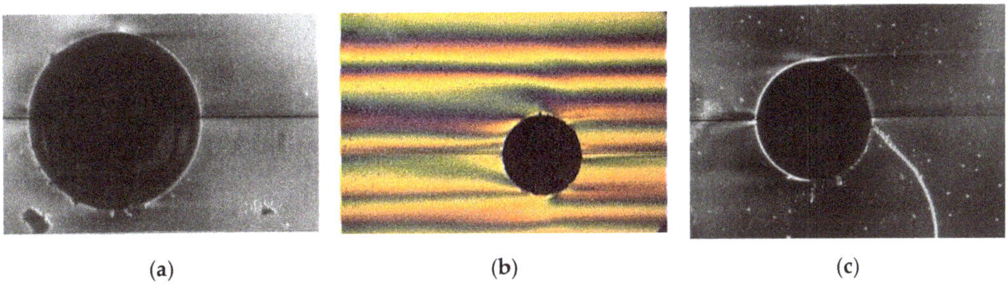

Figure 7. Schlieren images of the flow around a sphere moving slowly from right to left in CSF: (a,b)—$T_b = 8.0$ s, $N = 0.785$ s^{-1}, $D = 4$ cm, $U = 0.04$; 0.06 cm/c, $Fr = U^2/N^2D^2 = 1.6 \times 10^{-4}$, Re = 16; 24, conventional and natural rainbow schlieren methods; (c)—$T_b = 4.5$ s, $N = 1.4$ s^{-1}, $D = 5$ cm, $U = 0.08$ cm/s, Re = 40, $Fr = 1.3 \times 10^{-4}$ (vertical slit—Foucault knife).

DiF are transformed into the inclined beams of non-stationary internal waves, which are oriented in the direction of the sphere motion in front of the body. Light bands near the upper and the lower poles ahead of the sphere in Figure 7a,c, outline the blocked fluid moving together with the body. The blocked fluid is characterized with a more uniform density profile than the initial one. The boundaries of the blocking zone are better expressed in the color schlieren image of the flow in Figure 7b ("natural rainbow" method). Here they are represented by an inclined blue strip, extending from the lower pole of the sphere and pale from the upper pole, oriented in the direction of the body's velocity; it is converging with distance.

Contractions and expansions of the colored bands in a schlieren image of above the sphere flow (Figure 7b) visualize the crests and troughs of the attached internal waves. On the whole, according to linear theory, the calculations of the phase surfaces of the internal wave shapes [30,31,68] are consistent with the visualization images, even at the smallest values of the body velocity [69]. Here and below, the ligaments are presented by thin interfaces separating the internal waves and density wake.

The images of the electrolytic precipitation suspension, shown in Figure 8, illustrate the spatial geometry of the density wake past a slowly moving sphere (and small value of the Froude number, respectively). Fine particles are formed on the transverse tin belt.

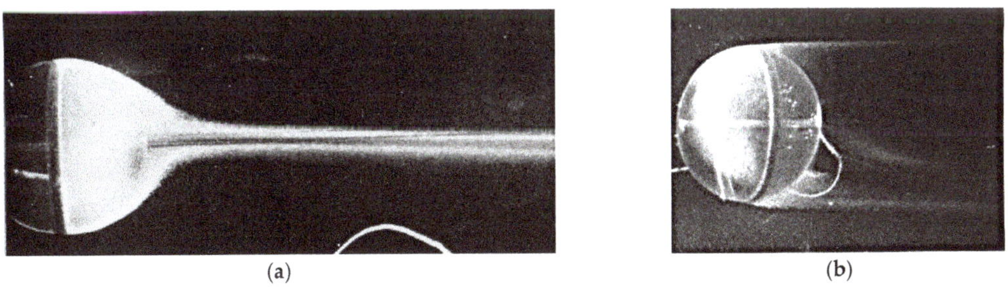

Figure 8. Patterns of electrolytic precipitation suspension past a slowly moving sphere in CSF, $T_b = 4.5$ s, $N = 1.4$ s^{-1}, $D = 5$ cm, $U = 0.03$ cm/s, $Re = UD/\nu = 15$, $Fr = 1.8 \times 10^{-5}$; (a,b) top and side view.

In the top view, the dense suspension is distributed evenly over the surface of the sphere and the plane wake. In the side view, only the edges of the wake are colored. They are in contact with the sphere poles. The sharpness of the separation lines on the sphere surface indicates the planar geometry of the shells bounding the wake. They are formed by converging flows, which sharpen DiF on the impermeable sphere surface. The general shape and the narrowness of the colored wake in the horizontal plane as well as the high altitude in the vertical plane determine the selection of a rectangle for approximating the cross-sectional shape of the density wake [69,70]. The maximum suspension concentration is observed on the edges of the rectangular flow section.

The discrete symmetry of the cross-section shape of the wake with respect to the horizontal and vertical planes, which passes through the motion line of the body center, indicates the dominance of the buoyancy effects over inertial and nonlinear effects. Their symmetry reflects the perfect continuous symmetry of the body. The initial density gradient intensification in the vicinity of the body enhances the consequences of buoyancy effects in the creeping flow mode. At low velocities of the sphere, the liquid mainly flows around an obstacle along isopycnal (close to horizontal) lines.

Structural differences in the geometry of the flow components become more distinct with an increasing sphere velocity [70]. In the schlieren flow pattern in Figure 9a, the blocked fluid is visualized by dark and light spots in front of the body. They are bounded by tilted rays of non-stationary internal waves extending from the poles of the sphere.

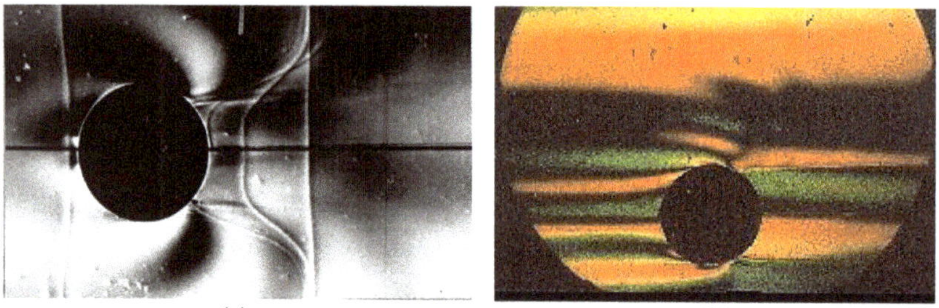

(**a**) (**b**)

Figure 9. Schlieren images of a stratified flow around a uniformly moving sphere at small Froude numbers: (**a**) $T_b = 12.0$ s, $N = 0.52$ s^{-1}, $D = 4$ cm, $U = 0.43$ cm/s, Re = 132, Fr = 0.04, (**b**)$T_b = 3.0$ s, $N = 2.09$ s^{-1}, $D = 5$ cm, $U = 0.5$ cm/s, Re = 250, Fr = 0.002.

Curvilinear bands extending from the separation points of sharp almost horizontal interfaces on the rear part of the sphere represent attached internal waves. The fine interfaces outline the density wake. The sharpness of the boundaries separating the domains with different kinds of a flow demonstrates strong interaction between the fine ligaments with more long internal waves [71] in different flow regimes past a sphere [69,72].

The three successive curved vertical lines in Figure 9a are deformed markers, which create the arising gas bubbles [59]. The vertical line is the initial bubble wake deformed by the shear flow with time. A profile with an almost uniform central part surrounded by two inclined sections visualizes the constant velocity in the center of the wake inside two wide shear layers. The third marker line directly adjacent to the surface of the body illustrates the patterns of velocity distribution on the boundaries of the wake. Here, the heights of the velocity shear layers and the thickness of the ligaments submerged in their central parts differ in more than an order of magnitude, as in the nonstationary DiF (Formulas (2) and (3)).

With a further increase in the velocity of the body, the length and amplitude of the attached waves grow. A group of five attached waves is presented in Figure 9b. The shapes of the color bands ahead and past a sphere reflect the difference in upstream and downstream density profiles. A more homogeneous domain of a blocking fluid is bounded by non-stationary waves. Behind the body, the ligaments forming the density wake envelopes are expressed. The linear theory for the uniform motion of the sphere satisfactorily describes the phase surfaces of the attached waves, both along a horizontal [31,68] and an inclined trajectory [32]. The selection of the position and the intensity of the model sources and sinks allows achieving satisfactory agreement between the numerical and measured displacement amplitudes [32,68].

With an increase in the body velocity at Fr << 1, the length and amplitude of all types of internal waves increase. It concerns non-stationary upstream waves, attached downstream, and trapped waves inside a density wake with their own profiles of velocity and density. The trapped waves inside the wake correspond to a set of tilted antisymmetric dark and light bands past the body, limited by high-gradient diverging interfaces in Figure 10a.

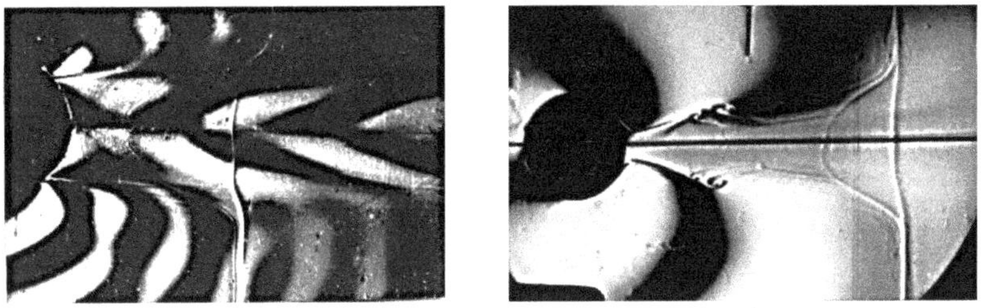

(a) (b)

Figure 10. Schlieren images of a stratified flow around a uniformly moving sphere: (**a**) $T_b = 3.0$ s, $N = 2.09$ s^{-1}, $D = 5$ cm, $U = 0.76$ cm/s, Fr $= 0.005$, Re $= 384$; (**b**) $T_b = 12.0$ s, $N = 0.52$ s^{-1}, $D = 4$ cm, $U = 0.8$ cm/s, Re $= 256$, Fr $= 0.15$.

The curvilinear phase surfaces of the attached waves adjoin the wake boundaries almost vertically. Their contour is distorted by the shear flow at the wake boundary, the height of which noticeably exceeds the thickness of the interfaces.

A further increase in the body velocity enhances the generation of waves, sharpens the gradients in the wake envelope, and activates the interaction between the various components of the flow. At first the wake expands uniformly with distance from the body and then contracts in accordance with the wave pattern (Figure 10b). The wake envelopes become wavy with sharpened crests and flattened troughs.

The deformation of the density marker indicates that the flow velocity profile in the wake changes from a stepped profile (in Figure 9a) to a profile of Poiseuille's type (in Figure 10b). At the same time, the retained sharpness of the envelopes maintains a generally rectangular cross-sectional shape of the density wake. The shear flow at the top of the wake expansion folds its high-gradient envelopes into pronounced flat vortices with a horizontal axis (Figure 10b).

Although the thickness of the ligaments, which confine the wave field, is much smaller than the length of waves, they limit the propagation of waves rather efficiently. The sharpness of the boundary indicates the effectiveness of the interaction of the different length-scale components of the flows. They are waves and ligaments consisting of a complete solution of the system of fundamental equations [8,27,71].

The spatial structure of the flow in this range of parameters is illustrated by patterns of electrolytic precipitation, shown in Figure 11. Plane convex separation lines on the surface of the sphere outline the flatness of the upper horizontal "lids" of the density wake. Two vertical convex separation lines, indicating the flatness of the lateral border of the density wake, are also distinct.

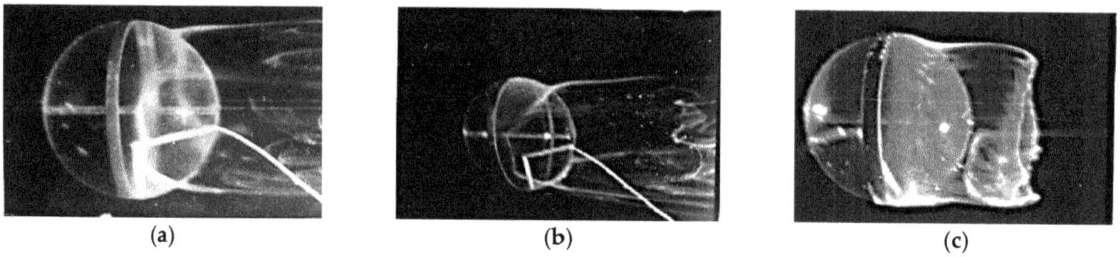

Figure 11. Electrolytic precipitation of a stratified flow past a uniformly moving sphere at small Froude numbers: (a) $T_b = 4.5$ s, $N = 1.4$ s^{-1}, $D = 5$ cm, $U = 0.7$ cm/s, Re $= 350$, Fr $= 0.01$; (b) $T_b = 5.0$ s, $N = 1.3$ s^{-1}, $D = 5$ cm, $U = 0.76$ cm/s, Re $= 380$, Fr $= 0.013$; (c) a homogeneous fluid, $N = 0$, $D = 5$ cm, $U = 0.70$ cm/s, Re $= 350$.

As the sphere velocity increases, the horizontal planes of the wake boundary approach each other, while the vertical ones diverge (Figure 10b). Here, the shape of the wake cross-section is close to a squared one. Four extended edges are formed by contacting flat upper horizontal and lateral envelopes of the wake. Their convex lines of intersection with the surface of the sphere and the straightness of the extended edges allow us to consider the shape of the density wake as a vertically compressed prism.

At the right boundary of the wake, the leading parts of the vertical "vortex columns", separated from the sidewalls of the density wake, are visible.

The electrolytic precipitation visualization of the spatial structure of the flow past a uniformly moving sphere in a homogeneous fluid at close values of the Reynolds number is shown for comparison in Figure 11c.

Here, the suspension carried by the flows from the transverse equatorial annular belt is uniformly distributed over the surface of the sphere and forms an annular separation line. In the vicinity of the separation line, the uniformity of the suspension is disturbed; it is redistributed in the form of more densely colored line segments and empty intervals between them. We have to study further whether the flow pattern indicates the formation of a "grooved" shape of the separation line of the cylindrical density wake, or whether such an impression is created by the play of light and shadows on the unevenly distributed density of the suspension, and thus the cylindrical shell itself is smooth. The circular contraction of the right edge of the suspension in the wake specifies the action of an axisymmetric toroidal vortex behind the sphere, which was repeatedly observed in the visualization of flows in experiments [12] and numerical simulations [44].

The prismatic wake in this flow regime is filled with pairs of "vertical vortex columns" (Figure 12a). Vortices are cut from below and above by the ligaments, which are the horizontal interfaces shown in the schlieren images (Figure 10). The flow in the wake has a complex three-dimensional character: individual, brightly colored fibers and surfaces with a complex shape and variable curvature, twisting into spirals, are distinguished in a distribution pattern in the visualized suspension. The vortex columns with flattened inner walls are separated by a faster narrow jet in the central part of the wake (see Figure 12b).

Successive frames of the flow pattern show that the entire vortex structure is stretched in a longitudinal direction and slowly oscillates in a transverse direction (Figure 13). We observe the proper structuralization of the suspension, which is initially uniformly distributed along the surface of the sphere. Here, dense threads are pronounced on weakly colored curvilinear surfaces.

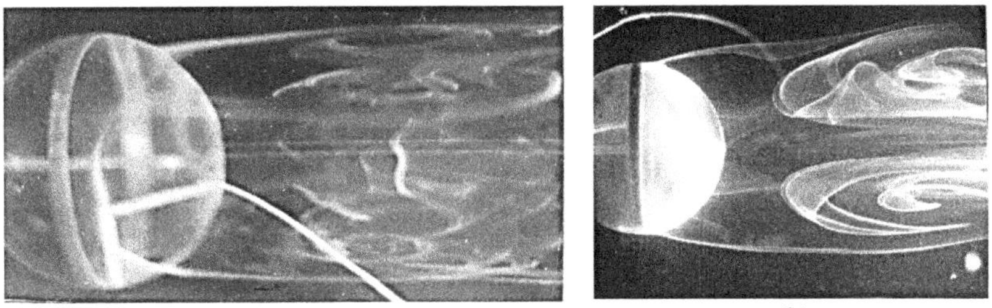

(**a**) (**b**)

Figure 12. Pattern of electrolytic precipitation suspension in "vertical vortex columns" in the wake past a sphere ($T_b = 4.5$ s; $N = 1.4\ \text{s}^{-1}$; $D = 5$ cm) : (**a**) side view $U = 0.7$ cm/s, Re = 350; Fr = 0.01; (**b**) top view $U = 0.62$ cm/s; Re = 310; Fr = 0.008.

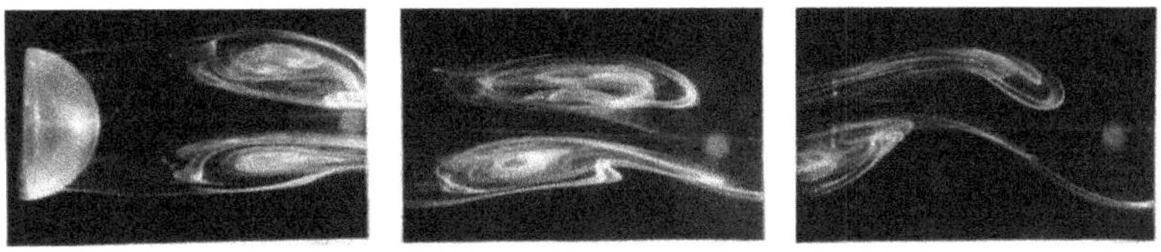

Figure 13. Evolution of an electrolytic precipitation suspension pattern in the wake past a sphere in CSF (top view): $T_b = 8.0$ s, $N = 0.785\ \text{s}^{-1}$, $D = 5$ cm, $U = 0.65$ cm/s, Re = 325; Fr = 0.027; time interval between frames $\Delta t = 1$ s.

The flow pattern loses regularity with a further increase in the flow velocity. The vertical separation line of the suspension from the surface of the sphere begins to elongate and approaches the separation point of the edges from above, forming the characteristic intersection of the dyed lines (Figure 14).

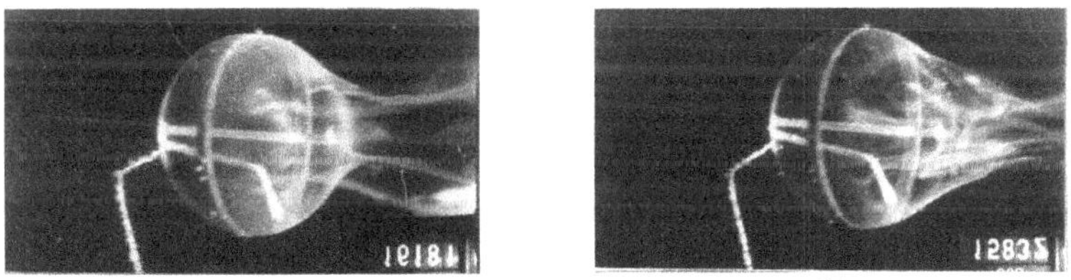

(**a**) (**b**)

Figure 14. The shape of the separation lines of the suspension from the sphere and the edges of the wake in the "expanding rectangle" stage in transient regime: (**a**,**b**) $T_b = 4.5$; 9.0 s, $N = 1.4$; $0.7\ \text{s}^{-1}$, $D = 5$ cm, U =1.08; 1.96 cm/s, Re = 540; 980 Fr = 0.02; 0.32.

The structure of the vortices in the wake is reconstructed with an increase in the sphere velocity and a decrease in the flow separation region height. In this case, a thin layer at the upper and lower envelopes of the wake expansion areas is firstly twisted into a spiral, the position of which is synchronized with the phase of the attached internal wave (crest in the upper hemi-space and trough in the lower one in Figures 10b and 15a).

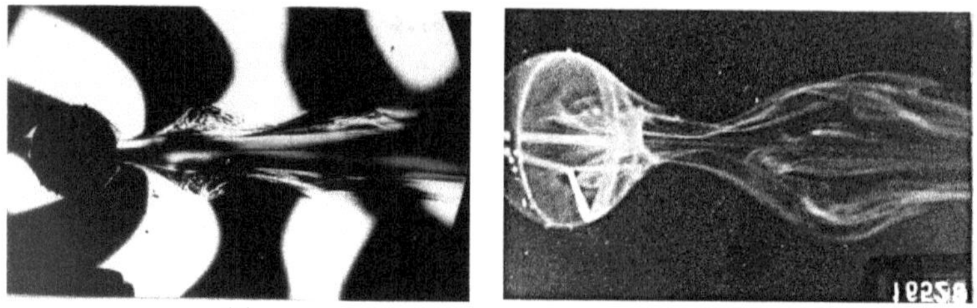

(a) (b)

Figure 15. Formation of a "vortex bubble" frozen in the field of attached internal waves by a pair of vortices with a horizontal axis: (**a**) conventional schlieren image with Foucault knife: $T_b = 3.0$ s; $N = 2.09$ s^{-1}; $D = 4$ cm; $U = 1.37$ cm/s, Re = 750, Fr $= U^2/N^2D^2 = 0.03$; (**b**) electrolytic precipitation $T_b = 9$ s, $N = 0.7$ s^{-1}; $D = 5$ cm; $U = 1.38$ cm/s, Re = 690, Fr = 0.15.

As the sphere velocity and Froude number increase, the vortex with a horizontal axis thickens and captures an increasing part of the "vortex bubble" (Figure 15a). Pronounced edges and individual lines emphasize the flatness of the wake boundaries. With distance, the flow degenerates with the formation of elongated horizontal interfaces and fibers throughout the volume of the wake. The inner structure and outer contour of the wake are visualized by the suspension of electrical precipitation in Figure 15b.

The black-and-white replica of the color schlieren image (Figure 16) shows that after the collapse of the "vortex bubble", the surface of the wake remains wavy [72]. The flow degenerates with the formation of elongated horizontal interfaces and filaments throughout the volume of the wake.

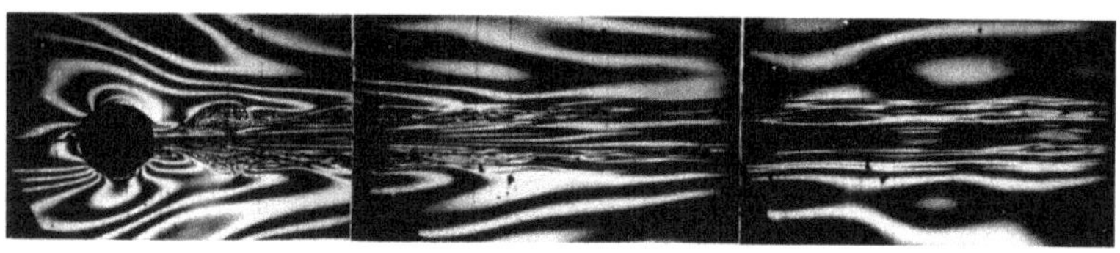

Figure 16. Black and white replica of the color schlieren image of the flow in the wake of the sphere: $T_b = 3.0$ s, $N = 2.09$ s^{-1}, $D = 4$ cm, $U = 1.45$ cm/s, Re = 780, Fr = 0.03.

The shapes of waves and vortices in the colored schlieren images differ significantly from the traditional schlieren patterns with the Foucault knife (Figure 15). It allows estimating the position of the high-gradient interfaces and the general deformation of the density profile.

8. Cylindrical Shape of the Wake Past a Sphere at High Froude Number

A further increase in the flow velocity is accompanied by a qualitative change in the flow structure. The inertial and nonlinear effects gradually become more pronounced and dominate over the effects of stratification, and they even enhance with the formation of high-gradient interfaces. The large length of the internal wave, which exceeds the diameter of the sphere in Figure 17 at almost an order of magnitude, indicates a violation of the spatial synchronism condition, which is necessary for the effective generation of waves. The attached wave amplitude decreases rapidly with a further-increasing Froude number. The weakening of the attached wave does not ensure the fulfillment of the condition of their spatial synchronization with the wake boundary geometry, which is defined by the positons of the vortices.

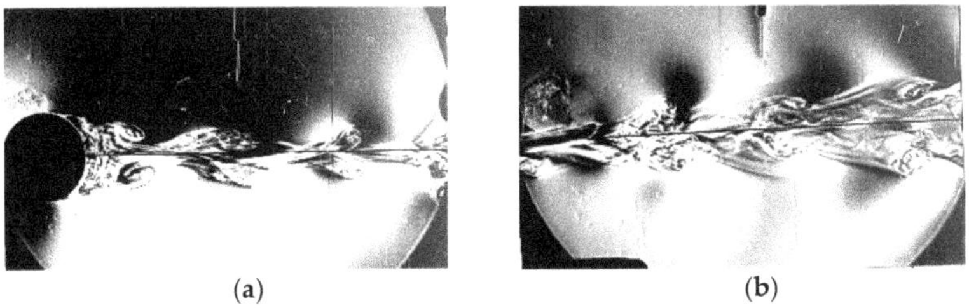

Figure 17. Schlieren images of periodically separating axisymmetric vortices from the flow at the rear part of the sphere in a weakly CSF: $T_b = 12$ s, $N = 0.52$ s^{-1}, $D = 4$ cm, $U = 2.1$ cm/s, Re = 840, Fr = 1.02; (**a**,**b**) $t = 0$; 4.5 s.

The length of the region of increased vorticity in the rear part of the sphere increases. Having reached a critical size, the bottom vortex splits up and its outer part is thrown into the wake in the form of a toroidal ring with a conical front part (Figure 17). Here, the position of the inner vortex structures is determined predominantly by dynamic conditions. Moving separated vortices covered by sharp envelopes emit internal waves similar to the waves of a solid body. Bands, visualized in black-and-white images, of the internal waves around the wake are tilted in the direction of the body motion (the geometry and dynamics of a free laminar vortex ring motion in a CSF were studied in detail in [73]).

As the density gradient increases, so does the frequency of vortex shedding, the intensity of short internal waves generated by the separate vortices, and the degree of manifestation of fine structures. The resulting ligaments, which correspond to the fibers and interfaces in the schlieren image (Figure 18), extend in the direction of body motion.

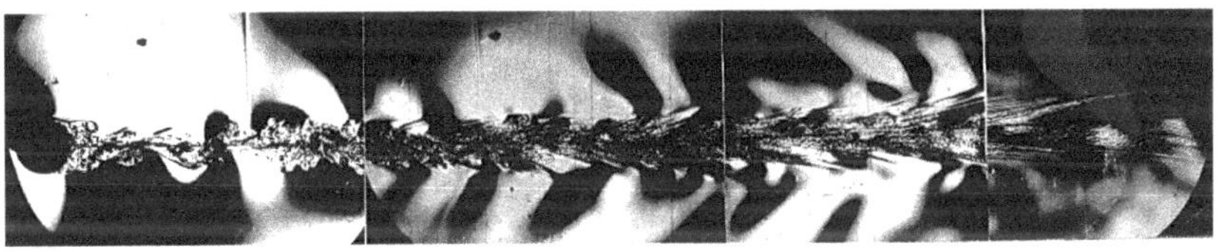

Figure 18. Coalescence of separated vortices into density wake with irregular envelope: $T_b = 3.5$ s, $N = 1.8$ s^{-1}, $D = 3$ cm, $U = 6.8$ cm/s, Re = 2040, Fr = 1.6.

The difference between the attached internal waves generated by the body itself [68], with the length proportional to its velocity $\lambda = UT_b$, and the short non-stationary waves emitted by vortices in the wake, are clearly expressed in the flow pattern in Figures 17 and 18

With a further increase in velocity, the structure of the wake boundary changes and the vortices are merged into a single wake with a continuous envelope (Figure 19). However, the existence and the periodicity of the location of radiating vortices in the wake indicate the regularity of the pattern of short internal waves.

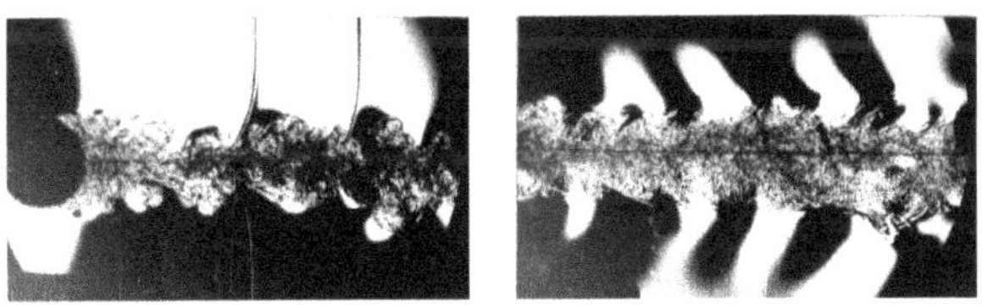

Figure 19. Irregular axisymmetric vortex wake past a sphere: $T_b = 3$ s, $N = 2.09$ s^{-1}, $D = 4$ cm, $U = 8.7$ cm/s, Re = 3488, Fr = 1.1, $\Delta t = 1.5$ s.

An oscillating wake is observed behind a uniformly moving sphere along with an axisymmetric pulsating wake. In this regime, vortices periodically separate from a flow directly past a sphere and propagate along a sloping trajectory at the separation region (Figure 19). Perfect axial symmetry of the wake is lost. When the excess lateral momentum is carried away by the radiated internal waves, the detached vortex rings return to the center of the flow. In this case, as well as in previously considered flow patterns, the ligaments, which are high-gradient envelopes, form a boundary separating the vortices from the radiated waves. Here, the mechanical action of the ligaments is equivalent to the surface of a solid body impact.

Consideration of the data presented in Figures 7–20 shows that two types of symmetries are distinguished in the wake flow past a uniformly moving sphere. A rectangular wake with a discrete symmetry is observed at small Froude numbers (Fr $<<$ 1). A round wake is formed at large Froude numbers (Fr $>$ 1).

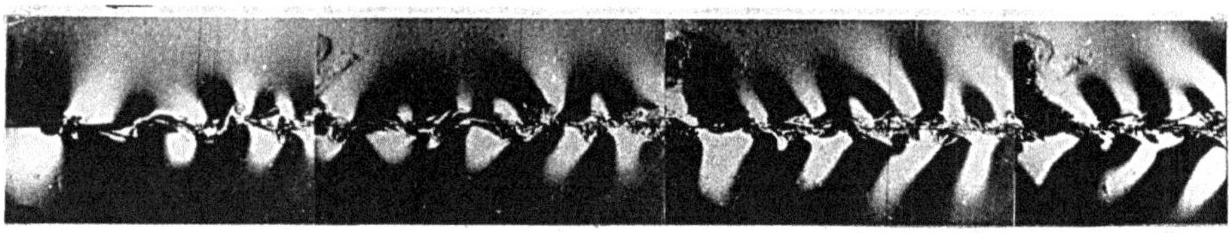

Figure 20. Oscillating vortex wake behind a sphere of a small diameter in CSF: $T_b = 3.5$ s, $N = 1.8\,s^{-1}$, $D = 1$ cm, $U = 7.4$ cm/s, Re = 742, Fr = 17.

The scheme of the flow past a sphere at intermediate Froude and Reynolds numbers is close to a rectangular form. A transverse cross-sectional view of the density wake is shown in Figure 21. The position of the long side depends on the value of the Froude number (or velocity for a constant body diameter).

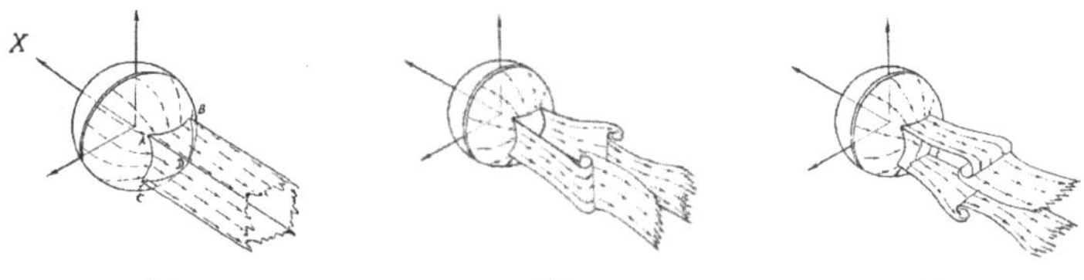

Figure 21. Scheme of the vortex flow past a sphere, (**a**) rectangular density wake, (**b**) wake with submerged vertical vortex columns, (**c**) wake with horizontal rotors.

At the smallest values of the sphere velocity, when buoyancy effects dominate, the rectangle is elongated vertically. If the generation of vorticity in a shear flow at the vertical boundaries of the wake is supplemented by the action of the baroclinic mechanism due to the multidirectional pressure and density gradient vectors [13], it leads to the formation of vortex pairs.

In this flow, the particles move mainly along isopycnal trajectories. The axes of the downstream symmetric "vortex columns" are oriented vertically.

Flat inner boundaries of the vortices in Figures 12b and 13 visualize the central back jet in the wake. With increasing velocity, inertial effects pull the upper and lower separation lines to the center and push the side faces apart (a wake scheme with a square section is shown in Figure 2a). Gradually, slides of the vertical envelopes are separated and twisted in vertical vortex columns, as shown in Figure 2b.

Increased stratification on the horizontal wake "lids" hinders the formation of large eddies with a horizontal axis that moves fluid to the horizons of neutral buoyancy. Deviations in the position of the boundary from the horizontal axis practically stop the removal of the vorticity from the wake. The vorticity turns out to be "frozen" in the wake, as in case of the particles of the suspension being transferred.

A gradual increase in velocity and Froude number leads to further contraction of the separation region height and "twisting" into a thin vortex curl of high-gradient interfaces at the horizontal boundaries of the wake. Vorticity in horizontal rotors is higher than in the "vertical vortex columns". Large vertical vortices are destroyed and replaced by intensive horizontal rotors (Figure 21c). A more intense vortex motion leads to the destruction of corners and edges of the wake and the formation of a toroidal bottom vortex behind the sphere, as in a homogeneous liquid.

The dependence of the angular position of the horizontal separation lines on the surface of a sphere with a diameter of 4 cm in a weakly CSF with a buoyancy period of = 12 s on the Froude number is shown in Figure 22 (the angle is measured from the horizontal position). Measurements of the separation angles of the wake envelopes in the central vertical plane were made according to three types of schlieren visualization: traditional (vertical slot—Foucault knife), slot-thread set at an angle of 45°, and color schlieren method. On the decreasing and increasing parts of the curve, the data of the independent measurements agree with each other.

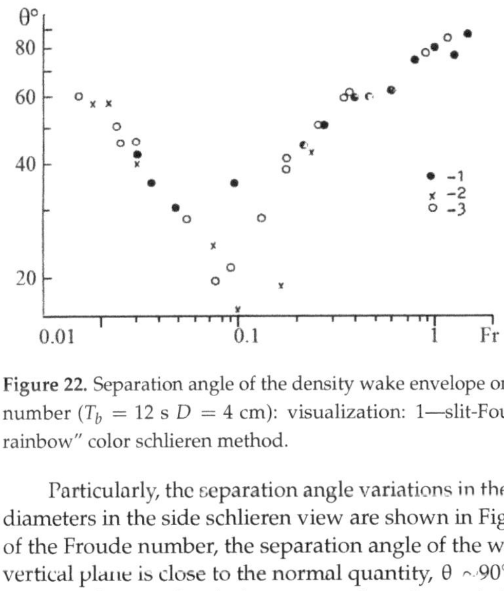

Figure 22. Separation angle of the density wake envelope on a sphere in a weakly CSF versus Froude number (T_b = 12 s D = 4 cm): visualization: 1—slit-Foucault knife; 2—slit-thread; 3—"natural rainbow" color schlieren method.

Particularly, the separation angle variations in the density wake on spheres of different diameters in the side schlieren view are shown in Figures 23 and 24. At the lowest values of the Froude number, the separation angle of the wake envelope from the sphere in the vertical plane is close to the normal quantity, $\theta \sim 90°$, and in the horizontal plane it does not exceed $\varphi < 15°$, which corresponds to the vertical rectangle for the cross-section of the wake. As the body velocity increases, the separation angle in the vertical plane decreases, and in the horizontal plane it increases, and exceeds the vertical one, $\theta \sim 45°$, and $\varphi = 67°$ at Fr = 0.03. The shape of the wake cross-section is a rectangle, elongated horizontally.

The experimental data in these figures are approximated by power-law functions of the form $\theta = A_R(\mathrm{Re})^r$ and $\theta = A_F(\mathrm{Fr})^m$. Exponents and coefficients in these formulas depend on other parameters of the problem.

In the interval of the Froude number, when the height of the separation region decreases with the velocity increases, their values are distinguished and equal: $r = -1.25$, $m = -0.66$ (T_b = 12c, D = 4 cm); $r = -1.7$, $m = -077$ (T_b = 3 c, D = 2 cm); $r = -0.48$, $m = -0.35$ (T_b = 3 c, D = 4 cm).

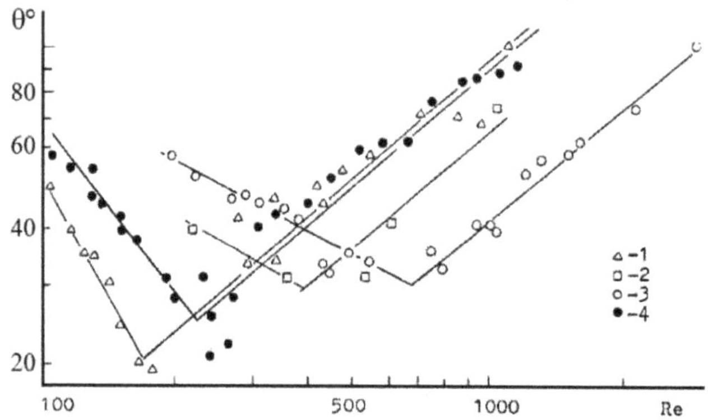

Figure 23. Separation angle of the density wake envelope on a sphere versus Reynolds number: curves (1–3)—$T_b = 3$ s, $D = 2, 3, 4$ cm; (4)—$T_b = 12$ s, $D = 4$ cm.

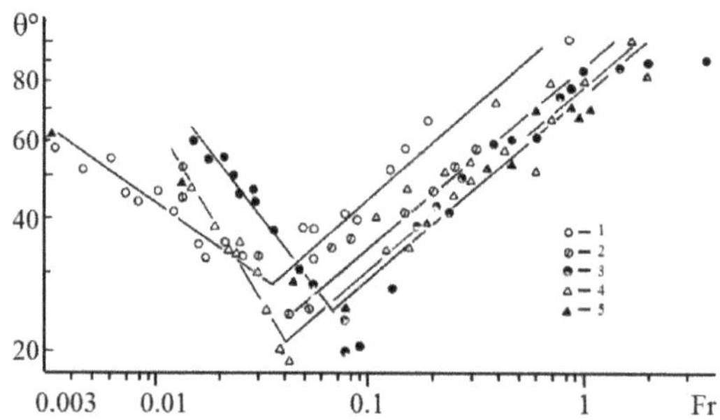

Figure 24. Separation angle of the density wake envelope on a sphere versus Froude number: curves 1 $T_b = 3.5$ s, $D = 4$ cm; 2—$T_b = 4.5$ s, $D = 4$ cm; 3—$T_b = 12.0$ s, $D = 4$ cm; 4—$T_b = 3.0$ s, $D = 2$ cm; 5—$T_b = 4.1$ s, $D = 2$ cm.

If the Froude number exceeds the critical value of the minimum wake height, the exponents is $m = 0.5$ in the approximation of the separation angle on the Reynolds and Froude numbers in all experiments performed at various values of the buoyancy period and sphere diameter.

At the lowest values of the Froude number, the separation angle of the wake envelope from the sphere in the vertical plane is close to normal, $\theta \sim 90°$, and in the horizontal plane, it does not exceed $\varphi < 15°$, which corresponds to the vertical rectangle for the cross-section of the wake. As the body velocity increases, the separation angle in the vertical plane decreases; in the horizontal plane it increases, and exceeds the value in vertical plane $\theta \sim 45°$ and $\varphi = 67°$ at $\mathrm{Fr} = 0.03$. The shape of the wake cross-section is a rectangle, which is elongated horizontally.

The given values of the separation angle of the wake from a sphere for different values of the Froude and Reynolds numbers are consistent with the results [40] obtained for a sphere with a diameter $D = 6.36$ cm/s.

The clarity of the shape is ensured by an increased concentration of the suspension on the edges of the wake and the small thickness of its boundaries (horizontally and vertically oriented high-gradient interfaces). The dependences of the boundary parameters and fine geometry of the flow on the properties of the medium and the conditions of motion of the body require a more detailed study.

At high body velocities, when the Froude number becomes greater than one, a cylindrical wake, filled with small-scale disturbances, is formed. The wake is separated from the annular toroidal vortex in the rear part of the sphere (Figure 19).

Having reached critical values, the vortex, contoured by high-gradient envelopes, breaks off and drifts in the wake. It remains connected to the flow past a sphere by a conical set of ligaments. Moving vortices radiate transient internal waves. Their rays, which represent phase surfaces, are oriented to the body motion. The vortices are gradually slowed down, deformed, and pulled out by the unsteady flow in the wake. The spiral interfaces of the vortex are elongated; they form line structures that connect the remains of the vortex rings into a single system.

9. Results and Discussion

The results of the first regular studies of the stratified fluid flows around 2D and 3D obstacles focused on visualizing the spatial structure of dispersive internal waves. They have been summarized in monographs [30,74], which maintained their cognitive value. An extensive series of experimental studies of the flow around a sphere uniformly moving in a pool filled with a stratified fluid were independently carried out in several countries: the USSR [34,68–72], the USA [75–78], France [79–82], and Great Britain [33,38]. In the experiments, flow patterns were visualized using high-resolution classical [3,4] and modified schlieren methods [25], including the density gradient registration instruments, modified for background continuous density distribution [34]; direct shadowgraph observations [75]; dyeing by common and fluorescent impurities; particle tracing (particle streak photograph) [76–78]; and electrolytic precipitation [69,70].

The experiments have shown that at low values of the Froude number, when the buoyancy effects dominate, the density wake has a prismatic shape. It is elongated vertically at first, then its cross-section becomes square [69], and, finally, it becomes flattened under the action of inertial effects at a range of Froude numbers, $0.03 < \mathrm{Fr} < 0.09$. The observed discrete symmetry of the wake is consistently reproduced in independent experiments.

The flat boundaries of such a density wake are formed by ligaments, i.e., high-gradient interfaces that arise in the flow in the vicinity of the sphere surface. The interfaces are located inside a thicker shear layer of velocity. The height of the shear layer in a liquid with a large Schmidt number ($\mathrm{Sc} = 700$ for an aqueous solution of sodium chloride, which is used in most experiments) is at least an order of magnitude greater than the ligaments thickness. Similar ratios of heights were observed in an unsteady DiF on the impermeable obstacle. The effects of amplification of the gradients at the wake boundaries have been registered in [34].

Classical schlieren methods have the highest sensitivity and spatial resolution among those listed above. These methods enable us to record disturbance patterns in the entire range of flow parameters (from the slowest ones induced by diffusion on topography to fast vortex flows) and to resolve all structural components in the studied flows. The low sensitivity of the direct shadow and moiré methods does not let us visualize weak internal waves and fine high-gradient interfaces, which was noted in [75]. The discrete nature of a number of flow visualization methods, in particular, using tracing particles for velocity measurement and the flow visualization (particle streak photograph) [78–80], does not allow us to register thin interfaces and assess their effect on the flow structure.

As the Froude number increases, the sidewalls of a rectangular wake begin to split, and form two and more vertical vortex columns [72] (lee-side eddies in the terminology of [75]) in the downstream wake.

The joint action of the velocity shear and pressure perturbations in the field of intense attached internal waves leads to the formation of "vortex bubbles" with a horizontal axis of symmetry. The sets of complete solutions of the system of fundamental Equation (1) contain functions that describe both periodic components—waves and vortices—as well as a large number of ligaments that characterize the thin high-gradient interfaces and fibers [8,27].

As vorticity accumulates in the rear part of the flow past the body, the wake faces become rounded and take a cylindrical shape. All the structural components, including ligaments, take part in nonlinear interactions with each other, with waves and mean flow. It provides a variety of available scenarios for the evolution of stratified flows.

The classification of the sequence of the vortex structures in the transitional flow regime was first presented in [75]. The pattern of the vortex structure near and far from the obstacle has been described in great detail in subsequent experiments [81,82]. The evolution of the fine structure of the density profiles downstream in the wake, revealed by a high-resolution electrical conductivity sensor, enabled studies of the dynamics of submerged vortices [83]. In a number of experiments, the pattern of internal waves [84] and the global structure of the vortex and turbulent wake were elucidated [85,86].

Analytical [87], numerical [88–91], and experimental studies [41,42] of the internal wave field and the generally round turbulent wake behind a rapidly moving sphere in a CSF are being actively pursued using various approaches.

10. Conclusions

Fluid stratification, even if it is weak, significantly affects the structure, symmetry, and dynamics of the flow past a uniformly towed sphere. At low values of the Froude number, ligaments that are high-gradient interfaces enhance the effect of stratification. Horizontal and vertical envelopes bound a narrow prismatic density wake past a slowly moving body.

With an increase in sphere velocity, the shape of the transverse cross-section of the wake transforms from a vertically elongated rectangle into a square and then into a narrow horizontally elongated rectangle. In a high wake, paired vertical vortex columns are formed. They are enclosed between the horizontal interfaces and form envelopes of the density wake.

Vortices with a horizontal axis, which are immersed in a horizontally stretched wake, lead to a periodic increase in its vertical size. The position of the wake expansion and compression regions is consistent with the phase pattern of the attached internal waves.

At large values of the Froude number, the accumulation of vorticity, accompanied by the formation of a toroidal vortex in the rear part of the sphere, causes a rearrangement of the discrete symmetry of the density wake into a continuous axisymmetric one. In the phase of flow restructuring, various forms of vortex components, contoured by thin high-gradient envelopes, are observed in the wake.

Funding: This research received no external funding.

Institutional Review Board Statement: Not applicable.

Informed Consent Statement: Not applicable.

Acknowledgments: The work was carried out as a state assignment (state registration No. AAAA-A20-120011690131-7). I thank E.Y. Sysoeva for assistance in conducting experiments.

Conflicts of Interest: The author declares no conflict of interest. The funders had no role in the design of the study; in the collection, analyses, or interpretation of data; in the writing of the manuscript, or in the decision to publish the results.

References

1. Popov, N.I.; Fedorov, K.N.; Orlov, V.M. *Sea Water*; Nauka: Moscow, Russia, 1979; 330p. (In Russian)
2. Franklin, B. Behavior of oil on water. Letter to J. Pringle. In *Experiments and Observations on Electricity*; R. Cole: London, UK, 1769; pp. 142–144.
3. Vasiliev, L.A. *Shadow Methods*; Nauka: Moscow, Russia, 1968; 400p. (In Russian)
4. Settles, G.S. *Schlieren and Shadowgraph Techniques: Visualizing Phenomena in Transparent Media*; Springer Science & Business Media: Berlin, Germany, 2001; 376p.
5. Chashechkin, Y.D.; Prokhorov, V.E. Visualization and echo sounding of stratified fluid disturbances in front of and behind a vertical plate. *Dokl. Phys.* **2020**, *65*, 178–182. [CrossRef]
6. Pao, H.-P.; Lai, R.Y.; Schemm, C.E. Vortex trails in stratified fluids. *J. Hopkins APL Techn. Digest* **1982**, *3*, 12–18.
7. Strutt, J.W.; Rayleigh, L. Investigation of the character of the equilibrium of an incompressible heavy fluid of variable density. *Proc. Lond. Math. Soc.* **1883**, *4*, 170–177.
8. Chashechkin, Y.D. Foundations of engineering mathematics applied for fluid flows. *Axioms* **2021**, *10*, 286. [CrossRef]
9. Prandtl, L. *The Essentials of Fluid Dynamics*; Blackie & Son Ltd.: London, UK, 1953; 452p.
10. Landau, L.D.; Lifshitz, E.M. *Fluid Mechanics. V.6. Course of Theoretical Physics*; Pergamon Press: Oxford, UK, 1987; 560p.

11. Mossa, M. The recent 500th anniversary of Leonardo da Vinci's death: A reminder of his contribution in the field of fluid mechanics. *Environ. Fluid Mech.* **2021**, *21*, 1–10. [CrossRef]
12. Van Dyke, M. *An Album of Fluid Motion*; Parabolic Press, Inc.: Stanford, CA, USA, 1982; 176p.
13. Kochin, N.E.; Kibel, I.A.; Rose, N.V. *Theoretical Hydromechanics. Part 1*; OGIZ. Gostekhizdat: Leningrad-Moscow, USSR, 1948; 728p. (In Russian)
14. Birkhoff, G. *Hydrodynamics. A Study in Logic, Fact and Similitude*, 2nd ed.; Princeton University Press: Princeton, NJ, USA, 1960; 184p.
15. Bernoulli, D. *Hydrodynamica, Sive de Viribus et Motibus Fluidorum Commentarii, Opus Academicum*; Sumptibus, J.R., Ed.; Argentorati: Dulseckeri, The Netherlands, 1738; 304p.
16. Robins, B. *New Principles of Gunnery*; J. Nourse: London, UK, 1742; 160p.
17. Euler, L. *Neue Grundsätze der Artillerie. Aus dem Englischen des Herrn Benjamin Robins Übersetzt und mit Vielen Anmerkungen Versehen*; Haude Königl. und der Academie der Wissenschaften: Berlin, Germany, 1753; 480p.
18. D'Alembert, J.-L.R.; la Marquis de Condorcet, J.M.A.; l'abbe Bossut, C. *Nouvelles Expériences sur la Résistance des Fluids*; C.-A. Jombert: Paris, France, 1777; 232p.
19. D'Alembert, J.-L.R. *Réflexions sur la Cause Générale des Vents*; David: Paris, France, 1747; 372p.
20. Euler, L. Principes généraux du mouvement des fluids. *Mémoires L'académie Des. Sci. Berl.* **1757**, *11*, 274–315.
21. Fourier, J.B.J. *Théorie Analytique de la Chaleur*; Firmin Didot Père et Fils: Paris, France, 1822; 639p.
22. Navier, C.-L.-M.-H. Mémoire sur les Lois du Mouvement des Fluids. *Mém. l'Acad. Sci.* **1822**, *6*, 389–417.
23. Stokes, G.G. On the theories of the internal friction of fluids in motion, and of the equilibrium and motion of elastic bodies. *Trans. Camb. Philos. Soc.* **1845**, *8*, 287–305.
24. Stokes, G.G. On the Effect of the Internal Friction of Fluids on the Motion of Pendulums. *Trans. Camb. Philos. Soc.* **1850**, *9*, 1–141.
25. Chashechkin, Y.D. Schlieren visualization of stratified flow around a cylinder. *J. Visualiz.* **1999**, *1*, 345–354. [CrossRef]
26. Müller, P. *The Equations of Oceanic Motions*; CUP: Cambridge, UK, 2006; 302p.
27. Chashechkin, Y.D.; Zagumennyi, I.V. 2D hydrodynamics of a plate: From creeping flow to transient vortex regimes. *Fluids* **2021**, *6*, 310. [CrossRef]
28. Monin, A.C.; Yaglom, A.M. *Statistical Fluid Mechanics–Vol. 1: Mechanics of Turbulence*, 1st ed.; The MIT Press: Boston, MA, USA, 1971; 782p.
29. Chashechkin, Y.D.; Kistovich, A.V. Classification of three-dimensional periodic fluid flows. *Dokl. Phys.* **2004**, *49*, 183–186. [CrossRef]
30. Lighthill, J. *Waves in Fluids*; CUP: Cambridge, UK, 2001; 524p.
31. Makarov, S.A.; Chashechkin, Y.D. Apparent internal waves in a fluid with exponential density distribution. *J. Appl. Mech. Tech. Phys.* **1981**, *22*, 772–779. [CrossRef]
32. Smirnov, S.A.; Chashechkin, Y.D. Internal lee(joined) waves at an arbitrary orientation of the incident flow. *Izv. Atmos. Ocean. Phys.* **1998**, *43*, 475–482.
33. Scase, M.; Dalziel, S. Internal wave fields generated by a translating body in a stratified fluid: An experimental comparison. *J. Fluid Mech.* **2006**, *564*, 305–331. [CrossRef]
34. Chashechkin, Y.D.; Gumennik, E.V.; Sysoeva, E.Y. Transformation of a density field by a three-dimensional body moving in a continuously stratified fluid. *J. Appl. Mech. Tech. Phys.* **1995**, *36*, 19–29. [CrossRef]
35. Phillips, O.M. On flows induced by diffusion in a stably stratified fluid. *Deep-Sea Res.* **1970**, *17*, 435–443. [CrossRef]
36. Chashechkin, Y.D. Hydrodynamics of a sphere in a stratified fluid. *Fluid Dyn.* **1989**, *24*, 1–7. [CrossRef]
37. Greenslade, M. Drag on a sphere moving horizontally in a stratified fluid. *J. Fluid Mech.* **2000**, *418*, 339–350. [CrossRef]
38. Scase, M.; Dalziel, S. Internal wave fields and drag generated by a translating body in a stratified fluid. *J Fluid Mech.* **2004**, *498*, 289–313. [CrossRef]
39. Mason, P.J. Forces on spheres moving horizontally in a rotating stratified fluid. *Geoph. Astrophys. Fluid Dyn.* **1977**, *8*, 137–151. [CrossRef]
40. Lofquist, K.; Purtell, L. Drag on a sphere moving horizontally through a stratified liquid. *J. Fluid Mech.* **1984**, *148*, 271–284. [CrossRef]
41. Cocetta, F.; Gillard, M.; Szmelter, J.; Smolarkiewicz, P.K. Stratified flow past a sphere at moderate Reynolds numbers. *Comp. Fluids* **2021**, *226*, 104998. [CrossRef]
42. Madison, T.; Xiang, X.; Spedding, G. Laboratory and numerical experiments on the near wake of a sphere in a stably stratified ambient. *J. Fluid Mech.* **2022**, *933*, A12. [CrossRef]
43. Shirayama, S.; Kuwahara, K. Patterns of Three-Dimensional Boundary Layer Separation. In Proceedings of the AIAA 25th Aerospace Sciences Meeting, Reno, NV, USA, 12–15 January 1987. Preprint AIAA-87-0461, 10p.
44. Gushchin, V.A.; Matyushin, P.V. Numerical simulation of separated flow past a sphere. *Comp. Math. Math. Phys.* **1997**, *37*, 1086–1100.
45. Li, J.; Zhou, B. The symmetry and stability of the flow separation around a sphere at low and moderate reynolds numbers. *Symmetry* **2021**, *13*, 2286. [CrossRef]
46. Nakamura, I. Steady wake behind a sphere. *Phys. Fluids* **1976**, *19*, 5–8. [CrossRef]

47. Karyagin, V.P.; Lopatkin, A.I.; Shvets, A.I.; Shilin, N.M. Experimental investigation of the separation of flow around a sphere. *Fluid Dyn.* **1991**, *26*, 126–129. [CrossRef]
48. Gerasimov, S.I.; Faikov, Y.I. *Shadow Photography in a Divergent Beam of Light*; FSUE RFSC-VNIIEF: Sarov, Russia, 2010; 344p. (In Russian)
49. Fortuin, J. Theory and application of two supplementary methods of constructing density gradient columns. *J. Polym. Sci.* **1960**, *44*, 505–515. [CrossRef]
50. Oster, G. Density gradients. *Sci. Am.* **1965**, *217*, 70–75. [CrossRef]
51. Mowbray, D. The use of schlieren and shadowgraph techniques in the study of flow patterns in density stratified liquids. *J. Fluid Mech.* **1967**, *27*, 595–608. [CrossRef]
52. Chashechkin, Y.D. Colour Schlieren method. In *Optical Methods in Dynamics of Fluids and Solids*; IUTAM Symposium; Springer-Verlag: Berlin, Germany, 1985; pp. 275–282.
53. Sakai, S. Visualization of internal gravity waves by Moire-method. *Kashika-Joho* **1990**, *10*, 65–68.
54. Dalziel, S.B.; Hughes, G.O.; Sutherland, B.R. Whole-field density measurements by "synthetic schlieren". *Exp. Fluid.* **2000**, *28*, 322–335. [CrossRef]
55. Honji, H. Near wakes of a sphere in a stratified flows. *Fluid Dyn. Res.* **1987**, *2*, 75–76. [CrossRef]
56. Honji, H. Vortex Motions in the Stratified Wake Flows. *Fluid Dyn. Res.* **1988**, *3*, 425–430. [CrossRef]
57. Hydrophysical Complex for Modeling Hydrodynamic Processes in the Environment and Their Impact on Underwater Technical Objects, as Well as the Distribution of Impurities in the Ocean and Atmosphere, Institute for Problems in Mechanics RAS. Available online: http://ipmnet.ru/uniqequip/gfk (accessed on 15 June 2022).
58. Smirnov, S.A.; Chashechkin, Y.D.; Il'inykh, Y.S. High-accuracy method for measuring profiles of buoyancy periods. *Meas. Techn.* **1998**, *41*, 514–519. [CrossRef]
59. Il'inykh, A.Y.; Il'inykh, Y.S.; Chashechkin, Y.D. Measuring transducers of electrical conductivity of liquids. *Meas. Techn.* **2016**, *58*, 1336–1340. [CrossRef]
60. Linden, P.F.; Weber, J.E. The formation of layers in a double-diffusive system with a sloping boundary. *J. Fluid Mech.* **1977**, *81*, 757–773. [CrossRef]
61. Kistovich, A.V.; Chashechkin, Y.D. The structure of transient boundary flow along an inclined plane in a continuously stratified medium. *J. Appl. Maths. Mech.* **1993**, *57*, 633–639. [CrossRef]
62. Zagumennyi, I.V.; Chashechkin, Y.D. Diffusion induced flow on a strip: Theoretical, numerical and laboratory modeling. *Procedia IUTAM* **2013**, *8*, 256–266. [CrossRef]
63. Baidulov, V.G.; Matyushin, P.V.; Chashechkin, Y.D. Evolution of the diffusion-induced flow over a sphere submerged in a continuously stratified fluid. *Fluid Dyn.* **2007**, *42*, 255–267. [CrossRef]
64. Kistovich, A.V.; Chashechkin, Y.D. Dissipative-gravity waves in subcritical regimes of multicomponent convection. *Izv. Atmosph. Ocean. Phys.* **2001**, *37*, 476–481.
65. Allshouse, M.; Barad, M.; Peacock, T. Propulsion generated by diffusion-driven flow. *Nat. Phys.* **2010**, *6*, 516–519. [CrossRef]
66. Dimitrieva, N.F.; Chashechkin, Y.D. Fine structure of stratified flow around a fixed and slow moving wedge. *Oceanology* **2018**, *58*, 340–349. [CrossRef]
67. Levitsky, V.V.; Dimitrieva, N.F.; Chashechkin, Y.D. Visualization of the self-motion of a free wedge of neutral buoyancy in a tank filled with a continuously stratified fluid and calculation of perturbations of the fields of physical quantities putting the body in motion. *Fluid Dyn.* **2019**, *54*, 948–957. [CrossRef]
68. Makarov, S.A.; Chashechkin, Y.D. Coupled internal waves in a viscous incompressible fluid. *Izv. Atmos. Ocean. Phys.* **1982**, *18*, 758–764.
69. Sysoeva, E.Y.; Chashechkin, Y.D. Spatial structure of a wake behind a sphere in a stratified liquid. *J. Appl. Mech. Tech. Phys.* **1988**, *29*, 655–660. [CrossRef]
70. Sysoeva, E.Y.; Chashechkin, Y.D. Vortex structure of a wake behind a sphere in a stratified fluid. *J. Appl. Mech. Tech. Phys.* **1986**, *27*, 190–196. [CrossRef]
71. Chashechkin, Y.D. Conventional partial and new complete solutions of the fundamental equations of fluid mechanics in the problem of periodic internal waves with accompanying ligaments generation. *Mathematics* **2021**, *9*, 586. [CrossRef]
72. Sysoeva, E.Y.; Chashechkin, Y.D. Vortex systems in the stratified wake of a sphere. *Fluid Dyn.* **1991**, *26*, 544–551. [CrossRef]
73. Belyaev, V.S.; Savinkov, A.M.; Chashechkin, Y.D. Dynamics of laminar vortex rings in a stratified liquid. *J. Appl. Mech. Tech. Phys.* **1987**, *28*, 34–43. [CrossRef]
74. Turner, J.S. *Buoyancy Effects in Fluids*; CUP: Cambridge, UK, 1980; 412p.
75. Lin, Q.; Lindberg, W.R.; Boyer, D.L.; Fernando, H.J.S. Stratified flow past a sphere. *J. Fluid Mech.* **1992**, *240*, 315–354. [CrossRef]
76. Lin, Q.; Boyer, D.L.; Fernando, H.J.S. Turbulent wakes of linearly stratified flow past a sphere. *Phys. Fluids* **1992**, *4*, 1687–1696. [CrossRef]
77. Lin, Q.; Boyer, D.L.; Fernando, H.J.S. Internal waves generated by the turbulent wake of a sphere. *Exps. Fluids* **1993**, *15*, 147–154. [CrossRef]
78. Lin, Q.; Boyer, D.L.; Fernando, H.J.S. The vortex shedding of a stream-wise-oscillating sphere translating through a linearly stratified fluid. *Phys. Fluids* **1994**, *6*, 239–252. [CrossRef]

79. Hopfinger, E.J.; Flor, J.B.; Chomaz, J.M.; Bonneton, P. Internal waves generated by a moving sphere and its wake in a stratified fluid. *Exps. Fluids* **1991**, *11*, 255–261. [CrossRef]
80. Chomaz, J.-M.; Bonneton, P.; Butet, A.; Perrier, M.; Hopfinger, E.J. Froude number dependence of the flow separation line on a sphere towed in a stratified fluid. *Phys. Fluids A Fluid Dyn.* **1992**, *4*, 254–258. [CrossRef]
81. Chomaz, J.-M.; Bonneton, P.; Hopfinger, E.J. The structure of the near wake of a sphere moving horizontally in a stratified fluid. *J. Fluid Mech.* **1993**, *254*, 1–21. [CrossRef]
82. Bonnier, M.; Bonneton, P.; Eiff, O. Far-wake of a sphere in a stably stratified fluid: Characterization of the vortex structures. *Appl. Sci. Res.* **1998**, *59*, 269–281. [CrossRef]
83. Bonnier, M.; Eiff, O.; Bonnetone, P. On the density structure of far-wake vortices in a stratified fluid. *Dyn. Atm. Oceans* **2000**, *31*, 117–137. [CrossRef]
84. Bonneton, P.; Chomaz, J.; Hopfinger, E. Internal waves produced by the turbulent wake of a sphere moving horizontally in a stratified. *J. Fluid Mech.* **1993**, *254*, 23–40. [CrossRef]
85. Spedding, G.R.; Browand, F.K.; Fincham, A.M. Turbulence, similarity scaling and vortex geometry in the wake of a towed sphere in a stably stratified fluid. *J. Fluid Mech.* **1996**, *314*, 53–103. [CrossRef]
86. Meunier, P.; Diamessis, P.; Spedding, G.R. Self-preservation in stratified momentum wakes. *Phys. Fluids* **2006**, *18*, 106601. [CrossRef]
87. Voisin, B. Lee waves from a sphere in a stratified flow. *J. Fluid Mech.* **2007**, *574*, 273–315. [CrossRef]
88. Gushchin, V.A.; Matyushin, P.V. Vortex formation mechanisms in the wake behind a sphere for 200 < Re < 380. *Fluid Dyn.* **2006**, *41*, 795–809. [CrossRef]
89. Gushchin, V.A.; Matyushin, P.V. Numerical simulation and visualization of vortical structure transformation in the flow past a sphere at an increasing degree of stratification. *Comput. Math. Math. Phys.* **2011**, *51*, 251–263. [CrossRef]
90. Gushchin, V.A.; Matyushin, P.V. Simulation and study of stratified flows around finite bodies. *Comput. Math. Math. Phys.* **2016**, *56*, 1034–1047. [CrossRef]
91. Orr, T.S.; Domaradzky, J.A.; Spedding, G.R.; Constantinescu, G.S. Numerical simulations of the near wake of a sphere moving in a steady, horizontal motion through a linearly stratified fluid at Re = 1000. *Phys. Fluids* **2015**, *27*, 035113. [CrossRef]

symmetry

MDPI

Article

Moving Forward by Shaking Sideways

Jean-Luc Thiffeault

Department of Mathematics, University of Wisconsin–Madison, 480 Lincoln Dr., Madison, WI 53706, USA; jeanluc@math.wisc.edu

Abstract: We investigate a simple model for a self-propelled swimmer, which consists of a fluctuating force acting at a point on a rigid body. The rigid body is subject to Newton's equations with linear friction, corresponding to drag in a viscous fluid. The force has zero time average, so net motion is challenging. We show that the swimmer can inch forward by shaking from side to side and exploiting friction coupled with nonlinearity. For large enough forcing amplitude it can reverse direction and swim backward.

Keywords: microswimmers; linear damping; swimming efficiency

1. Introduction

Microorganisms are all around us, and many self-propel through a fluid medium. Understanding their mechanism of locomotion is key to the understanding of biological processes, and is an interesting fundamental question in itself. As famously noted by [1], the net motion of a swimmer in a viscous fluid in the limit of zero Reynolds number is impossible without imposing a time-asymmetric forcing. Simple models of zero Reynolds number swimmers include Taylor's swimming sheet and Purcell's scallop [2]. But it is well-known that even a small amount of inertia can lead to a net motion for reciprocal swimmers, as can other effects such as viscoelasticity of the fluid [3–5] and fluid vibrations [6].

Here we investigate a very simple toy model of a swimmer subjected to a force with zero time average, so it is not a priori clear that the swimmer will experience any net motion. We will show that, somewhat paradoxically, the swimmer can make net progress by shaking from side to side, as long as the force is not acting at its center of mass. The motion will require a small amount of angular inertia, but no translational inertia is needed for it to occur (though, of course, it is also present). The mechanism also crucially depends on drag.

2. Model

Consider a disk of radius a of uniform density subjected to a mean-zero fluctuating force $f(t)$ (see Figure 1). We envisage the force is due to some propulsion mechanism, such as a flagellum for the case of a microswimmer. The line from the center of the disk to the point of application of the force defines a geometric symmetry line for the system. We assume the dynamics are governed by

$$\dot{\boldsymbol{x}} = \boldsymbol{u}, \qquad m\dot{\boldsymbol{u}} = -\sigma\boldsymbol{u} + \boldsymbol{f}; \tag{1a}$$

$$\dot{\phi} = \omega, \qquad I\dot{\omega} = -\kappa\omega + \tau, \tag{1b}$$

where $\boldsymbol{x}(t)$ is the position of the disk's center, $\boldsymbol{u}(t)$ its velocity, and m its mass. The force is acting in the plane of the disk ($\boldsymbol{f} \cdot \hat{\boldsymbol{z}} = 0$) at a fixed point in the disk's reference frame, a distance ℓ from its center. The disk's angular speed is ω, its moment of inertia is I, and the force creates a torque $\tau\hat{\boldsymbol{z}}$. We assume a linear damping in (1), with damping coefficient σ for translation and κ for rotation. The damping could come from the disk being immersed in a

Citation: Thiffeault, J.-L. Moving Forward by Shaking Sideways. *Symmetry* **2022**, *14*, 620. https://doi.org/10.3390/sym14030620

Academic Editors: Andrzej Herczyński and Roberto Zenit

Received: 24 February 2022
Accepted: 18 March 2022
Published: 20 March 2022

viscous fluid, or from friction with a surface. (In the former case, we neglect the fluid inertia, which is appropriate when the swimmer has a large density compared to the fluid [7,8]).

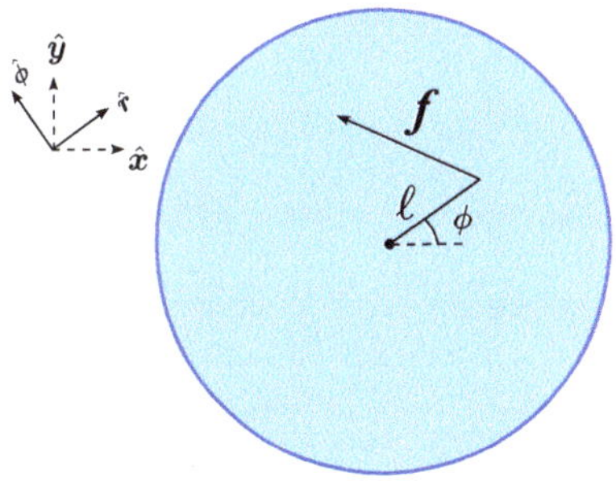

Figure 1. Disk of radius a with a force acting at a distance ℓ from its geometric center.

The force can be decomposed as

$$f(t) = f_\phi(t)\,\hat{\boldsymbol{\phi}} + f_r(t)\,\hat{\boldsymbol{r}}, \tag{2}$$

with the unit vectors

$$\hat{\boldsymbol{\phi}}(\phi) = \begin{pmatrix} -\sin\phi \\ \cos\phi \end{pmatrix}, \qquad \hat{\boldsymbol{r}}(\phi) = \begin{pmatrix} \cos\phi \\ \sin\phi \end{pmatrix}. \tag{3}$$

We refer to $\hat{\boldsymbol{\phi}}$ and $\hat{\boldsymbol{r}}$ as the lateral and longitudinal directions, respectively. The torque is then a vector out of the plane, with component

$$\tau(t) = \ell f_\phi(t) \tag{4}$$

where $0 \leq \ell \leq a$ is the distance from the center of the disk.

We define the long-time average of a bounded function $F(t)$ as

$$\overline{F} = \lim_{T\to\infty} \frac{1}{T} \int_{t_0}^{t_0+T} F(s)\,\mathrm{d}s. \tag{5}$$

The long-time average is independent of t_0. (For periodic $F(t)$ we can drop the limit $T \to \infty$ and take T to be the period.) The asymptotic net velocity of the disk is

$$\begin{pmatrix} U \\ V \end{pmatrix} = \overline{(\dot{x})} = \lim_{T\to\infty} \frac{1}{T} \int_0^T \dot{x}(s)\,\mathrm{d}s. \tag{6}$$

We take the imposed force to have components with zero long-time average:

$$\overline{f_r} = \overline{f_\phi} = \overline{\tau} = 0. \tag{7}$$

The question we wish to answer is whether the disk can 'swim', that is, can a translational motion in the longitudinal ($\hat{\boldsymbol{r}}$) direction be achieved by the fluctuating force. We will see that this is not possible with a purely longitudinal force ($f_r \neq 0$, $f_\phi = 0$), but it does occur for a purely lateral force ($f_r = 0$, $f_\phi \neq 0$). That is, the disk can have net forward motion by 'shaking' from side-to-side, as long as I, σ, κ, and ℓ are all nonzero. For small force amplitude, the motion is to the right in Figure 1, but for large enough amplitude the motion can reverse direction.

3. Longitudinal Forcing

First take

$$f(t) = f_r(t)\,\hat{\boldsymbol{r}}(\phi), \tag{8}$$

that is, the force is longitudinal along the symmetry direction. Then there is no torque ($\tau = 0$), and $\phi \to$ const. after a transient of duration I/κ. We take $\phi \equiv 0$ without loss of generality, and it follows from (1a) that $y \equiv 0$. The remaining equation is

$$m\ddot{x} = -\sigma\dot{x} + f_r(t). \tag{9}$$

Since $f_r(t)$ has zero mean, we can write $f_r(t) = mB'(t)$, where $B(t)$ also has zero mean. After a transient, $x(t)$ achieves the asymptotic solution

$$x(t) = \int_{-\infty}^{t} \mathrm{e}^{-\sigma(t-s)/m}\, B(s)\,\mathrm{d}s = \int_{-\infty}^{0} \mathrm{e}^{\sigma s/m}\, B(t+s)\,\mathrm{d}s. \tag{10}$$

Clearly, $x(t)$ has zero mean:

$$\overline{x} = \int_{-\infty}^{0} \mathrm{e}^{\sigma s/m}\, \overline{B(\cdot + s)}\,\mathrm{d}s = 0, \tag{11}$$

which implies that the net speed $U = 0$. A purely longitudinal mean-zero forcing will thus not lead to net motion, perhaps supporting our intuition. To relate to the title of the paper: shaking front to back (with zero mean) will not cause forward motion, even with inertia.

4. Lateral Forcing: Dimensionless Scaling

We saw in Section 3 that a mean-zero longitudinal force will never lead to net propulsion. The next natural question is whether a purely *lateral* forcing can lead to propulsion along the longitudinal direction. We set

$$\boldsymbol{f}(t) = f_\phi(t)\,\hat{\boldsymbol{\phi}}(\phi). \tag{12}$$

It will be convenient to use dimensionless variables from now on. We choose the mass, length, and time scales respectively as

$$\mathcal{M} = I\sigma/\kappa, \qquad \mathcal{L} = \sqrt{\kappa/\sigma}, \qquad \mathcal{T} = I/\kappa. \tag{13}$$

The time scale is the rotational damping time, and the length scale is roughly the radius of the disk. We do not introduce new symbols for dimensionless variables; we simply treat all variables as dimensionless from now, with the definitions (13) implying

$$\kappa = \sigma = I = 1. \tag{14}$$

Equation (1) can then be rewritten as coupled second-order ODEs:

$$m\ddot{\boldsymbol{x}} = -\dot{\boldsymbol{x}} + \ell^{-1}A'(t)\hat{\boldsymbol{\phi}}; \tag{15a}$$

$$\ddot{\phi} = -\dot{\phi} + A'(t), \tag{15b}$$

where we defined for later convenience

$$f_\phi(t) = \ell^{-1}A'(t), \tag{16}$$

with

$$\ell A(t) = \int_0^t f_\phi(s)\,\mathrm{d}s - \overline{\int_0^t f_\phi(s)\,\mathrm{d}s} \tag{17}$$

such that $\overline{A} = 0$. We implicitly assumed that $A(t)$ is bounded, which requires

$$\left| \int_0^t f_\phi(s)\,\mathrm{d}s \right| < M, \qquad t \in \mathbb{R}, \tag{18}$$

for some positive constant M.

5. Temporal Evolution of the Angle

The solution to (15b) is

$$\phi(t) = \phi_0 + (1 - \mathrm{e}^{-(t-t_0)})\,\omega_0 + \int_{t_0}^{t} \mathrm{e}^{-(t-s)} A(s)\,\mathrm{d}s \tag{19}$$

where $\phi(t_0) = \phi_0$ and $\phi'(t_0) = \omega_0$. We are really only interested in what happens after an initial transient of $\mathrm{O}(1)$ duration (I/κ in dimensional variables), so we take the limit $t_0 \to -\infty$ in (19) and drop some constants to obtain

$$\phi(t) = \int_{-\infty}^{t} \mathrm{e}^{-(t-s)} A(s)\,\mathrm{d}s = \int_{-\infty}^{0} \mathrm{e}^{s} A(t+s)\,\mathrm{d}s. \tag{20}$$

The form (20) is such that that $\overline{\phi} = 0$, which follows from $\overline{A} = 0$ in the same manner as for Equation (11).

We shall work in the limit where the damping time $I/\kappa = 1$ is small compared to the typical time scale for changes in $A(t)$. This means that $A \gg A' \gg A'' \ldots$. In that case the integral in (20) can be evaluated as a series using Watson's lemma:

$$\phi(t) = \sum_{n=0}^{\infty} (-1)^n A^{(n)}(t). \tag{21}$$

The series converges for $A(t)$ smooth enough.

For the case where $f_\phi(t)$ is periodic with period T, the boundedness condition (18) requires

$$\int_0^T f_\phi(t)\,\mathrm{d}t = 0 \tag{22}$$

and it follows that $A(t)$ is also T-periodic, as well as $\phi(t)$.

6. Temporal Evolution of Position

Following a similar approach to Section 5, we see that the asymptotic solution to Equation (15a) is

$$\ell\dot{x}(t) = \int_{-\infty}^{t} \mathrm{e}^{-(t-s)/m}\, A'(s)\,\hat{\boldsymbol{\phi}}(s)\,\mathrm{d}s. \tag{23}$$

We substitute the unit vector (3) with the expansion (21) in the $\dot{x}$ integral (23), and again using Watson's lemma solve for the slow time evolution:

$$\ell\dot{x} = (A')^2 \cos A + \left(\cos A + mA' \sin A\right)' + \mathrm{O}(A'''), \tag{24a}$$

$$\ell\dot{y} = (A')^2 \sin A + \left(\sin A - mA' \cos A\right)' + \mathrm{O}(A'''). \tag{24b}$$

The second term on the right of both equations in (24) is an exact derivative, so its long-time average vanishes. Hence, the components in (6) are

$$\begin{pmatrix} U \\ V \end{pmatrix} = \lim_{T\to\infty} \frac{1}{\ell T} \left(\int_0^T (A'(t))^2 \begin{pmatrix} \cos A(t) \\ \sin A(t) \end{pmatrix} \mathrm{d}t + \mathrm{O}(A''') \right). \tag{25}$$

The simplest case is to take the forcing $A(t)$ to be a periodic function:

$$A(t) = A_0 \sin(\Omega t). \tag{26}$$

The integral (25) can be taken over a period $T = 2\pi/\Omega$ and the limit dropped, to obtain

$$U = A_0 \Omega^2 \ell^{-1} J_1(A_0) + \mathrm{O}(\Omega^3), \qquad V = 0, \tag{27}$$

where J_1 is a Bessel function of the first kind. (V is zero to all orders, by symmetry, since $A(t)$ is odd.) Restoring dimensional variables,

$$\begin{aligned} U &= \frac{A_0 I \Omega^2}{\sigma \ell} J_1(A_0) + O((\Omega I/\kappa)^3) \\ &= \frac{I \Omega f_{\phi 0}}{\sigma \kappa} J_1(\ell f_{\phi 0}/\Omega \kappa) + O((\Omega I/\kappa)^3). \end{aligned} \tag{28}$$

where $f_{\phi 0} = A_0 \Omega / \kappa \ell$ is the amplitude of the fluctuating force (see Equation (16)). The dimensional form makes it clear that the net velocity goes to zero for $\ell = 0$ (no torque).

Figure 2 shows some numerical trajectories for the forcing (26), for a few different amplitudes A_0.

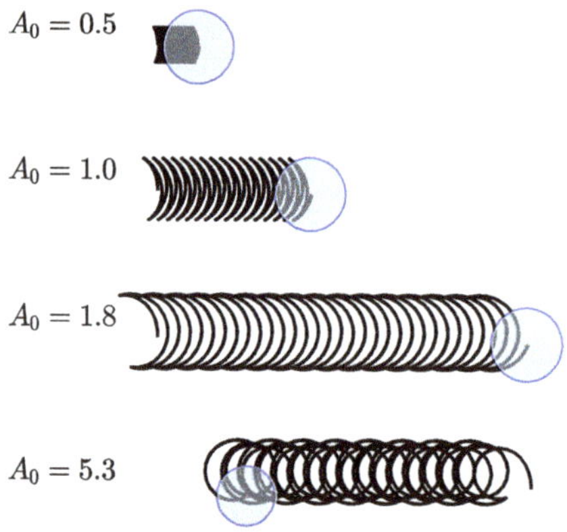

Figure 2. Trajectories for $A(t) = A_0 \cos(\Omega t)$ and $\ell = 1, \Omega = 0.1$.

Note that the disk moves to the right (with the convention as in Figure 1) with a speed that increases with A_0. In Figure 3 we confirm Formula (27) by finding the asymptotic swimming speed as a function of A_0.

Note that the swimmer can reverse its direction of motion with a large enough 'shaking' amplitude. This corresponds to the swimmer shaking hard enough that it completely reverses its direction in one half-period (see Figure 2, bottom). Clearly, this regime is unrealistic, but is interesting to note nonetheless.

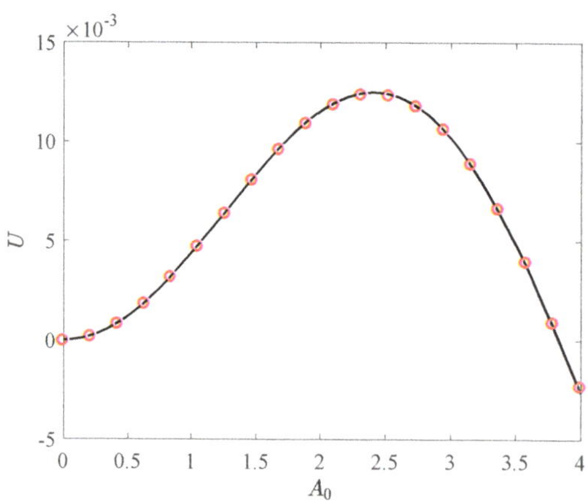

Figure 3. The mean speed U for numerical simulations with $\ell = 1$, $\Omega = 0.1$ (solid line) and the asymptotic form Equation (27).

7. Efficient Motion

A natural question to ask is: what is the optimal forcing to progress forward as rapidly as possible? This is a popular endeavor in the world of microswimmers; see for instance [9–17]. As we see below, we cannot completely answer this question here without some extra assumptions.

7.1. Power

From (15), we can get energy equations

$$\tfrac{1}{2}m\frac{\mathrm{d}}{\mathrm{d}t}\|\boldsymbol{u}\|^2 = -\|\boldsymbol{u}\|^2 + \boldsymbol{u}\cdot\boldsymbol{f}, \qquad \tfrac{1}{2}\frac{\mathrm{d}}{\mathrm{d}t}\omega^2 = -\omega^2 + \omega\tau. \tag{29}$$

The left-hand sides of (29) are the rate of change of kinetic energy; the right-hand sides include negative quadratic dissipation terms, and power expenditure terms

$$\mathcal{P}_{\boldsymbol{u}}(t) = \boldsymbol{u}\cdot\boldsymbol{f}, \qquad \mathcal{P}_\omega(t) = \omega\tau. \tag{30}$$

Averaging (29) gives the balance between power expended and dissipated:

$$\overline{\mathcal{P}_{\boldsymbol{u}}} = \overline{\boldsymbol{u}\cdot\boldsymbol{f}} = \overline{\|\boldsymbol{u}\|^2}, \qquad \overline{\mathcal{P}_\omega} = \overline{\omega\tau} = \overline{\omega^2}. \tag{31}$$

If we use the overdamped approximation (24a) for $\boldsymbol{u}$ and the asymptotic solution (21) for ϕ, we have

$$\mathcal{P}(t) = \left(1+\ell^{-2}\right)(A')^2 + \mathrm{O}(A'A''). \tag{32}$$

Taking the long-time average yields

$$\overline{\mathcal{P}} = (1+\ell^{-2})\overline{(A')^2} + \mathrm{O}(\overline{(A'')^2}). \tag{33}$$

For convenience we absorb the prefactor to define

$$P = \mathcal{P}/(1+\ell^{-2}). \tag{34}$$

7.2. Optimizing the Speed of Motion

From Equation (25), assuming a periodic forcing $A(t+T) = A(t)$,

$$\begin{aligned} U &= \overline{(A')^2\cos A} \\ &= P - \overline{(A')^2\,(1-\cos A)} \le P, \end{aligned} \tag{35}$$

where we neglected higher derivatives of A by assuming the period T is long. It is clear from (35) that the power P is an absolute limit to the net speed of motion. A natural question is whether there is an optimal periodic $A(t)$ that leads to the fastest speed. For simplicity we take $A(t)$ to be an odd function of t, so that $V = 0$ in Equation (25). Consider the particular choice of a triangular function

$$A(t) = \begin{cases} \sqrt{P}\,t, & -\tfrac{1}{4}T \le t \le \tfrac{1}{4}T, \\ \sqrt{P}\,(\tfrac{1}{2}T - t), & \tfrac{1}{4}T \le t \le \tfrac{3}{4}T, \end{cases} \tag{36}$$

defined on $[-\tfrac{1}{4}T, \tfrac{3}{4}T]$ and extended periodically (Figure 4). This piecewise linear function has constant power $P = (A')^2$, so $\overline{P} = P$. The speed for this periodic forcing is

$$U = P\,\frac{4}{T}\int_0^{T/4} \cos(\sqrt{P}\,t)\,\mathrm{d}t = P\,\mathrm{sinc}(\sqrt{P}\,T/4) \tag{37}$$

where $\operatorname{sinc} x := x^{-1} \sin x$. But now we replace T by a sub-period $T/2^k$, and we can see that $U \to P$ as $k \to \infty$. The maximum net speed $U = P$ is thus achieved by a constant-power forcing that 'switches' infinitely rapidly with a very small amplitude (Figure 4, bottom). This is not very natural, but still suggests what kinds of motions lead to most efficient propulsion. A follow-up investigation could involve a constraint with a second derivative of A in order to regularize the problem, but this requires introducing additional physics or biology.

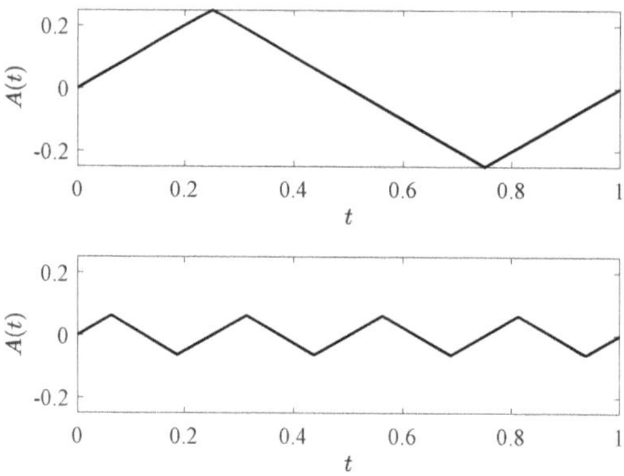

Figure 4. Top: The periodically-extended triangular function (36) for $P = T = 1$. **Bottom**: The same function with T replaced by $T/4$. The two functions have the same constant power $P = (A')^2$, but the bottom function has smaller amplitude.

8. Discussion

In this paper, we introduced a very simple mechanical model where it's not initially clear that a swimmer can make forward progress. A mean-zero the forward-backward motion does not lead to net motion, but somewhat surprisingly a sideways motion can do so, as long as it is driven by a force that also exerts a torque. A small amount of rotational inertia is necessary, as is some damping mechanism such as linear drag or friction with a surface. It is likely that the mechanism would still exist in a more realistic model, though this has yet to be verified.

In practice, the forward progress made by the swimmer is extremely slow, so this toy model is probably not a practical means of motion. Nevertheless, it is instructive to explore the full range of possibilities when it comes to self-propelled motion since nature often surprises us. Moreover, the setup described here could probably be easily realized with a simple mechanical robot.

The choice of a disk for the shape is fairly immaterial: the only consequence of the disk shape that was used here is the isotropy of the drag. A nonisotropic drag law (with a resistance matrix) could be used, but would complicate the math. In closing, we note that it is also possible to solve a stochastic version of this problem [18].

Funding: This work was supported by EPSRC grant no EP/R014604/1.

Institutional Review Board Statement: Not applicable.

Informed Consent Statement: Not applicable.

Acknowledgments: The author thanks Jiajia Guo, Giovanni Fantuzzi, and Andrzej Herczynski for helpful discussions. The author also thanks the Isaac Newton Institute for Mathematical Sciences, Cambridge, for support and hospitality during the program *Mathematical aspects of turbulence: where do we stand?* where work on this paper was completed.

Conflicts of Interest: The author declares no conflict of interest.

References

1. Purcell, E.M. Life at low Reynolds number. *Am. J. Phys.* **1977**, *45*, 3–11. [CrossRef]
2. Lauga, E. *The Fluid Dynamics of Cell Motility*; Cambridge University Press: Cambridge, UK, 2020.
3. Datt, C.; Nasouri, B.; Elfring, G.J. Two-sphere swimmers in viscoelastic fluids. *Phys. Rev. Fluids* **2018**, *3*, 123301. [CrossRef]
4. Pak, O.S.; Zhu, L.; Brandt, L.; Lauga, E. Micropropulsion and microrheology in complex fluids via symmetry breaking. *Phys. Fluids* **2012**, *24*, 103102. [CrossRef]
5. Zhu, L.; Lauga, E.; Brandt, L. Self-propulsion in viscoelastic fluids: Pushers vs. pullers. *Phys. Fluids* **2012**, *24*, 051902. [CrossRef]
6. Klotsa, D.; Baldwin, K.A.; Hill, R.J.; Bowley, R.M.; Swift, M.R. Propulsion of a two-sphere swimmer. *Phys. Rev. Lett.* **2015**, *115*, 248102. [CrossRef] [PubMed]
7. Bocquet, L.; Piasecki, J. Microscopic derivation of non-Markovian thermalization of a Brownian particle. *J. Stat. Phys.* **1997**, *87*, 1005–1035. [CrossRef]
8. Hinch, E.J. Application of the Langevin equation to fluid suspensions. *J. Fluid Mech.* **1975**, *72*, 499. [CrossRef]
9. Childress, S. *Mechanics of Swimming and Flying*; Cambridge University Press: Cambridge, UK, 1981.
10. Gueron, S.; Levit-Gurevich, K. Energetic considerations of ciliary beating and the advantage of metachronal coordination. *Proc. Natl. Acad. Sci. USA* **1999**, *96*, 12240–12245. [CrossRef] [PubMed]
11. Lauga, E.; Powers, T.R. The hydrodynamics of swimming micro-organisms. *Rep. Prog. Phys.* **2009**, *72*, 096601. [CrossRef]
12. Michelin, S.; Lauga, E. Efficiency optimization and symmetry-breaking in a model of ciliary locomotion. *Phys. Fluids* **2010**, *22*, 111901. [CrossRef]
13. Michelin, S.; Lauga, E. Optimal feeding is optimal swimming for all Péclet numbers. *Phys. Fluids* **2011**, *23*, 101901. [CrossRef]
14. Michelin, S.; Lauga, E. Unsteady feeding and optimal strokes of model ciliates. *J. Fluid Mech.* **2013**, *715*, 1–31. [CrossRef]
15. Pironneau, O.; Katz, D.F. Optimal swimming of flagellated micro-organisms. *J. Fluid Mech.* **1974**, *66*, 391–415. [CrossRef]
16. Spagnolie, S.E.; Lauga, E. The optimal elastic flagellum. *Phys. Fluids* **2010**, *22*, 031901. [CrossRef]
17. Tam, D.; Hosoi, A.E. Optimal stroke patterns for Purcell's three-link swimmer. *Phys. Rev. Lett.* **2007**, *98*, 068105. [CrossRef] [PubMed]
18. Thiffeault, J.-L.; Guo, J. Shake your hips: An anisotropic active Brownian particle with a fluctuating propulsion force. *arXiv* **2021**, arXiv:2102.11758.

symmetry

Article

Numerical Investigation of the Automatic Air Intake Drag Reduction Strut Based on the Venturi Effect

Hai An *, Zhenyu Hu, Haozhe Pan and Po Yang

College of Aerospace and Civil Engineering, Harbin Engineering University, Harbin 150001, China; huzhenyu@hrbeu.edu.cn (Z.H.); panhaozhe@hrbeu.edu.cn (H.P.); yangpo@hrbeu.edu.cn (P.Y.)
* Correspondence: anhai@hrbeu.edu.cn

Abstract: Drag reduction by injecting air is a promising engineering method for improving ship performance. A novel automatic air intake drag reduction strut structure based on the Venturi effect is proposed for the high-speed small water-plane area twin hull vessels in the present study. The drag reduction strut can achieve the function of automatic air intake when the vehicle is moving at high speed, and the air inhaled and the incoming flow form bubbly flows to cover the strut surface, effectively reducing the drag of the strut. Considering the longitudinal symmetry of the strut structure, a two-dimensional single-chip drag reduction strut structure is designed to facilitate analysis and a solution. The volume of fluid model is combined with the k-ω SST turbulence model, and a numerical simulation is carried out to investigate the variation of the air inflow, the air volume fraction in the bubbly flows of the strut and the drag reduction rate of the strut for different sailing speeds. The analysis result shows that when the proposed model reaches a certain speed, the external air is inhaled by the strut intake duct, and the bubbly flows are formed with the incoming flow covering the surface of the strut, thereby reducing the drag coefficient. Meanwhile, it is found that as the sailing speed increases, the drag reduction rate of the strut gradually rises and its maximum value reaches about 30%. For high sailing speeds, the drag reduction rate is affected by wave-making resistance so that it gradually declines.

Keywords: symmetry; numerical investigation; strut drag reduction; bubbly flows; Venturi effect; air inflow

Citation: An, H.; Hu, Z.; Pan, H.; Yang, P. Numerical Investigation of the Automatic Air Intake Drag Reduction Strut Based on the Venturi Effect. *Symmetry* **2022**, *14*, 367. https://doi.org/10.3390/sym14020367

Academic Editors: Andrzej Herczyński and Roberto Zenit

Received: 18 January 2022
Accepted: 9 February 2022
Published: 12 February 2022

1. Introduction

A new type of high-speed surface vehicle similar to the small water-plane area twin hull (SWATH) [1] has attracted people's attention. Forming supercavities on the SWATH vessels via artificial ventilation can lower resistance by reducing the wetting surface of the submerged bodies by applying supercavitation techniques [2]. However, as the strut passes through the water, supercavities cannot be formed on its surface. The skin friction drag of the strut surface can account for almost half of the total drag of the ship. Considering the effects of skin friction drag of the strut on the total resistance of the vehicle, drag reduction by bubbly flows is an effective way of strut drag reduction.

Various methods have been explored to reduce the skin friction drag in the past. One such method—drag reduction by bubbly flows—has recently become a focus in the expectation that it might be applicable to ships. Early reporting of drag reduction by bubbly flows was by Mccormick [3]. They observed viscous drag reduction of a fully submerged body by creating hydrogen gas on the hull by electrolysis. Murai [4] developed the study of the mechanism of drag reduction using relatively large air bubbles compared to the boundary layer thickness in a horizontal turbulent channel flow and indicated that there is a negative correlation between the local skin friction and the local void fraction.

A common method to form bubbly flows is gas injection into the liquid boundary layer. Such injection results in the formation of bubbles that produce drag reduction. Drag reduction by injecting air can be categorized into microbubble drag reduction (MBDR) and air layer drag reduction (ALDR), according to the shape and distribution of air bubbles.

In MBDR, a number of experiments and numerical studies have been carried out to investigate. Madavan et al. [5] indicated that microbubbles can cause a reduction of high-frequency shear-stress fluctuations and a destruction of some of the turbulence in the near-wall region by spectral measurements. Kodama et al. [6] carried out the microbubble experiments using a circulating water tunnel specially designed for microbubble experiments, and skin friction reduction by microbubbles of up to 40% was obtained. Sayyaadi et al. [7] estimated a simple formulation for calculating an efficient injection rate by considering the main parameters of the ship by the model test results of a 70 cm catamaran model. The test results showed that excessive air injection decreases the drag reduction effect, while suitable injection reduces total drag by about 5–8%. Paik et al. [8] developed the study of a bubbly turbulent boundary layer. The behaviors of the microbubbles were visualized quantitatively by using the conventional PIV technique with a field-of-view of 200 mm^2. The velocity fields of the bubbles showed that the PIV technique is highly effective in reducing skin friction by decreasing Reynolds stress. Because the direct numerical simulation for bubbly flows can provide insight into the physics of bubbly boundary layer flows, numerical simulations have been widely performed to predict the observed MBDR. Kanai et al. [9] clarified the structure of the turbulent boundary layer containing microbubbles and the mechanism of frictional drag reduction by conducting direct numerical simulation (DNS) for bubbly flows. Kawamura et al. [10] presented a new computational method for investigating interactions between bubbles and turbulence, which is applied to a direct numerical simulation of a fully developed turbulent channel flow containing bubbles. Xu et al. [11] conducted a series of numerical simulations of small bubbles seeded in a turbulent channel flow at average volume fractions of up to 8%. These results showed that an initial transient drag reduction can occur as bubbles disperse into the flow and that small spherical bubbles will produce a sustained level of drag reduction over time. Ferrante et al. [12] conducted numerical simulations for a microbubble-laden spatially developing turbulent boundary layer and compared skin friction due to the presence of the bubbles for two Reynolds numbers: $Re\theta = 1430$ and $Re\theta = 2900$. The results showed that increasing the Reynolds number decreases the percentage of drag reduction. Mohanarangam et al. [13] investigated the phenomenon of drag reduction by the injection of microbubbles into the turbulent boundary layer by using an Eulerian–Eulerian two-fluid model. Feng et al. [14] investigated effects of the parameters of microbubbles, including the gravity, the injection height and the volume fraction, on friction drag in a turbulent boundary layer by using large-eddy simulation. Yanuar et al. [15] developed a comparison between MBDR and ALDR for a self-propelled barge. The results showed that the ship model using the air layer has greater drag reduction than microbubbles.

ALDR is another form of air injection drag reduction. As opposed to MBDR, ALDR occurs when a continuous or nearly continuous layer of air is formed between a solid surface and the outer liquid flow [16]. Shen et al. [17] and Murai et al. [18] studied the possibility and mechanism of drag reduction using relatively large air bubbles compared to the boundary layer thickness in a horizontal turbulent channel flow. Choi et al. [19] described a numerical method based on the boundary element method, which is used to establish trends of the total resistance and its dependence on the Froude number. On this basis, Choi et al. [20] addressed the issue of unsteadiness of the air layer free surface with regards to the ship motion and the upstream conditions for future practical hulls with air layers. Elbing et al. [16] conducted the ALDR experiments, and it was concluded that (i) drag reduction with injection of air can be divided into three distinct regions: a BDR zone, a "transition zone" and an ALDR zone; (ii) the "critical" gas injection rate required to form a persistent air layer is approximately proportional to the square of the free-stream liquid velocity and (iii) ALDR may have persistence lengths much greater than the lengths of the current test model. Kim et al. [21] performed direct numerical simulations in order to examine the stability and mechanism of ALDR for different air injection rates and investigated the stability of the air layer theoretically by solving the Orr–Sommerfeld equations in both phases in order to find the stabilizing parameters and stability conditions for ALDR. Zhao et al. [22] investigated the differences between MBDR and ALDR by using the Eulerian—Eulerian two-fluid model and the volume of fluid (VOF) model. It was concluded that the Eulerian—Eulerian two-fluid model and the VOF model are suitable for MBDR and ALDR, respectively. Kim et al. [23] investigated flow change of the horizontal channel flow including large bubbles and presented the numerical procedure of how to inject large bubbles into turbulent channel flow. Zhao et al. [24] investigated the ALDR of an axisymmetric body in oscillatory motions. The results showed that the variation of the drag reduction is related to the morphological change of the air layer, and the heave motion is more likely to reduce the effects of the ALDR than the pitch motion.

ALDR can be realized on the strut surface by artificial ventilation. Furthermore, a large amount of ventilation power is also needed. In the present study, an approach to drag reduction by using natural air flow to form bubbly flows of the strut is explored for the SWATH vessel. Based on the Venturi effect, the strut drag reduction structure is designed, and a numerical analysis is carried out to study the variation law of the air inflow amount, the air volume fraction in bubbly flows as well as the strut drag reduction rate with the sailing speed.

2. Description of the Physical Model

The Venturi effect is defined as the flow through the cross-section of the passage, where the flow velocity increases continuously. The flow velocity is inversely proportional to the passage cross-section. Moreover, Bernoulli's law indicates that the increment in the flow velocity is accompanied by a decrease in the fluid pressure. Therefore, when the fluid flows through a nozzle with an expanding cross-section, the pressure of the trailing edge is lower than that of the incoming flow. Figure 1 displays the process of the strut's automatic air intake. In Figure 1, the strut of the SWATH vessel is a longitudinal symmetrical structure. An intake duct is designed on the strut. The opening at one end of the intake duct is connected to the surrounding atmosphere, while that at the other end is below the waterline. During the vehicle sailing, since the pressure at the opening below the waterline is lower than the external atmospheric pressure, the automatic air intake can be realized based on the Venturi effect. The implementation of Venturi effect in the design scheme can be theoretically verified in the following.

Figure 2 shows the planform of the strut's automatic air intake. As can be seen from Figure 2, the bubbly flows form on the strut surface in the state of the strut's automatic air intake when the vehicle is moving at high speed. Consider the portion from the leading edge to the air outlet of the strut as the control body, as shown in Figure 3. In the control body of Figure 3, water flows through the control plane $\sum_1$ and the control plane $\sum_2$. The Bernoulli equation is established as follows:

$$Z_1 + \frac{p_1}{\rho g} + \frac{v_1^2}{2g} = Z_2 + \frac{p_2}{\rho g} + \frac{v_2^2}{2g} \tag{1}$$

where p_1 and p_2 are the pressure of the control plane $\sum_1$ and the control plane $\sum_2$, respectively; Z_1 and Z_2 are the depth of the control plane $\sum_1$ and the control plane $\sum_2$, respectively; v_1 and v_2 are the velocity of the control plane $\sum_1$ and the control plane $\sum_2$, respectively; ρ is the density of water and g is acceleration of gravity.

Considering the equal depths of control planes $\sum_1$ and $\sum_2$ and the continuity of the fluid, it can be obtained that

$$\begin{cases} Z_1 = Z_2 \\ v_1 A_{\sum_1} = v_2 A_{\sum_2}. \end{cases} \tag{2}$$

where $A_{\sum_1}$ and $A_{\sum_2}$ are the area of the control plane $\sum_1$ and the control plane $\sum_2$, respectively.

Substituting Equation (2) into Equation (1), it can be obtained that

$$p_2 = p_1 + \frac{\rho v_1^2}{2}\left[1 - (\frac{A_{\Sigma_1}}{A_{\Sigma_2}})^2\right] < p_1 \tag{3}$$

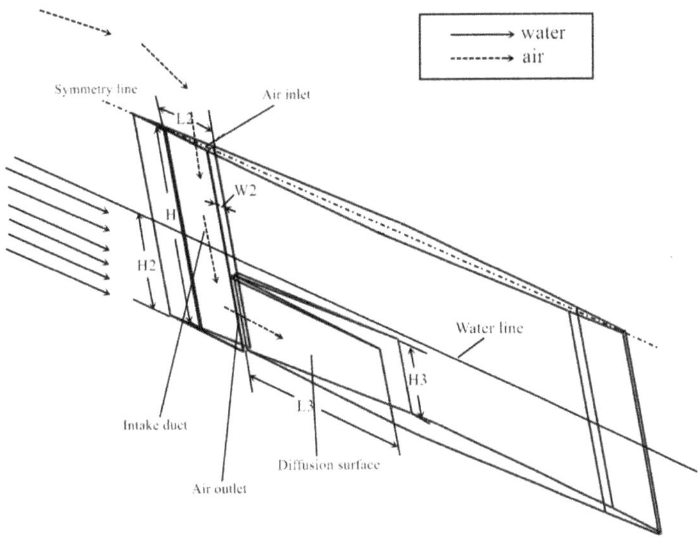

Figure 1. Process of the strut's automatic air intake.

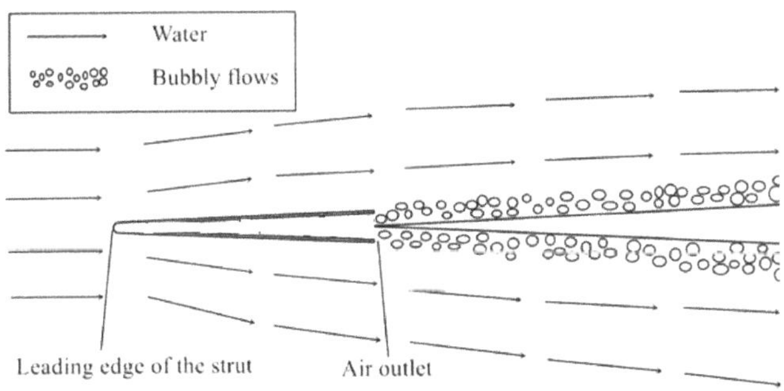

Figure 2. Planform of the strut's automatic air intake.

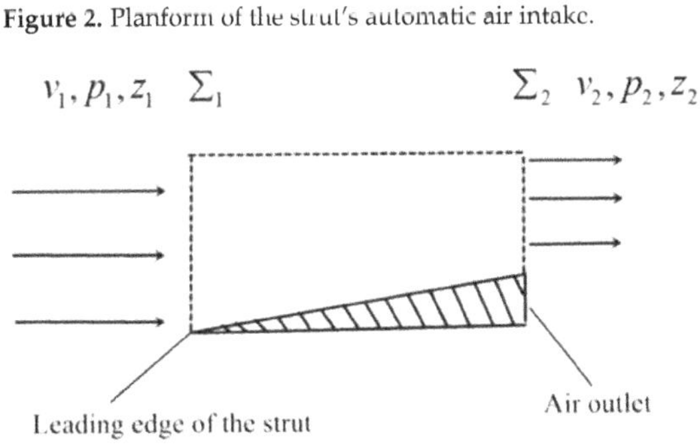

Figure 3. Control body.

According to Equation (3), the pressure at the expansion control plane $\sum_2$ is lower than that at the control plane $\sum_1$ of the incoming flow. In the design of this paper, the end of the control plane $\sum_2$ is connected with the external atmosphere through an intake duct. When the vehicle reaches a certain speed, once the pressure p_2 of the control plane $\sum_2$ is lower than atmospheric pressure, automatic air intake can be realized.

Because of the longitudinal symmetry of the SWATH vessel's struts, the bubbly flows that form on the surface of the struts are also symmetrical during navigation. Adopting a two-dimensional single-chip connection strut model provides a better understanding as well as simpler analytical and numerical solutions for studying ALDR of the struts.

Figure 4 shows the layout of the strut in the proposed model. As can be seen from Figure 4, the proposed model consists of the strut and the submerged vehicle. A groove is designed in the underwater part of the rear side of the leading edge of the strut. The process of the strut's automatic air intake is shown in Figure 1. When the vehicle moves, a low-pressure zone appears in the section with the groove, and the air outlet of the intake duct is placed in this low-pressure zone. The air inlet of the intake duct is designed on the top of the strut and connected to the external atmosphere. In this case, due to the Venturi effect, the external air is blended with the incoming flow through the air outlet of the intake duct. Moreover, the resultant air–water mixture propagates backwards along the diffusion surface, covering the underwater surface of the strut and having the effect of drag reduction by bubbly flows.

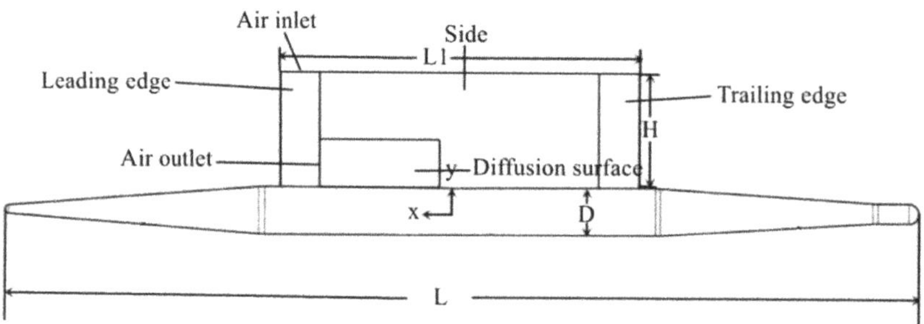

Figure 4. Layout of the strut in the proposed model.

The structural dimensions of the proposed model are L = 2320 mm, L_1 = 1000 mm, L_2 = 90 mm, L_3 = 300 mm, H_2 = 290 mm, H = 140 mm, H_3 = 120 mm, D = 120 mm and W_2 = 10 mm.

In order to evaluate the drag reduction effect of the proposed model, a control model is designed for comparison. The appearance and the dimension of the control model are consistent with those of the proposed model. The only difference between these two models is that the intake duct and the diffusion surface are absent in the control model so that the leading edge is smoothly joined to the side surface.

3. Numerical Model

3.1. Basic Governing Equation

Both water and air phases are treated as incompressible fluids, and the continuity of stress is implemented at the interface. According to the homogenous equalized multi-phase theory, the continuity equation and the momentum conservation equation of the mixed media can be written as follows [25]:

$$\frac{\partial \rho_m}{\partial t} + \frac{\partial}{\partial x_i}(\rho_m u_i) = 0. \tag{4}$$

$$\frac{\partial}{\partial t}(\rho_m u_j) + \frac{\partial}{\partial x_i}(\rho_m u_i u_j) = -\frac{\partial p}{\partial x_j} + \rho_m g + \frac{\partial}{\partial x_i}\left[(\mu_m + \mu_t)\left(\frac{\partial u_i}{\partial x_j} + \frac{\partial u_j}{\partial x_i}\right)\right] \tag{5}$$

where t is time; p is pressure; g is acceleration of gravity; u_i is the velocity component and ρ_m and μ_m are the density and the dynamic viscosity of the mixed media, respectively. They can be obtained by weighted averaging the volume components as follows:

$$\rho_m = \alpha_l \rho_l + \alpha_g \rho_g,\ \mu_m = \alpha_l \mu_l + \alpha_g \mu_g \tag{6}$$

where μ_t is the dynamic viscosity of turbulent flow, ρ_l is the liquid density, ρ_g is the gas density, α_l is the liquid fraction, α_g is the gas volume fraction, μ_l is the liquid dynamic viscosity and μ_g is the gas dynamic viscosity.

The volume components of various phases should satisfy:

$$\alpha_l + \alpha_g = 1 \tag{7}$$

In this study, numerical simulation is performed on both the bubbly flows covering the surface of struts and the wave-making when the strut of the vehicle passes through the water surface. While the VOF model [26] is suitable for solving the motion rules of the multiphase intersection interface, when the air flow ratio is high, it is suitable for ALDR [22]. Therefore, the VOF model is selected for conducting numerical simulation on the change observed in the bubbly flows' formation.

In the VOF model, the method of tracing the inter-phase boundary is achieved by solving the volume fraction continuity equation. As for the qth phase, the volume fraction equation is [26]:

$$\frac{\partial \alpha_q}{\partial t} + \frac{\partial}{\partial x_i}(\alpha_q u_i) = 0. \tag{8}$$

where the subscript $q = l, g$ denotes the liquid phase and the gas phase, respectively.

In addition, due to the speed ($\leq$14 m/s) of the proposed model in the study, although there is local low pressure in the outlet area of the air intake duct, the pressure is not enough to vaporize water, so the natural cavitation phenomenon is ignored.

3.2. Turbulence Model

It should be emphasized that for high air flow rate, a stable air layer is formed on the solid surface, whereas, in the case of low air flow rate, the air layer breaks up and ALDR is not achieved [27]. Due to the sailing speed ($\geq$5 m/s) of the proposed model in this study during the automatic air intake of the strut, the pressure difference between the air outlet and the air inlet of the intake duct is enough to ensure high air flow rate, so the air volume fraction of the formed bubbly flows is above 60%. Therefore, it can be considered that drag reduction by bubbly flows adopted in this study conforms to the ALDR model.

In handling turbulent bubbly flows, the SST $k-\omega$ model developed by Menter [28] is employed in the present study. The SST $k-\omega$ model is the combination of the $k-\varepsilon$ and $k-\omega$ models, which eliminates errors arising from the empirical wall function and thus provides high simulation precision for the ALDR model [22].

The turbulent kinetic energy and the specific dissipation rate can be calculated by [28]:

$$\frac{\partial(\rho k)}{\partial t} + \frac{\partial}{\partial x_j}(\rho k u_j) = \tau_{ij}\frac{\partial u_i}{\partial x_j} - \beta^* \rho k \omega + \frac{\partial}{\partial x_j}\left[\left(\mu + \frac{\mu_t}{\sigma_{k3}}\right)\frac{\partial k}{\partial x_j}\right] \tag{9}$$

$$\frac{\partial(\rho\omega)}{\partial t} + \frac{\partial(\rho\omega u_j)}{\partial x_j} = \frac{\omega}{k}(\alpha_3 \tau_{ij}\frac{\partial u_i}{\partial x_j}) - \beta_3 \rho \omega^2 + \frac{\partial}{\partial x_j}\left[\left(\mu + \frac{\mu_t}{\sigma_{\omega 3}}\right)\frac{\partial \omega}{\partial x_j}\right] + 2(1-F_1)\frac{\rho}{\sigma_{\omega_2}\omega}\frac{\partial k}{\partial x_j}\frac{\partial \omega}{\partial x_j} \tag{10}$$

$$\tau_{ij} = \mu_t(2s_{ij} - \frac{2}{3}\frac{\partial u_k}{\partial x_k}\delta_{ij}) - \frac{2}{3}\rho k \delta_{ij}. \tag{11}$$

$$s_{ij} = \frac{1}{2}(\frac{\partial u_i}{\partial x_j} + \frac{\partial u_j}{\partial x_i}). \tag{12}$$

where the model parameters such as α_3, β_3, σ_{k3} and $\sigma_{\omega 3}$ are the linear combinations of the corresponding coefficients in $k-\omega$ and the modified $k-\varepsilon$ turbulence model as follows:

$$\psi = F_1\psi_{k\omega} + (1-F_1)\psi_{k\varepsilon},\ \alpha_3 = F_1\alpha_1 + (1-F_1)\alpha_2. \tag{13}$$

$$k-\omega:\ \alpha_1 = 5/9,\ \beta_1 = 3/40,\ \sigma_{k1} = 2,\ \sigma_{\omega 1} = 2,\ \beta^* = 9/100 \tag{14}$$

$$k-\varepsilon:\ \alpha_2 = 0.44,\ \beta_2 = 0.0828,\ \sigma_{k2} = 1,\ \sigma_{\omega 2} = 1/0.856,\ C_\mu = 0.09 \tag{15}$$

μ_t is the vortex viscosity, which can be described as:

$$\mu_t = \rho\frac{k}{\max(\omega, SF_2)} \tag{16}$$

where S is the invariant measure of the strain rate and the values of the above coefficients come from [29]:

$$\begin{cases} F_1 = \tanh(\Gamma^4) \\ \Gamma = \min(\max(\frac{\sqrt{k}}{\beta^*\omega y}; \frac{500v}{\omega y^2}); \frac{4\rho\sigma_{\omega 2}k}{CD_{k\omega}y^2}) \\ CD_{k\omega} = \max\left(2\rho\sigma_{\omega 2}\frac{1}{\omega}\frac{\partial k}{\partial xj}\frac{\partial \omega}{\partial xj}, 10^{-20}\right) \\ F_2 = \tanh(\Gamma_2^2) \\ \Gamma_2 = \max(\frac{2\sqrt{k}}{\beta^*\omega y}, \frac{500v}{\omega y^2}) \end{cases} \tag{17}$$

where y is the distance to the nearest wall.

3.3. Computational Domain and Boundary Conditions

The study uses STAR-CCM+, a commercial computational fluid dynamics (CFD) software. The computing domain and boundary conditions are shown in Figure 5. Figure 5a displays the computational domain and the related grid division. The cross-sectional area is 50D × 21D and the length is 6 L. The inlet and outlet of the computational domain are defined as the velocity inlet and pressure outlet, respectively. The velocity inlet is 1.2 L away from the head of the vehicle, and the pressure outlet is located 3.8 L away from the tail of the vehicle. Parameter settings of the solver are shown in Table 1.

As part of the vehicle's strut is above the water surface, the influence of the water surface cannot be ignored. An appropriate two-phase flow region is created in STAR-CCM+ based on the VOF model. The upper and lower halves of the velocity inlet are air and water, respectively. The inlet pressure and the outlet pressure vary in accordance with the water depth variation rule $p = \rho_l gh$. A no-slip wall surface boundary condition is adopted for the wall of the vehicle. The grids on the wall surface of the vehicle are refined in Figure 5b, and mesh refinement is applied along the water-depth direction at the air–water multi-phase flow interface, as shown in Figure 5c.

3.4. Evaluation of Mesh Independence

The grid quality has an important influence on the calculation results, and the results will be more accurate if the grids are refined. In this study, three different sets of meshes with different sizes (namely, fine, moderate and coarse meshes) are generated on the surface of the control model for further mesh convergence analysis. By varying the basic mesh size, the time step is calculated, and the Courant number remains unchanged. Table 2 lists the detailed mesh parameters.

Under different grid conditions, the strut drag of the control model is calculated at the sailing speed of 8 m/s. Table 3 lists simulation results under different mesh conditions.

The mesh independence analysis results can be calculated [2] and listed in Table 4. It can thus be concluded that when mesh S_1 is adopted, $0 < R_k < 1$ and the monotone convergence condition is satisfied. The error of the strut drag of the control model is 3.3%. According to mesh independence analysis results, fine mesh (S_1) is used in the numerical simulation.

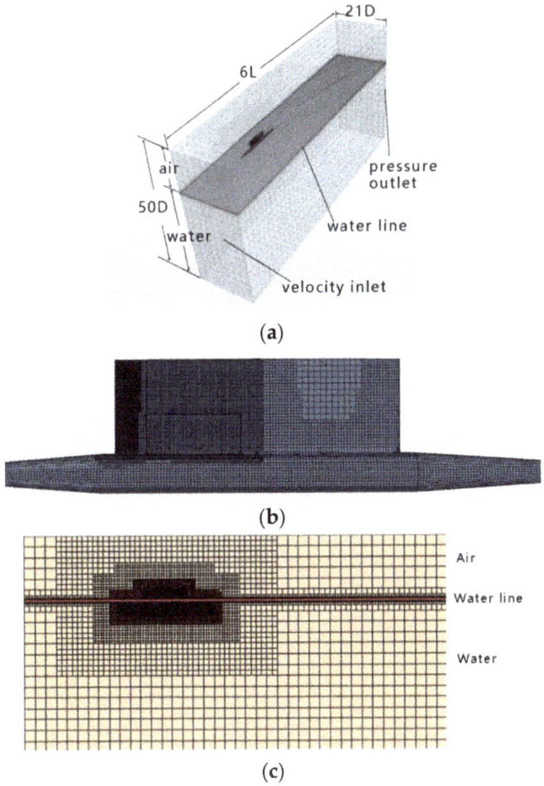

Figure 5. Computing domain and boundary conditions: (**a**) computation domain and grid division; (**b**) grids on the wall surface of the vehicle; (**c**) refined grids near waterline height.

Table 1. Parameter settings of the solver.

Parameter	Inlet Velocity (m/s)	Outlet Pressure (Pa)	Turbulence Intensity	Time Step (s)	Basic Pressure (Pa)
Setting value	5–14	Hydrostatic Pressure	0.01	0.0004	101,325

Table 2. Mesh parameters.

Mesh Generation	Meshing Size (mm)	Grid Number ($\times 10^4$)	Time Step (s)	Courant Number
Fine mesh (S_1)	3.2	140	0.0004	1
Moderate mesh (S_2)	4.8	90	0.0006	1
Coarse mesh (S_3)	6.4	50	0.0008	1

Table 3. Drag simulation results for the control model.

Mesh Conditions	S_1	S_2	S_3
Strut Drag (N)	43.56	44.98	51.02

Table 4. Mesh independence analysis results.

Analysis Parameter	R_k	C_k	δ^*_{k1} (%)	U_k (%)	U_{kC} (%)	S_C
Convergence	0.235	3.247	3.3%	5.5%	2.3%	44.98

4. Simulation Results and Analysis

4.1. Analysis of the Air Inflow Amount of the Strut Intake Duct

Since the sailing speed of the SWATH vessel in actual conditions is usually below 80 kn and considering the scale ratio of the model to be 1:9, the speed of the proposed model is set to be in the range of 1–14 m/s in the working conditions. Table 5 presents the air inflow amount of the outlet of the intake duct for each working condition.

Table 5. Air inflow amount for different sailing speeds.

Sailing Speed (m/s)	1–4	5	6	7	8	9	10	12	14
Air Inflow Amount (kg/h)	0	5.58	9.25	12.47	14.51	16.60	19.91	29.52	38.66

It is observed from Table 5 that when the sailing speed is 1–4 m/s the air inflow amount is 0. The reason for this is that at a lower sailing speed, the pressure difference between the outlet and the inlet of the duct induced by the Venturi effect is insufficient to overcome the hydrostatic pressure at the outlet as well as the pressure loss caused by the frictional drag when the air flows through the duct, with no air flowing out. However, when the sailing speed is above 4 m/s, the pressure difference between the outlet and the inlet of the duct can overcome the above-mentioned pressure loss. Therefore, the air starts flowing through the outlet of the duct. It should be indicated that higher sailing speed results in greater pressure difference and consequently more air inflow. Figure 6 illustrates the correlation between the air inflow amount and sailing speeds according to the data in Table 5. As can be seen from Figure 6, it is observed that the air inflow amount of the intake duct has a positive linear correlation with sailing speeds. Therefore, by adjusting the dimensions of the leading edge of the strut and appropriately designing the width of the intake duct, the optimal intake amount is obtained to realize the maximization of the strut drag reduction.

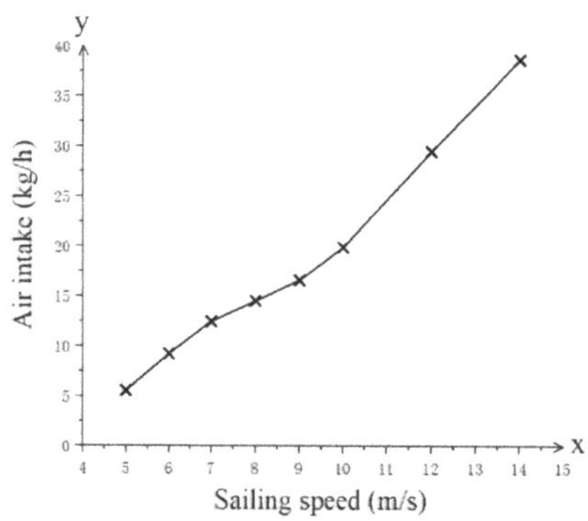

Figure 6. Correlation between the air inflow amount and sailing speeds.

4.2. Analysis for the Volume Content of Air on the Surface of the Strut

As the sailing speed changes, the varying intake amount of the intake duct of the strut leads to a variable air volume fraction in the bubbly flows covering the underwater strut surface. The air volume fraction cloud diagram of the bubbly flows on the surface of the strut of the proposed model at different sailing speeds is shown in Figure 7.

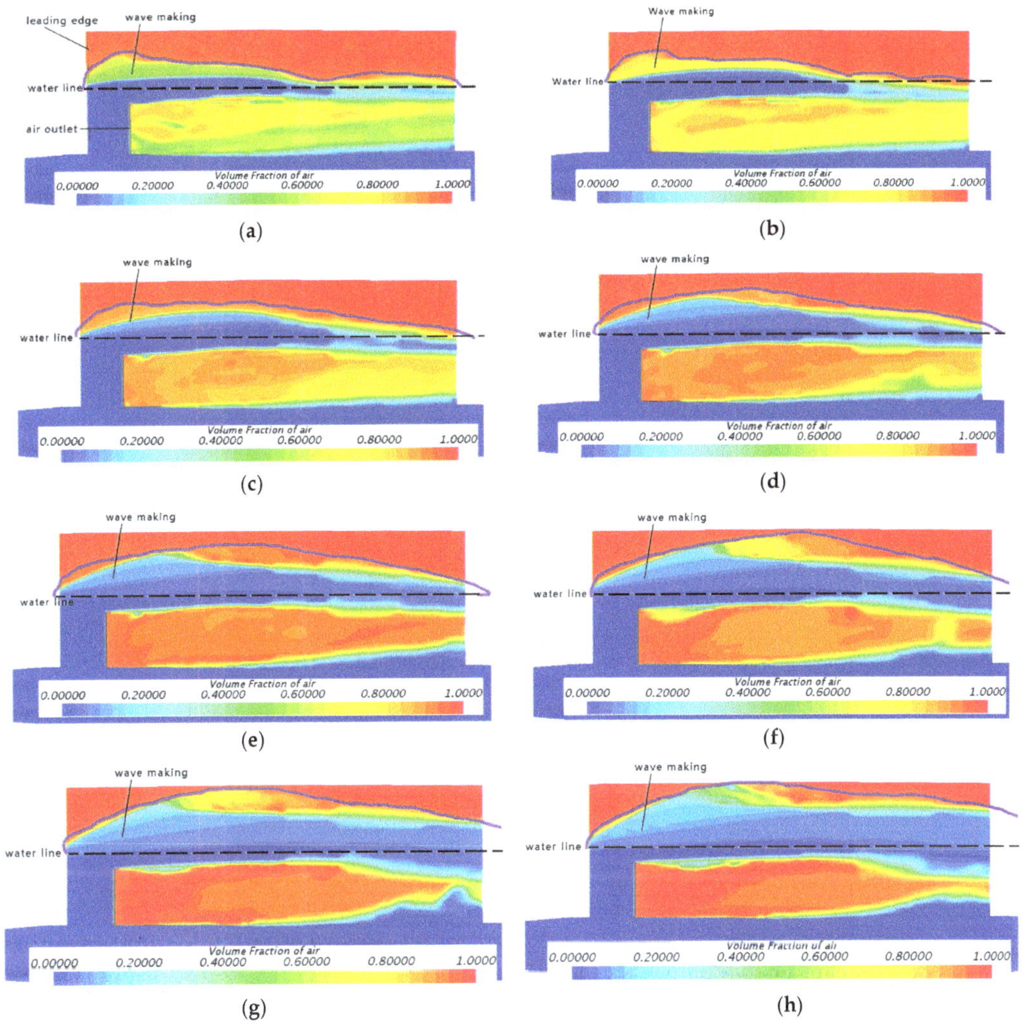

Figure 7. Volume Fraction of air cloud diagram of the bubbly flows on the surface of the strut at different sailing speeds: (**a**) at the sailing speed 5 m/s; (**b**) at the sailing speed 6 m/s; (**c**) at the sailing speed 7 m/s; (**d**) at the sailing speed 8 m/s; (**e**) at the sailing speed 9 m/s; (**f**) at the sailing speed 10 m/s; (**g**) at the sailing speed 12 m/s; (**h**) at the sailing speed 14 m/s.

It is found from Figure 7a–h that the external air is inhaled by the intake duct and blended with the incoming flow, and the resultant bubbly flows cover the underwater strut surface. The air volume fraction of the formed bubbly flows in the front of the strut is relatively higher. However, around the medium and the rear of the strut, the air volume fraction gradually declines as the inhaled air is persistently blended with water. As the sailing speed increases, the air amount of the intake duct suction increases and the air volume fraction of the formed bubbly flows increases gradually. When the sailing speed reaches 10 m/s, the air volume fraction of the bubbly flows around the front of the strut in Figure 7f is close to 1. As the speed increases further, the air volume fraction of the bubbly flows on the entire strut surface in Figure 7h is close to 1. Then, the covering effect on the strut is at its strongest.

Meanwhile, Figure 7a–h show that as the sailing speed increases, the wave-making is extended. It should be indicated that the part between the red part and the waterline in Figure 7 is the wave-making. It is worth noting that at a lower sailing speed, the wave-making is small. In Figure 7a, the wave-making above the waterline has a relatively considerable distance from the top of the strut. Therefore, the wet area of the strut above the waterline is relatively limited. As the sailing speed increases, the wave-making continuously increases. When the sailing speed reaches 14 m/s, the wave-making is considerable, and the wave-making above the waterline in Figure 7h almost reaches the top of the strut so that the part of the strut above the waterline is mostly wet.

4.3. Analysis for the Strut Drag

In this section, in order to evaluate the effect of drag reduction of the proposed model, comparative analysis is conducted on strut drags of the control model and the proposed model at different sailing speeds. Table 6 presents the strut drags at different sailing speeds.

Table 6. Strut drags at different sailing speeds.

Sailing Speed (m/s)	Model Type	Strut Drag (N)	Drag Reduction Rate (%)
1	Control model Proposed model	0.60 0.72	−20.00
2	Control model Proposed model	3.35 3.66	−9.25
3	Control model Proposed model	6.79 7.30	−7.51
4	Control model Proposed model	11.67 11.98	−2.66
5	Control model Proposed model	17.36 14.37	17.22
6	Control model Proposed model	24.16 18.82	22.10
7	Control model Proposed model	33.03 24.46	25.95
8	Control model Proposed model	43.56 31.27	28.21
9	Control model Proposed model	56.30 40.37	28.29
10	Control model Proposed model	70.47 50.72	28.03
12	Control model Proposed model	104.79 76.55	26.95
14	Control model Proposed model	144.55 107.05	25.94

It is observed from Table 4 that when the sailing speed is 1–4 m/s, the strut structure of the proposed model does not only reduce the drag but also enhances the drag. This is because, on one hand, the intake duct does not inhale air and the bubbly flows are not formed on the strut surface. On the other hand, the design of the duct and diffusion surface in the proposed model extend the wet surface of the strut in comparison to the appearance of the control model so that the frictional drag is increased. When the sailing speed is above 4 m/s, the intake duct initiates inhaling air and the bubbly flows are formed on the surface of the strut. The existence of the bubbly flows leads to reduction in the frictional drag. Moreover, as the speed increases, the air content in the bubbly flows and the drag reduction rate gradually increases. When the speed is 9 m/s, the drag reduction rate reaches its maximum at 28.29%. Figure 8 shows the variation of the drag reduction rate of the strut at different sailing speeds according to the data in Table 6.

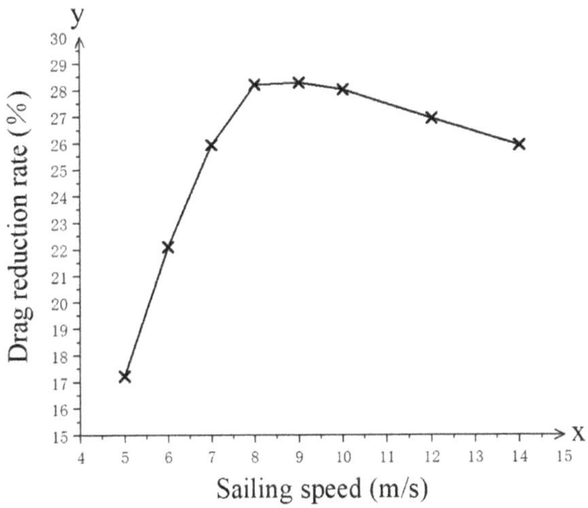

Figure 8. Variation of the drag reduction rate of the strut at different sailing speeds.

It is observed from Figure 8 that before the sailing speed reaches 9 m/s, the drag reduction rate tends to rise linearly. However, it gradually declines when the speed reaches 9 m/s. This is because, as can be seen in Figure 5, although the increment of the air content in the bubbly flows causes the frictional drag of the strut to gradually decline as the sailing speed increases, the drag of the wave-making of the strut gradually rises. When the sailing speed is below 8 m/s, the drag reduced by formation of the bubbly flows is greater than the drag caused by the wave-making, and the drag reduction rate of the strut gradually grows. When the sailing speed reaches 8–9 m/s, the drag reduced by the bubbly flows is balanced with the drag caused by the wave-making, so the drag reduction rate of the strut remains constant. When the sailing speed is greater than 9 m/s, the drag reduced by the bubbly flows is lower than the drag caused by the wave-making. Therefore, the drag reduction rate of the strut gradually declines as the speed increases. It should be indicated that the strut drag reduction structure should be designed according to the range of the SWATH vessel speeds to achieve the best drag reduction effect.

5. Conclusions

A novel drag reduction approach to an automatic air intake strut structure is proposed in the present study. Based on the Venturi effect, a drag reduction strut structure for the SWATH vessel is designed, and a numerical study is carried out to investigate the variation of the air inflow amount, the air volume fraction in the bubbly flows and the drag reduction rate of the strut for different sailing speeds. Through analysis, the following conclusions are drawn:

(1) As the sailing speed increases, the external air flows through the air outlet of the intake duct and it is blended with the incoming flow, forming bubbly flows on the surface of the strut and reducing the frictional drag of the strut.

(2) The air volume content of the bubbly flows on the surface of the strut increases as the sailing speed increases, which leads to the continuous increment of the drag reduction rate of the strut. The maximum drag reduction rate can reach about 30%, demonstrating a favorable drag reduction effect.

(3) As the sailing speed increases, the drag of the wave-making of the strut gradually increases. When a certain speed range is reached, the drag reduces due to the formation of bubbly flows being gradually balanced with the drag caused by the wave-making,

and the drag reduction rate does not increase anymore. Further increment of the speed results in a gradual decrease in the drag reduction rate.

Compared with the existing drag reduction approaches by injecting air, the proposed scheme is based on the Venturi effect with the utilization of the pressure difference between the air outlet and the air inlet of the intake duct to inhale air, not requiring artificial ventilation, which saves considerable energy and demonstrates a favorable drag reduction effect. In the future, we will continue to study the influence of the immersion depth and the shape of the automatic air intake strut structure on the air inflow amount and the drag reduction rate of the strut, and we will further design the test model of the drag reduction strut to carry out experimental verification. In a word, the proposed the automatic air intake drag reduction strut has good application prospects for achieving the targets of ship energy conservation and emission reduction. The approach of drag reduction by bubbly flows proposed in the present study provides a useful reference for the study of drag reduction of similar types of ships.

Author Contributions: Conceptualization, H.A. and Z.H.; methodology, H.A.; software, Z.H.; validation, Z.H. and H.P.; formal analysis, H.A.; investigation, Z.H.; resources, P.Y.; data curation, Z.H.; writing—original draft preparation, Z.H.; writing—review and editing, H.A. All authors have read and agreed to the published version of the manuscript.

Funding: The International Cooperation Special Program of the State Ministry Science and Technology of China, grant number: 2016DFR00070.

Acknowledgments: This work is supported by the international cooperation special program of the State Ministry Science and Technology of China (Grant No. 2016DFR00070). The financial contributions are gratefully acknowledged.

Conflicts of Interest: The authors declare no conflict of interest.

References

1. Begovic, E.; Bertorello, C.; Mancini, S. Hydrodynamic performances of small size SWATH craft. *Brodogradnja* **2015**, *66*, 1–22.
2. An, H.; Sun, P.; Ren, H.; Hu, Z. CFD-based numerical study on the ventilated supercavitating flow of the surface vehicle. *Ocean Eng.* **2020**, *214*, 107726. [CrossRef]
3. Mccormick, M.E.; Bhattacharyya, R. Drag reduction of a submersible hull by electrolysis. *Nav. Eng. J.* **1973**, *85*, 11–16. [CrossRef]
4. Murai, Y.; Fukuda, H.; Oishi, Y.; Kodama, Y.; Yamamoto, F. Skin friction reduction by large air bubbles in a horizontal channel flow. *Int. J. Multiph. Flow* **2007**, *33*, 147–163. [CrossRef]
5. Madavan, N.K.; Deutsch, S.; Merkle, C.L. Measurements of local skin friction in a microbubble-modified turbulent boundary layer. *J. Fluid Mech.* **1985**, *156*, 237–256. [CrossRef]
6. Kodama, Y.; Kakugawa, A.; Takahashi, T.; Kawashima, H. Experimental study on microbubbles and their applicability to ships for skin friction reduction. *Int. J. Heat Fluid Flow* **2000**, *21*, 582–588. [CrossRef]
7. Sayyaadi, H.; Nematollahi, M. Determination of optimum injection flow rate to achieve maximum micro bubble drag reduction in ships; an experimental approach. *Sci. Iran.* **2013**, *20*, 535–541.
8. Paik, B.G.; Yim, G.T.; Kim, K.Y.; Kim, K.S. The effects of microbubbles on skin friction in a turbulent boundary layer flow. *Int. J. Multiph. Flow* **2016**, *80*, 164–175. [CrossRef]
9. Kanai, A.; Miyata, H. Direct numerical simulation of wall turbulent flows with microbubbles. *Int. J. Numer. Methods Fluids* **2001**, *35*, 593–615. [CrossRef]
10. Kawamura, T.; Kodama, Y. Numerical simulation method to resolve interactions between bubbles and turbulence. *Int. J. Heat Fluid Flow* **2002**, *23*, 627–638. [CrossRef]
11. Xu, J.; Maxey, M.R.; Karniadakis, G.E. Numerical simulation of turbulent drag reduction using micro-bubbles. *J. Fluid Mech.* **2002**, *468*, 271–281. [CrossRef]
12. Ferrante, A.; Elghobashi, S. Reynolds number effect on drag reduction in a microbubble-laden spatially developing turbulent boundary layer. *J. Fluid Mech.* **2005**, *543*, 93–106. [CrossRef]
13. Mohanarangam, K.; Cheung, S.C.P.; Tu, J.Y.; Chen, L. Numerical simulation of micro-bubble drag reduction using population balance model. *Ocean Eng.* **2009**, *36*, 863–872. [CrossRef]
14. Feng, Y.-Y.; Hu, H.; Peng, G.-Y.; Zhou, Y. Microbubble effect on friction drag reduction in a turbulent boundary layer. *Ocean Eng.* **2020**, *211*, 107583. [CrossRef]
15. Haryanto, Y.; Waskito, K.T.; Pratama, S.Y.; Candra, B.D.; Rahmat, B.A. Comparison of Microbubble and Air Layer Injection with Porous Media for Drag Reduction on a Self-propelled Barge Ship Model. *J. Mar. Sci. Appl.* **2018**, *17*, 165–172.

16. Elbing, B.R.; Winkel, E.S.; Lay, K.A.; Dowling, D.R.; Perlin, M. Bubble-induced skin-friction drag reduction and the abrupt transition to air-layer drag reduction. *J. Fluid Mech.* **2008**, *612*, 201–236. [CrossRef]
17. Shen, X.; Ceccio, S.L.; Perlin, M. Influence of bubble size on micro-bubble drag reduction. *Exp. Fluids* **2006**, *41*, 415–424. [CrossRef]
18. Murai, Y. Frictional drag reduction by bubble injection. *Exp. Fluids* **2014**, *55*, 1773. [CrossRef]
19. Choi, J.K.; Hsiao, C.T.; Chahine, G.L. Numerical Studies on the Hydrodynamic Performance and the Startup Stability of High Speed Ship Hulls with Air Plenums and Air Tunnels. In Proceedings of the 9th International Conference on Fast Sea Transportation, FAST 2007, Shanghai, China, 23–27 September 2007; pp. 286–294.
20. Choi, J.K.; Georges, L.; Chahine, G.L. Numerical Study on the Behavior of Air Layers Used for Drag Reduction. In Proceedings of the 28th Symposium on Naval Hydrodynamics, Pasadena, CA, USA, 12–17 September 2010; pp. 1–15.
21. Kim, D.; Moin, P. *Direct Numerical Study of Air Layer Drag Reduction Phenomenon Over a Backward-Facing Step*; 63rd Annual Meeting of the APS Division of Fluid Dynamics; Center for Turbulence Research: Stanford, CA, USA, 2010; pp. 351–362.
22. Zhao, X.; Zong, Z.; Jiang, Y.; Sun, T. A numerical investigation of the mechanism of air-injection drag reduction. *Appl. Ocean Res.* **2020**, *94*, 101978. [CrossRef]
23. Kim, S.; Oshima, N. Numerical prediction of a large bubble behavior in wall turbulent flow. In Proceedings of the 14th World Congress in Computational Mechanics and ECCOMAS Congress 600, Paris, France, 19–24 July 2020; pp. 1–7.
24. Zhao, X.-J.; Zong, Z.; Jiang, Y.-C. Numerical study of air layer drag reduction of an axisymmetric body in oscillatory motions. *J. Hydrodyn.* **2021**, *33*, 1007–1018. [CrossRef]
25. Zhang, Y.J. *The Daqo of Fluid Mechanics*; Beijing University of Aeronautics and Astronautics Press: Beijing, China, 1991. (In Chinese)
26. Hirt, C.W.; Nichols, B.D. Volume of fluid (VOF) method for the dynamics of free boundaries. *Comput. Phys.* **1981**, *39*, 201–225. [CrossRef]
27. Kim, D. Direct numerical simulation of two-phase flows with application to air layer drag reduction. Ph.D. Thesis, Standford University, Stanford, CA, USA, June 2011.
28. Menter, F.R. Two-equation eddy-viscosity turbulence models for engineering applications. *AIAA J.* **1994**, *32*, 1598–1605. [CrossRef]
29. Pendar, M.R.; Roohi, E. Investigation of cavity around 3D hemispherical head-form body and conical cavitators using different turbulence and cavity models. *Ocean Eng.* **2016**, *112*, 287–306. [CrossRef]

MDPI AG
Grosspeteranlage 5
4052 Basel
Switzerland
Tel.: +41 61 683 77 34

Symmetry Editorial Office
E-mail: symmetry@mdpi.com
www.mdpi.com/journal/symmetry

www.ingramcontent.com/pod-product-compliance
Lightning Source LLC
Chambersburg PA
CBHW061608210926
51850CB00009B/2255
* 9 7 8 3 7 2 5 8 1 8 6 4 8 *